C0-DWJ-135

WITHDRAWN
UTSA Libraries

Innovative Capability of Chinese Enterprises

Innovative Capability
of Chinese Enterprises

Yi Zhihong

Enrich Professional Publishing

Published by

Enrich Professional Publishing (S) Private Limited
16L, Enterprise Road,
Singapore 627660
Website: www.enrichprofessional.com
A Member of Enrich Culture Group Limited

Hong Kong Head Office:
2/F, Rays Industrial Building, 71 Hung To Road, Kwun Tong, Kowloon, Hong Kong, China

China Office:
Rm 1800, Building C, Central Valley, 16 Hai Dian Zhong Jie, Haidian District, Beijing

United States Office:
PO Box 30812, Honolulu, HI 96820, USA

English edition © 2013 by Enrich Professional Publishing (S) Private Limited
Chinese original edition © 2008 China Renmin University Press

Translated by Li Ling and Li Yanmin

Edited by Barbara Cao and Angela McGaffin

All rights reserved. This book, or parts thereof, may not be reproduced in any form or by any means, electronic or mechanical, including photocopying, recording or any information storage and retrieval system now known or to be invented, without prior written permission from the Publisher.

ISBN (Hardback)   978-981-4298-36-0
ISBN (ebook)   978-981-4298-37-7 (pdf)
                       978-981-4298-63-6 (epub)

This publication is designed to provide accurate and authoritative information in regard to the subject matter covered. It is sold with the understanding that the publisher is not engaged in rendering legal, accounting, or other professional service. If legal advice or other expert assistance is required, the services of a competent professional person should be sought.

Enrich Professional Publishing is an independent globally-minded publisher focusing on the economic and financial developments that have revolutionized new China. We aim to serve the needs of advanced degree students, researchers, and business professionals who are looking for authoritative, accurate, and engaging information on China.

Library
University of Texas
at San Antonio

Contents

Contents

Preface

One of China's basic strategies is to enhance its independent innovative capability so as to build an innovation-oriented country where enterprise is the main force. The key for a business to seize opportunities and take the initiative amid a fierce international competition lies in the sustainable development of its independent innovative capability which represents the core competitiveness of an enterprise.

Since the reform and opening-up in 1978, China has witnessed continuous economic growth and become a major economy in terms of size and economic aggregate. In terms of GDP, China's economic ranking has moved from 11th in 1990 to 4th in 2006, and from 15th to 3rd in terms of the total export and import volume. During this period, Chinese enterprises have become the leading drive for China's economic growth under the market economic system. Many enterprises transformed from quantitative to qualitative changes and saw significant development in many aspects, such as scientific and technological innovation, independent research and development capability, brand innovation, business mode innovation and management innovation. Some even became competitive in the international market. However, China still has a long way to go before it becomes an economic giant, and most Chinese enterprises are still a far cry from their international counterparts.

A thorough study on the characteristics of the Chinese enterprises in independent innovation and the environment and organization system which influence the independent innovative capability of enterprises is expected to help improve the policy environment and strengthen their innovative capability. For this matter, when setting up Key Projects of Renmin University of China to implement Project 985, we brought together professors from the School of Business in relevant disciplines to collaborate and embark on "research for sustainable development of the independent innovative capability of Chinese enterprises." Since there are already many works relating to independent innovation of Chinese enterprises, we tried to make breakthroughs in two aspects. Our first methodology combines empirical research and case studies to explore the effects of environmental influences on independent innovation of Chinese enterprises and their characteristics. Second, we made an all-round analysis and research of the sustainable development mode for the independent innovation of Chinese businesses from three aspects, namely institutional environment, innovative behavior, and innovation governance and organization. In order to access first-hand information, many professors carried out investigations and surveys inside enterprises and some have even made follow-up studies for many years. Since there are still many new challenges for the study of business independent innovation, we continue to research and evaluate this issue.

Abbreviations

CRSB	China Resources Snow Breweries
CRM	Customer Relationship Management
FP	Financial Performance
FOF	Fund of Funds
GEM	Growth Enterprise Market
HR	Human Resources
IPO	Initial Public Offerings
IPR	Intellectual Property Rights
M&A	Merger and Acquisition
NBTT	Net Barter Terms of Trade
OECD	Organization for Economic Co-operation and Development
OEM	Original Equipment Manufacturer
R&D	Research and Development
ROI	Return on Investment
S&OP	Sales and Operation Planning
SDM	Service Delivery Management
SSC	Service Supply Chain
SFTT	Single Factor Terms of Trade
SMEs	Small- and Medium-Sized Enterprises
SOEs	State-Owned Enterprises
SRM	Supplier Relationship Management
TFP	Total Factor Productivity
VEC	Vector Error Correction
VC	Venture Capital
WTO	World Trade Organization

1

Chapter

Introduction and Overview

Yi Zhihong

Research Subject

Since the 1990s, the way of business transactions and the economic environment have changed dramatically, and the surging development of technology and the formation of the global market have greatly changed the way of competition. We are now entering into a new era—the era of the knowledge-based economy. The knowledge-based economy as a new economic form, compared with the traditional industry-based economy, refers to an economy that is built on production, allocation, and utilization with knowledge at the core. This calls for enterprises to make corresponding changes in line with the requirements of the new era. In the era of the knowledge-based economy, the economy of scope has become more important than the economy of scale over time, and the business philosophy of simply relying on quantity to gain competitive edge has fallen behind the times. The emergence of new technologies and new knowledge requires enterprises to make a swift response to production and marketing-related information processing. In this sense, new requirements have been set for businesses in management mode and organizational structure. Peter F. Drucker, an American expert in management, pointed out: "A typical enterprise in the future shall be known as the information-based organization. It is knowledge-based and consists of all kinds of experts…Its organizational structure will also be affected as a result." Business innovation with knowledge at the core has become an important feature of corporate behaviors in the era of knowledge-based economy. This kind of innovation does not just happen to material products or to the simple processing of objective information. Rather, it is the capability to realize sustainable development or the secret of success in operations which are formed on the basis of fully tapping into people's potential ideas, intuition and inspiration. It is because of this that the rapid expansion of innovation propelled enterprises to become a learning organization so that they can meet the requirements of the knowledge-based economy.

As early as in 1930s, the American economist Joseph A. Schumpeter pointed out that innovation of various kinds based on the recombination of existing knowledge had made a great difference in boosting economic development and changes. In the theoretical system of Schumpeter, "innovation" is a strict economics category that is creatively included in the theoretical system of economy. After that, the widespread application of econometrics and the emergence of property right theory have provided a new theoretical support for the development of the innovation theory in two aspects. First, there have been in-depth studies on technological innovation and diffusion on the one hand and technological innovation and market structure on the other, which have been integrated into the technological innovation economics with technological change and technology transfer as objectives. Second, by introducing the property right theory and the transaction cost concept into the analysis of institutional formation and changes, Coase et al. formed institutional innovation economics. As the research deepens, the research field of innovation keeps witnessing subdivision and the application of innovation concepts becomes widespread. For instance, the concepts of management innovation and organization innovation have been proposed apart from technological innovation and institutional innovation. These innovation concepts have promoted and enriched the studies on innovation theory.

It is fair to say that the concept of "independent innovation" is creatively proposed by China. Both the central government and provincial and municipal governments have put independent innovation at the top of their agendas in recent years and during the "11th Five-Year Plan" period. This was made a priority in the hope it would make independent innovation a key measure to implement the Scientific Outlook on Development and transform the economic growth pattern. Chinese economic development and growth mode have for long focused on external expansion, and the innovative capability is weak and undervalued. The concept of independent innovation was just put forward to address this vulnerability. Competition between countries and regions is, in effect, the competition in science and technology between countries with the independent innovative capability at the core. Only by improving independent innovative capability, focusing on making breakthroughs in core technologies and possessing more proprietary intellectual property rights in key areas, could China seize the commanding height of strategy in the international industrial division of labor and in the world economic structure, and shape sustainable competitiveness to firmly keep the initiative in economic development. The decade-long reform shows that only by enhancing independent innovative capability of business could China keep upgrading the industrial structure, improving resource utilization ratio and market competitiveness of products in a real sense and finally keep the economy sound, strong and growing. In this connection, China will undoubtedly highlight and strengthen the position of business' independent innovative capability in the reform of Chinese modern business management.

However, we need to be clear that emphasizing innovation is different from setting up innovation mechanism. In the era of knowledge-based economy, innovation could become the driving force for business growth only by building on a systematic mechanism. Specifically speaking, the establishment of innovation mechanism must balance the relation between knowledge-based innovation and technological innovation on the one hand and better integrate innovative capability and control on the other. Technological innovation is only meaningful when it is built on the production, dissemination and application of knowledge. If we say the rising economy of Japan and East Asia in 1970s and 1980s was attributed to the strong national technological innovation system, then the rejuvenation of the European and American economy, especially the economy of the United States in 1990s, was attributed to the national innovation system valuing both knowledge innovation and technological innovation. Another issue that needs to be addressed by innovation mechanism is how to unify the innovative capability and the control. Compared with the innovative capability, the control has yet to be valued by the business community in China. As a matter of fact, the failure of some companies in operation is not just because of the lack of innovative capability. More importantly, they lost the control, which has resulted in great losses of business assets. In this sense, a major issue that merits serious studies amid innovation of knowledge management in China is how to integrate the incentive mechanism and the restraint mechanism in institutional arrangement and realize the unification of innovative capability and control.

We believe, out of this consideration, that the connotation of independent innovation is not just limited to a single technological innovation. Rather, it is an integrated innovation system with technological innovation at the core, the institutional and financial environment as the background, the organization management and operation model as the form, and the governance and organizational structure as the support. We classify this type of innovation mechanism as the "independent innovation system with sustainable development." The concept of "sustainable development" not only refers to our relationship and respect for the natural environment but also to a harmonious development of the human society. More importantly, it lies more in the formation of an operation mechanism for facilitating a sustainable development of independent innovation and an continuous increase of benefit through the building of a comprehensive economic environment.

Literature Review and Research Objectives

Traditional foreign innovation theories have developed for decades. Of these, technological innovation, institutional innovation and management innovation are the themes of modern studies on innovation theories. *The*

Theory of Economic Development, published in 1912, and *Capitalism, Socialism and Democracy* (Schumpeter), published in 1942, represent the pioneering monographs about studies on innovation. Beyond that, *Institutions, Institutional Change and Economic Performance* (Coase), published in 1990, and *The Visible Hand: The Managerial Revolution in American Business* (Chandler), published in 1977, are also works about the development of innovation research. In general, research from foreign scholars on innovation can be divided into three categories:

1. Taking technological innovation as a key element for driving economic development and business growth, whose meaning is to realize "revolutionary" changes and "breakthroughs" during the whole process of material production by introducing a new mode of production. Schumpeter (1921) broke innovation down into new product, new mode of production, new market, controlling new source of raw materials supply and realizing new business organization. Hicks (1963) later divided innovation into three categories, namely labor-saving innovation, capital-saving innovation, and neutral innovation. According to Mensch (1979), innovation falls under two categories, namely fundamental innovation and second innovation. Based on this classification, different scholars respectively probed into the different aspects and relevant issues of technological innovation, such as technological innovation diffusion (Mansfield Model, 1968), technological innovation and market structure (Kamien and Schwartz Model, 1975), technological innovation and business size (Shimshoni Model, 1970; Freeman Model, 1982), technological innovation and business entry (Lerin Model) and technological innovation and uncertainties (Nelson, Winter and Freeman Theory).

2. Research must continue on institutional innovation. Schumpeter's Theory of Innovation also puts forward the innovation concept of "realizing a new organization in any industry" while elaborating on technological innovation, but he did not conduct specialized studies on this issue. Davis and North took the lead in substantive research of institutional innovation. Institutional innovation and technological innovation have some similarities. For example, neither would happen until the expected returns on innovation surpass the anticipated costs for innovation. However, their differences are not only in the contents of innovation but also in the different durations of different contents. The time of technological innovation depends on the lifetime of physical capital while the time of institutional innovation does not rest with materialized factors. This is what is behind the observation of the theory of institutional innovation–that institutional innovation has to go through five stages. The five stages are: a. Form "the first action independent innovation"; b. "The first action

independent innovation" proposes plans on institutional innovation, and there is a need to wait for a new invention in institution if feasible, ready plans do not exist; c. When some alternative plans on institutional innovation are available, "the first action independent innovation" will make comparisons and selections in line with the principle of profit maximization; d. Form "the second action independent innovation" to help "the first action independent innovation"; e. The two independent innovations work together to make institutional innovation a reality. According to the theory of institutional innovation, institutional innovation happens at three levels, i.e. individual, group and government levels (North, 1991).

3. Foreign researchers focus on the uniqueness of innovation as an emerging management element from different perspectives, further explore ways to include and integrate innovation behavior into the management environment and strategy, try to provide a theoretical basis and practical guidance for business innovation activities, and blaze a new trial for business innovation.

First, against the current background of internationalization and networking of competition, business performance depends on various factors: the first group includes the track, type and characteristic of innovation, and these different factors have different influences on innovation performance (Teece and Pisano, 1994; Nelson, 1995); the second group includes the new mode of production, management style and resources adopted for realizing innovation. Some researchers carried out studies on innovation behaviors and models. As cases in point, Green et al. studied the differences between sustainable innovation and the capability to promote innovation (Green et al., 1995; Anderson and Tushman, 1990); Christensen et al. researched the differences between structural innovation and disruptive innovation (Christensen, 1998; Henderson and Clark, 1990); in particular, Gatignon et al. believed that the way of innovation formation (internal or external formation), the utility mode of innovation (capability improvement or disruption), innovation form (core or outer circle) and innovation path (continuous or fundamental) are totally different standards, and they have respectively affected the ultimate performance and competitiveness of enterprises (Gatignon et al., 2002). Some researchers put forward that the customized mass production (Pine II, 2001; Duray, 2002), postponed production (Suetal, 2004) and modular production (Tu, 2004) are all major manifestations of the realization of innovation when it comes to the production and management mode and channel for innovation. On top of that, just-in-time management and supply chain logistics also represent the essential channels for management innovation (Flint, 2005).

Second, a number of scholars represented by Simpson (1999), including

Tidd (2001), Greve and Taylor (2000), Stuart et al. (2000), Powell et al. (1996), Goes et al. (1997), Jones (2005) and Frambach et al. (2001) incorporated the theory of organization to research innovation. They observed that only more flexible organization forms and organization types with fewer bureaucratic characteristics could promote the development of innovation and adapt to uncertainties of innovation and changes. In particular, the flat management, the process-oriented organization, and even the sound network and relationship between enterprises represent the enabling environment resulting from innovation.

Third, Corso et al. (2001), Majchrzak (2004), Tsai (2001), Johnson et al. (2002), Hall et al. (2002, 2003), and Almedia (2004), among other scholars, studied the knowledge transformation and transfer during the process of innovation. They raised the question of the influence of knowledge transformation and transfer on innovation performance. Skilton et al. (2003) conducted empirical research on knowledge maturity, innovation and production efficiency.

Finally, innovation has created a new type of relationship between businesses and customers and a new type of relation inside and outside enterprises on the one hand, and resulted in the fundamental changes and model transformation of business strategy on the other (Govindarajan et al., 2004). Drucker (1984) observed that innovation comes from conscious and purposeful research in most cases. Innovation has seven sources: unexpected events, incongruous incidents, process needs, industrial and market changes, changes in population statistics, changes in views, and new knowledge. These seven sources are overlapped, thus the potential for innovation is not just in one aspect in some time point. Leonard (1997) proposed that enterprises need to identify and meet the needs that might not be realized by consumers during the process of improving and innovating products that are not familiar to consumers. To make that happen, enterprises need to fulfil this innovative mission by applying "customized technologies." In particular, Kim et al. (1997) noted that the innovation of modern businesses stresses value-added innovation and requires enterprises to give up the traditional competition mode based on physical products as a way to realize the yet-to-be-realized value via innovative services and operations.

Chinese scholars are latecomers to research on innovation compared with their foreign counterparts. Some earlier works in this regard include: Chang Xiuze's *On Modern Business Innovation* (1994), Li Baoshan's *Integrated Management: Management Innovation in High-tech Era* (1998), Wu Guisheng's *The Management of Technological Innovation* (2004), *Theory and Practice of Business Innovation Management* of Hou Xianrong et al. (2003), and Zhang Pinghua's *Chinese Business Innovation Management* (2004). However, studies on independent innovation only started in recent years. Zhang Shouzheng (2006) put forward that independent innovation is the means to enhance the core competence. To do this, there is a need to change ideas, value talents, integrate production, learning and research, and improve intellectual property rights protection. Mei Yonghong (2006) elaborated on the role of independent innovation in China's long-term development from the perspectives of independent innovation and national interests. Cui Jinhua (2006) came up with three types of independent innovation–namely original innovation, integrated innovation and re-innovation after assimilation–and pointed out that China's independent innovation needs to be realized in the aspects of strength, institution and talent. Cai Bing (2006) discussed the obstacles for China's independent innovation from cultural and institutional perspectives and presented that only these two aspects can really fuel the development of China's independent innovation. Chen Jinhua (2006) illustrated the factors for independent innovation from six aspects: subjectivity of enterprises, the building of innovative culture, market orientation, utilization of domestic and foreign social resources, talent incentive, and the innovative spirit of entrepreneurs.

Scholars outside of China mainly base their studies on innovation on the context of the Western World. Their studies are typically limited to partial or individual innovations and have neglected institutional and characteristic analysis during the innovation and changing process of developing countries amid economic transition. In contrast, relevant research on independent innovation within the Chinese mainland mostly focuses on relevant philosophies, methods and applications and few have touched upon innovation issues. In particular, there are insufficient and less in-depth research on key elements for innovation and specific measures for development. When it comes to the research method, they have mainly employed theoretical research and case study, and the methods of integrating qualitative and quantitative research are insufficient and need to be promoted. This book plans to adopt a systematic and integrated angle and regulate the research methods of the combination of research and empirical research as well as case study, as well as the combination of qualitative and quantitative studies, to conduct independent innovation research under the background of economic transition in China.

Research Framework

Studies in this book are mainly carried out in three areas: first, the institutional and financial environment for business independent innovation; second, the functional activities for business independent innovation; third, the governance structure and organizational structure of business independent innovation. We make this arrangement out of the following considerations; these parts are the three interrelatedlevels of business sustainable development: the institutional and financial environment comes as the restraining factor for corporate independent innovation and has guided the innovation behaviors of enterprises; functional activities are the specific demonstration of innovation, and business innovation is reflected in the process of a series of correlative activities; and governance structure and organizational structure are the guarantee for business innovation. It is the interaction of these three parts that has made a difference in the overall business innovation performance (see Fig. 1.1). The research topic for each part of this book is as follows:

Fig.1.1.   Research framework

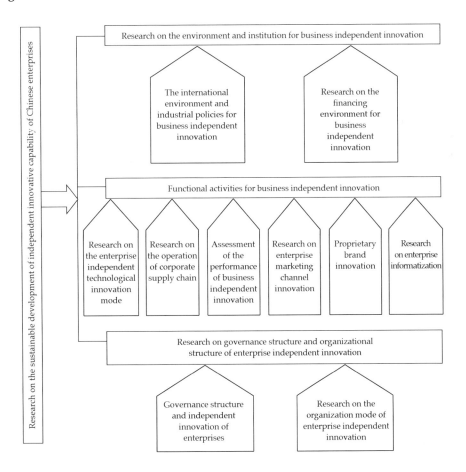

The international environment and industrial policies for the independent innovation of Chinese enterprises

The socialization of production and circulation is facilitating the reform of organizational forms. In the era when the surging development of production and circulation socialization becomes internationalized, the development of enterprise independent innovation closely relies on the international environment with globalization as the major theme. The contents of this subject research include:

(1) The dependency relationship between the sustainable development and restructuring of the world economy and the business organization evolution. The research in this book gives priority to the discussion about the practice of the world industrial revolution and technological revolution, especially about the influence of the upsurge of the information technological revolution and the development of the information technology industry in the 1990s on the development and organizational structure of enterprise independent innovation.

(2) The major impact of the cross-border circulation of trade, investment, labor and service, and the evolution and status quo of global governance on the development of enterprise independent innovation. This book makes an in-depth study on the changes of the multilateral trading system represented by the World Trade Organization (WTO) and that of global trade policies and their influences on the formation and expansion of enterprise independent innovation scale; the influence mechanism of global governance structure of international activities, including bilateral investment system and foreign investment policies of different countries, international conventions on intellectual property rights, policies on technology transfer and on labor regulation, for the development of enterprise independent innovation; and relevant analysis of the international environment and the strategic choice for the development of Chinese enterprise independent innovation.

(3) Hypothesis and testing of the enterprise's globalized operation mode in the international trade model; the American mode, European mode, Japanese mode, and "the circumstances in small countries" mode and their lessons for the development of Chinese enterprise independent innovation.

(4) The role of the new form and new progress of globalization in boosting the development of the independent innovation of modern enterprises. The discussions mainly focus on the carrying capacity and its changes in enterprise independent innovation under the conditions of drastically changing the modes of global manufacturing and distribution as well as international division of labor.

(5) The international environment and generative mechanism for the formation of enterprises organizational capital in the modern enterprise and capital theories; the building of the optimal balance mechanism model for enterprise organizational capital and the international environment for business operation.

(6) The interaction between the setup and development of Chinese enterprise independent innovation and the international environment; the study focuses on the mechanism influencing China's industrial layout from the dual viewing angles of China's independent opening up and external changes in the international environment, to establish a theoretical description on the experiment environment for the development of Chinese enterprise independent innovation.

Studies on the financing environment for the independent innovation of Chinese enterprises

The independent innovation of enterprises, especially the innovation in product and technology, requires financial support. But international experiences and China's practices signify that most enterprises may face financing difficulties during the process of independent innovation. This is especially true for small and medium-sized enterprises that are the foundation of independent innovation because the capital they obtained from financial institutions and their contributions to economic growth are seriously unbalanced. If the bottleneck for enterprise innovation in financing cannot be effectively broken, then the motivation and capability of enterprises for innovation will become constrained, and affect the innovative capability of the state. This section of the study takes small and medium-sized innovative businesses as research objectives. It primarily studies the financing demand and the financing characteristics of these enterprises at different stages of the corporate lifecycle, from the perspective of information asymmetry according to research results of relevant disciplines, including economics, finance and business theory. This section also probes into the financing environment and policy support for independent innovation of Chinese enterprises through investigation and surveys of the practicecompiled in theBeijing Zhongguancun Science and Technology Park, for example. The main contents of this research include:

(1) Characteristics of the financing demand of enterprises for independent innovation at different stages of the corporate lifecycle;

(2) Innovative business financing and government's institutional support;

(3) Innovative business financing and venture capital mechanism;

(4) Innovative business financing and the credit supply of commercial banks;

(5) Innovative business financing and the operation of the capital market.

Studies on the independent technological innovation of Chinese enterprises

This section takes the technological innovation mode of China Shenhua Group as a case study. The purpose is to analyze the establishment of the main body of the independent technological innovation of Chinese businesses, the driving force for technological innovation, and business technological innovation under the networking and integrated conditions. This section also makes an analysis of the management mechanism and innovation elements for business technological innovation.

The major issues that are studied and discussed in this section include:

(1) The main body and driving force for business independent technological innovation. The focus of this study is on clearly stating that the enterprise is the main body in technological innovation and analyzing the driving force for innovation on this basis.

(2) The system and channel of business independent technological innovation. Based on networked, knowledgeable and integrated social economy, the formation and channel of business technological innovation have differentiated themselves from previous modes. This mode is typically reflected by the fact that modern technological innovation is no longer the partial innovation of a single technology. Rather, it is the innovation system integrating wheel radiation, integrated innovation and aggregated innovation.

(3) The management mechanism for business technological innovation. The performance of business technological innovation depends on the project management and organization mode of the innovation. The research is to explore a unique way of project management for business technological innovation and the organization mode of innovation based on cooperation and coordination.

Studies on the operation of the innovation-oriented service supply chain of Chinese enterprises

The innovation of modern enterprises is not only the innovation of an individual enterprise but also the innovation of the network and supply chain. In addition, innovation does not just happen in production and manufacturing; more importantly, it happens to the integrated service supply chain. Based on this logic, this section explores the operation mode of the innovation-oriented service supply chain of Chinese enterprises and relevant innovation performance.

The contents of the research in this section include:

(1) Similarities and differences between the service supply chain and the manufacturing supply chain, and the connotation and model structure of the service supply chain;

(2) The operation mode of the service supply chain of Chinese enterprises, key points and law of mode innovation, and the influence of the service supply chain on business innovation;

(3) The preconditions and key elements for the establishment of operation model of the innovation-oriented service supply chain.

Studies on the enterprise's informatization performance under institutional innovation

With the development of Chinese society and economy, industries are experiencing institutional changes and innovation, and informatization plays a pivotal role in this process. Given this context, how to assess the implication of informatization for the performance of business changes in an all-round and rational manner becomes a focus of the study in this section.

The main points of the study in this section include:

(1) Major manifestations of business informatization;

(2) The influences and effects of business informatization on the internal management and operation;

(3) Performance assessment and measurement of business informatization.

Studies on the marketing channel innovation of Chinese enterprises

Compared with the technology, brand, and management of multinational enterprises, the competitive edge of Chinese local businesses is primarily demonstrated in the marketing channel, especially in the marketing, logistics and service network resources built in the secondary and tertiary markets. However, this advantage of domestic manufacturers has been held back by the lagging development of domestic market structure (large retailers dominate the market) and market regulations. Additionally, domestic manufacturers have little influence and a reduced profit space in the market.

Given this background, a multitude of local companies represented by Gree Electric Appliance have focused on marketing channel innovation by adopting the strategy of building independent marketing channels. Gree created the "joint-stock regional sales company" mode. This mode has made Gree internalized and systematized over time through effectively integrating the marketing resources of external dealers and leveraging the management approaches of holding shares and granting franchise. As a result, this has become Gree's largest competitive resource in the market. At the same time, this mode has also added to Gree's ability to compete with giant retailers like Gome and Suning, and has effectively eliminated the price war among dealers and stopped their short-sighted actions in brand management and administration.

Through empirical research, this section reveals the mode and characteristics of the marketing channel of the Chinese household appliance industry with the retailers dominating the market. On this basis, this section summarizes the causes for the continuous profit increase from the channel innovation perspective by analyzing how Gree built the independent marketing channels. In addition, the section also discovers and discusses the characteristics and success stories of the Chinese household appliance industry in marketing channel innovation based on case studies.

Studies on the brand innovation and performance of Chinese enterprises

Brand innovation and management is part and parcel of Chinese enterprise independent innovation, but it is also a weak point compared to international corporations. This section takes China Resources Snow Breweries as an example to study business brand innovation and brand operation and management.

Since its founding in 1993, China Resources Snow Breweries has become the largest beer group in mainland China after a series of merger and acquisition activities. In 2006, the group realized a sales volume of 5.3 million tons and the product Snow Beer alone recorded 3 million tons, ranking first nationwide; Snow Beer enjoys a market share of 15.1% of the entire industry, more than that of Tsingtao Beer (another influential beer producer with the longest history in mainland China). The main purpose of this section is to analyze the causes for the rapid growth of Snow Beer and explore the general features of brand innovation.

This section, on the basis of literature review, prioritizes the analysis of the brand innovation and growth history of Snow Beer, including capital management and brand operation, product innovation and brand integration, and leveraging opportunities of brand promotion under the "Mushroom Strategy." This section also comprehensively evaluates the brand image of Snow

Beer based on investigation and surveys, and the results based on customers' evaluation indicate that it ranks among leaders in terms of its image.

Through case study, this section generalizes the best practices of brand innovation and brand operation and management: the first is to strengthen product innovation and improve product quality; the second is to disseminate resources in a concentrated way and integrate marketing communications; the third is to implement effective channel management.

Studies on the building of evaluation indicators for the continuous technological innovative capability of enterprises

This section takes soft financial theory as the basis to describe the formation mechanism for enterprise's continuous technological innovative capability that is embedded in enterprise's value network from the three dimensions, namely element dimension, space dimension and time dimension. Considering that information might not be equally important in the system, some indicators are even redundant; therefore, this section uses Rough Set to reduce indicators and offers corresponding calculations.

The main contents of the research in this section include:

(1) The three-dimensional system and elements for continuous technological innovation of enterprises;

(2) The building of the indicator system for assessing the continuous technological innovative capability of enterprises based on Rough Set;

(3) Calculation method for the evaluation of enterprises' continuous technological innovative capability.

Governance structure and independent innovation of Chinese enterprises

The influence of governance structure on the strategic choice of enterprises for innovation is not just limited to resources. Enterprises must take into consideration its legitimacy in the society while seeking economic benefits. Differences in governance structure have generated costs and returns on gaining resources through two different channels: relational network and market competition. Business owners and board members affected by relying on a historical path will impose different institutional pressures on the enterprise. Governance structure will also exert influences on the decision-making of enterprises in innovation.

The main contents of the research in this section include:

(1) Understanding the complex relationship between corporate governance and enterprise innovation in economic transition, especially the complex relations resulting from the multifunction of the state and government;

(2) Research on agency-related problems that might arise from state-owned shares of listed companies;

(3) Investment of individual investors, especially institutional investors in cooperative innovation and its influence on strengthening the competitiveness of enterprises.

Studies on the organizational model of enterprise independent innovation

New product development as an important form of enterprise independent innovation is the process whereby intellectual enquiry and knowledge application are integrated. However, intellectual enquiry and knowledge application are more often than not regarded as a pair of contradictions. Based on the systematic literature review of the paradox relationship and response strategy of the two, this text extends the definition of the ambidextrous organization from the "dual structure" and "dual-character structure" to the "ambidextrous capability" in an effort to answer the "hows" at a rather lower organization level, such as how to dialectically solve the paradox or contradictory relations between intellectual enquiry and knowledge application.

The main contents of the research in this section include:

(1) Paradox in innovation and several handling strategies;

(2) Research from dual structure to ambidextrous capability;

(3) Changes of the organization mode and practices of different enterprises in whole-staff innovation.

2 Chapter

The International Environment and Industrial Policies for Independent Innovation

Gu Kejian and Li Gang

Introduction

No matter what path a state might choose for its national innovation system, it will ultimately be the national innovation system with enterprises playing a leading role. The identification of the international environment by enterprise innovation system, however, must be realized through state-to-state economic and technological exchanges. Trade flow carries abundant information about market, profit and cost, and it not only underlies the decisions of enterprises in short-term production and operation but also exerts great influence on their long-term development strategy. At the new stage of globalization and the comprehensive development of the Chinese open economy, the international environment has and will continue to be a profound, broad, yet complex influence and even determine the technological innovation that is an enterprise's major long-term development strategy.

The further development of economic globalization and global economic informatization have directly strengthened the characteristic of technological dependency that has shaped after the second technological revolution, and promoted countries, especially developing countries, to pay more attention to technical progress and keep trying to improve their technologies in all sectors as they seek to reap more trade benefits. Under the conditions of the accelerated economic globalization, the technical progress driven by enterprise innovation has influenced and even determined the changes in terms of trade. These changes have shaped the international environment for enterprise independent innovation of a nation from multi-angles, including the actual cost and the opportunity cost, long-term decision-making and current decision-making, and have made it necessary to hammer out relevant industrial policies in reference to terms of trade effects of technical progress.

Thanks to the reform and opening-up since the 1980s, Chinese economic scale and structure and the openness to the outside world have witnessed a historic leapfrog, and China's economic structure is undergoing profound changes. However, the structure of China's commodity trade and its share in the national economy have differed vastly from those in most developing countries. Additionally, the technical progress and independent innovation mechanism of Chinese enterprises have a large gap between that of developed countries and are similar to the typical features of enterprises in developing countries due to the status quo of the global technology structure and the long-term evolution of China's advantageous factor supply in the international division of labor pattern.

Since the World War II, the developing world has always sounded the alarm of "trade immiserizing growth." Statistics show that the terms of trade index of developing countries began worsening around 1980 and has dropped dramatically since 1997, showing a brief improvement in the mid-1990s; in the meantime, industrialized countries have witnessed continuous improvement in this regard.

While actively moving forward with the reform and opening-up, China, as a large developing country, has been trying to implement technological innovation strategy, promote technical progress, and upgrade and increase technology contents in foreign trade, and has achieved corresponding progress. China's foreign trade volume has also expanded rapidly, making it the third largest trader in the world. To some extent, technological factors affect China's international trade activities. Yet, China's overall terms of trade and the terms of trade of different sectors have demonstrated remarkable changes during this process.

With regard to the overall terms of trade and the terms of trade of leading export industries in China, between 1995 and 2005, the fluctuations of China's finished products were basically identical to that of China's overall terms of trade. Before 1997, the terms of trade improved little. After 1997, the terms of trade kept deteriorating; before 2000, the terms of trade of the two leading industries, namely the machinery and electronic industry and the textile and garment industry, initially improved by a large margin but deteriorated after 2001, and have since improved in the recent two years.

Once we classify manufactured goods based on technology intensity, we will find that terms of trade of labor-intensive products were better than that of technology-intensive industries before 2003. Since then, the former has worsened more seriously than the latter; but the technology industry emerged with the strongest changing trend among technology-intensive industries. Its terms of trade increased in the late 1990s and have kept improving during the past two years. The high- and low-tech industries have shown a "unilateral downslide" trend, and the deterioration degree of low-tech industries is worse than that of high-tech industries.

This phenomenon goes against the original plan of improving terms of trade through technical progress, and makes developing economies doubt that they have dragged themselves into the trap. What role can technical progress play in the changes of terms of trade in a country? How does technical progress affect the changes of a nation's terms of trade, and why does the continuous technical progress of countries, including developing ones, worsen the terms of trade? All these questions need to be answered and also represent the motivation of this research project.

Literature Review

The international economic community continually introduces new variables to rectify and develop basic theoretical propositions. It also makes reference to other theories to analyze international trade, economic growth and national welfare, which have made the theoretical research on terms of trade and its changes become increasingly mature. The technical progress factor is just one of these variables. But the analyses can still be divided into three stages: the classical trade theory, the neoclassical trade theory, and the new trade theory.

Classical analysis—introduction of technical progress in Ricardo-Viner framework

Although a generally accepted viewpoint is that John Stuart Mill first analyzed the influence of technical progress on terms of trade in 1848,[1] in fact many scholars had begun to analyze this issue primitively before then.

Analyses of technical progress and terms of trade before John Stuart Mill

Ricardo (1817) used to imply that the addition of export and import price ratio might lead to the increase of enjoyment based on the theoretical assumption of trade benefits, including the increase of commodity categories and its resulting increase of enjoyment. He failed to propose taking this ratio as a term of trade. However, his theoretical assumption of "two states, two kinds of products, one production factor (labor)" laid a basic framework for the entire classical analysis. Follow-up analyses also focused on the changes of unit labor productivity, which is unit labor requirement, to analyze the influence of technical progress on the ratio of exchanging goods. Many scholars have followed these two paths to make analysis simple in this stage.

(1) The path from the perspective of the actual conditions for barter trade. Malthus (1817) initially made a clear-cut analysis of the implications of growth for terms of trade based on the barter trade framework. He focused on the analysis that terms of trade of a technologically advanced country would deteriorate if the increase of exportable products did not match the corresponding increase of foreign demand, under the conditions that the country sees increased productivity triggered by the levelling-up of technologies and the adoption of machineries. Scrope (1831) further developed this argument in 1930s and presented the rudiment of "immiserizing growth" in his analysis. However, he failed to explain

the causes by which the deterioration of terms of trade in innovation countries is bound to surpass the effect of productivity increase. The analyses and conclusions of Ellis (1825) and Sleeman (1829) during this period also belong to this category.

(2) The path from the perspective of the pure monetary conditions. Pennington (1840) focused on the analysis of the subject that mutual demand determines the terms of trade between Britain and Portugal, from a purely financial perspective. But, he neither told us the ultimate impact of technical progress on terms of trade nor examined the circumstance when the elasticity of the demand of Portugal for cloth from Britain is less than one.

It is easy to see that these two kinds of analyses have yet to clearly classify the different types of technical progress and the different industries involved[2] and differentiate the welfare effect of terms of trade changes sparked by technical progress. Apart from the brief discussion of the Single Factor Terms of Trade (SFTT) made by Malthus, other researchers only focused on the commodity terms of trade.

Analysis of the classical trade theories of John Stuart Mill and his followers

The analysis of this stage is primarily to keep correcting the imperfection in the analysis of the previous stage. The evolving path can be divided into three categories.

(1) Analyzing the terms of trade effects from technical progress itself. John Stuart Mill (1848) firstly introduced technical progress into the analysis of terms of trade in two circumstances. The first circumstance is that the technical progress of a nation creates a new product for export, and this product can replace the product manufactured by other countries at home. Driven by this increased foreign demand, the terms of trade of technologically-advanced countries has been improved. At this time, terms of trade are commodity terms of trade. The other circumstance is that the technical progress of a country cuts down the costs of existing export products. Technical progress may worsen domestic terms of trade. The differences between the changes in the ratio of domestic terms of trade and that of product costdepend on the elasticity of foreign demand for domestic products. Analysis at the moment focuses on SFTT. Edgeworth and Bastable have since proven the second circumstance.

(2) Differentiating the changes in terms of trade triggered by technical progress in different industries. Marshall is one of the pioneer researchers in this aspect. In 1903, he utilized "leading industries" and "lagging industries" to respectively represent "the export industry" and "the import competing industry" He

differentiated through case studies the changes in terms of trade sparked by two kinds of technical progress in the two different industries respectively.[3] He believed that the technical progress in leading industries would expand trade volume and the terms of trade would become worse; the technical progress in lagging industries would decrease trade volume and the terms of trade would improve.

(3) Preliminary definition of terms of trade effects of technical progress. This theoretical investigation emerged from the clarification of the definition of welfare, analyzed by Edgeworth (1903). He argued that the welfare effects at this time include not only the effects under the conditions of terms of trade remain unchanged, but also the effects related to terms of trade changes. Technologically-advanced countries are still likely to be damaged due to changes when their residents consume their own export commodities. What needs to be pointed out is that Viner (1937) clearly called for the use of SFTT in cross-time welfare analysis. But all classical scholars, including Haberler (1950), took the actual national income changes as the welfare effect of terms of trade changes in their investigations. They believed that the improvement or deterioration of terms of trade would directly increase or decrease the actual national income, that is, to increase or decrease the national welfare.

It is easy to see that all classical analyses by the end of 1950s were still made based on the Ricardo-Viner barter trade framework. The terms of trade at that time were determined by mutual demand. Technical progress was still marginalized compared with capital accumulation and population increase, and a set of systematic technical progress theories was yet to take shape.[4] This job was accomplished by Dombush, Fisher and Samuelson until 1977. They expanded the Ricardo Model to multi-commodities circumstance, and took into account identical technical progress between two countries and the international transfer of technologies with minimum cost. They argued that the former may lead to the falling of domestic relative wage and the improvement of terms of trade; the latter can also be called technological synergy that could improve the terms of trade and increase national welfare in low wage countries while high wage countries could be damaged.

Neoclassical analysis—introduction, expansion and application of the H-O model

The neoclassical trade theory which arose from the H-O Model points out that terms of trade are determined by production technology, factor scarcity and family and individual preferences. This has directly made technical progress a core of analysis and the H-O Model has become the starting point of the first stage research.

Introduction—examinations under the traditional factor proportions theory framework

Hicks (1953) is one of the forerunners of the analysis in this regard. He, "under the H-O Model assumption,"[5] broadened the feasible range of terms of trade effect of growth and brought it into the modern analysis stage. Follow-up neoclassical scholars in trade have made progress in the following four aspects:

(1) In-depth analysis of technical progress. Works in this aspect include two parts. The first part is to further differentiate technical progress and its terms of trade effects from the industry perspective. Hicks put forward that importers would benefit if improvement of productivity is export-oriented. But if technical progress happens to the import-oriented industries–technical progress happens in the industries where the importing countries are fiercely competing with other countries–it would definitely harm the factor terms of trade in exporters. Kemp (1955) and Corden et al. (1956) further perfected and evidenced this judgement. Quite a number of Chinese researchers believe, on this basis, that the deterioration of terms of trade in China is directly linked to the export-biased technical progress. Researches in this aspect include those made by Zeng Zheng, Hu Xiaohuan (2004), Wang Ping, and Qian Xuefeng (2007) et al.

The other part is to further enrich the definition of technical progress. Hicks drew on the analysis of operational cost by Mill and clearly defined the export-biased technical progress as not only the improvement in productivity but also the transportation costs and commercial technologies. The technical progress at this time was deemed by later generations as "Hicks Neutral Technical Progress."[6] Findly and Grubert (1959) presented another interpretation of neutral technical progress,[7] and pointed out that the neutral technical progress in export-based industries would result in a kind of output expansion that is "highly conducive to trade tendency," with clear-cut terms of trade effects. But Batra (1969) re-defined technical progress by following extremely biased studies of technical progress, broke it down into neutral- and intensive-factor-using and intensive-factor-saving categories,[8] and compared the different results triggered by the three kinds of technical progress through algebraic and geometric analyses.

(2) The conductive effect of the introduction of technical progress. Neoclassical scholars in trade have examined two kinds of conductive mechanisms. One

is the technical progress-product manufacturing conductive mechanism. Asimakopulos (1957) observed that the different results of terms of trade effects triggered by identical productivity changes rely on the quantity and variety of the commodities actually produced. The identical productivity growth in one country could benefit the terms of trade of a country without changes only when there are no difficulties in monetary adjustment. The other is the conduction of technical progress between trade sectors and non-trade sectors. Harrod, (1939), Balassa (1964) and Samuelson (1964) introduced non-trade sectors into the two-state model. Chinese scholar Gu Kejian (1998) improved the traditional HBS deduction and employed it into China's economic and trade practices. He used both the national sector and the foreign capital sector to replace the trade sector and the non-trade sector, and defined the former as the capital-intensive sector and the latter as the labor-intensive sector. He also presented the prices of tradable goods in each sector and designed a wage coefficient. His basic theory is that the relative price between two sectors hinges on wage changes based on productivity, and the improvement of the competitiveness of each sector, i.e. the improvement of terms of trade, has a direct positive correlation with the wage coefficient.

(3) In-depth examination of terms of trade effects triggered by technical progress. Relevant analyses are demonstrated in two aspects. The first aspect is to define the welfare effects of terms of trade changes triggered by technical progress. Its definition adopts the production theory and the utility theory. The production theory directly links the terms of trade effect to production or supply, while the utility theory directly connects the terms of trade effect to consumer surplus and consumer utility; the latter is less frequently adopted. The other aspect is to examine the conditions for the realization of the terms of trade effects. Kempe (1955) inferred that the terms of trade effects resulting from technical progress depend on the preference of residents of the nation importing and exporting goods, and the changes in price. Lerdau (1962) observed that terms of trade effects depend on consumers' response while response relies on policy, resource abundance, market scale and technological conditions, among others. This has aroused the interest of their followers to examine the domestic economic landscape. Bhagwati (1963) took the lead in studying the welfare effects of the terms of trade changes sparked by domestic distorted growth, and put forward the concept of immiserizing growth.[9]

(4) Preliminary analysis of the South-North Trade. The analysis of this issue can be traced back to that of Prebisch (1950) and Singer (1950). They observed that the allocation mechanism for the fruits of technical progress is asymmetric

between producers and consumers of the South and North. Technical progress in the manufacturing industry has increased income but cut down the price of food and raw materials in underdeveloped countries. Sarkar (2001) further evidenced and improved this standpoint under the dual economic framework.

Expansion—further interpretation under the neo-factor proportions theory framework

Above examinations seem in vain before the "Leontief Paradox." In order to simplify the analyses, one mustgo back to a single factor assumption, which seems rather unrealistic for the economy. For that matter, Jones (1970)[10] directly utilized technological differences to adjust the traditional H-O framework. Scholars, represented by Keesing (1965) and Waehrer (1966), have shaped a new theory system: the neo-factor proportions theory. This has simplified the analysis of the terms of trade effects of technical progress. As a result, an array of in-depth explanations of this issue has emerged.

(1) Analysis of the terms of trade effects of technical progress under the conditions that the technological factor differences exist. The leading analyst in this aspect is Jones (1970). He examined the influences of technological differences on the terms of trade in 1970, and believed that the labor-intensive products in a nation, even in a low-wages country, would be relatively expensive if its trade partner enjoyed relatively strong technological advantages in making labor-intensive products; it would make the country comparatively disadvantageous in the manufacturing of labor-intensive products. The technical progress in the production of export-oriented products would lead to improvement of the terms of trade.

(2) New interpretation of the issue of South-North trade. The emergence of the neo-factor proportions theory has offered a new explanation for South-North trade. Prebisch (1959) was the first to consider the differences of "technological densities."[11] He observed that, with these differences, presuming that identical technical progress or productivity enhancement happen to two countries, the improvement in the central nation would be fully reflected by the increase of the wage rate, while some part of the improvement in the periphery countries would be transferred to the central nation by decreasing their export prices. In this situation, the terms of trade in the periphery countries would deteriorate. Jones (1970) even directly utilized Vernon's "Product Cycle" theory to bring in the third kind of production factors, i.e. a host of specific technologies in line with parts of labor or capital demand.

Application I—mathematical model building

Probing into the theoretical basis for the model

(1) Complexity and diversity of theoretical examinations. The analyses in these two stages continue to use the H-O framework or its expansion, and make corrections in a more complex way. Additionally, it had been focusing on the "biased-ness" concept and had never gained a sufficient condition for the terms of trade changes driven by growth. During the analysis process, some contradictions and vagueness also exist (Takayama, 1964). To this end, economists have started to use the mathematical method to conduct model-building.

(2) Two-factor mathematical model. Meade (1951) used the H-O model framework to build a clear-cut, two-factor model mathematically, and employed it in the analysis of the international trade theory. The successful introduction of production factors has laid a building block for the introduction of technical progress factors.

Mathematical model-building

Neo-classical scholars in trade have always depended on general equilibrium theory framework in building the mathematical model and exploring the relations between technical progress, terms of trade and terms of trade effects from the demand-supply perspective.

(1) Demand equation and supply equation. The demand equation is built both directly and indirectly. Direct building of the equation is represented by Takayama (1964) while the indirect one is represented by Jones (1970). The supply equation is built strictly based on the H-O model with its major changes in the variety of production factors. The manifestation of the traditional factor proportions theory is the equation of output changes under the "two states, two factors, two products" model. Under the neo-factor proportions theory, different input and output coefficients are used to derive their respective output.

(2) Analysis of terms of trade changes triggered by technical progress and the corresponding effects. The analysis of this issue has two expressions: the expression of import-export price margin, and the mode of differential. Bhagwati is the trailblazer in this regard.

Application II—quantitative analysis

Quantitative analyses of the integrity of the three variables—technical progress, terms of trade, and terms of trade effect—are rarely seen in literature on neo-classical trade theory. They have only progressed in the following groundwork:

Definition and measurement of relevant variables

Either Paasche index or Laspeyres index is used in actual analyses of the terms of trade. The latter prevails in most cases.

Form of quantity of technical progress is represented by the overall labor productivity, marginal productivity or the Total Factor Productivity (TFP) calculated by the Solow residual.

Effects of the terms of trade changes are typically represented three different ways. The first is to attribute part of trade balance changes to the terms of trade. Stvel (1956, 1958) is the representative for this method. The second way is to measure the "national income in real terms" based on a standard that is different from actual GDP and to attribute the differences between the two to the terms of trade effects. Geary (1955) and Nicholson (1960) successively used similar formulas to define the terms of trade effects. However, this method can only make comparisons of static welfare under the conditions that technical progress does not exist. The third way is to measure and integrate actual national income, actual GDP and the terms of trade effects into a continuous accounting system for actual national income. Geary and Burge (1967) as well as Stvel (1967) offered an integrated measuring system.

Quantitative analysis of the correlations between variables

At that time, the quantitative analysis was mainly reflected by factor terms of trade. Lerdau (1962) was the first to successfully calculate productivity index, commodity terms of trade, and single factor terms of trade, and demonstrate the consistency of the changes in the productivity index and those in single factor terms of trade.

Objectively speaking, under the framework of the entire neo-classical trade theory, the theoretical examinations and quantitative analyses of technical progress are basically carried out according to the neo-classical growth theory, and have overlooked domestic demand and international transmission. They have focused on whether the effect of technical progress on production is "extremely biased" and defined technical progress as an exogenous given factor. They are also limited to certain one-off improvements of labor productivity (or TFP) and have concentrated in the meaning of wage changes resulting from productivity changes with respect to the terms of trade, but neglected other forms of sustainable technical progress. These include the growth effects of technology learning and technology spillover. As a result, the new trade theory rapidly applies some advances of new (endogenous) growth theory in the analyses of the terms of trade, and drives the closed economy to an open economy.

New trade theory analysis—the introduction and application of the endogenous technical progress theory and the macroeconomics framework of open economy

New trade theory starts to focus on the examinations of the terms of trade effects of endogenous technical progress, especially the utilization of New Keynesianism framework, or the macroeconomic framework of open economy, to study the long-term dynamic effects under the open conditions.

Theoretical progress I— expansion and endogenesis of technical progress theory

(1) Examination of product innovation-oriented technical progress. Krugman (1979) capitalized on the "product cycle" theory of Vernon (1966) to develop a South-North Trade Model and prioritized the examination of the influences of technological innovation by creating new products on terms of trade. Krugman (1989) extended his examination from the simple South-North trade to global general circumstances and made a similar conclusion. Yet, these analyses only paid attention to the influence of innovation and technology transfer and all presumed exogenous formation. Grossman and Helpman (1991) developed it into the endogenous product cycle theory and made an opposite conclusion to that of Krugman. Ruhl (2003) made attempts to build a model including the supply-demand elasticity and creating a new product, based on the analysis framework of Krugman. Gorsetti, Martin and Resenti (2005) directly incorporated the "spillover effect" of productivity into their analyses.

(2) Examinations of "technological innovation-oriented" technical progress. Schleifer (1986) and Krugman (1990) discussed the process of continuous cost innovation. Segerstrom (1990), and Aghion and Howitt (1990) creatively delivered the repetitive quality innovation theory that preliminarily shaped the framework of innovation-oriented technical progress theory. They assumed that technological innovation is carried out in order among products or that it is carried out at an economic level, which have inevitably led to difficulties in analysis. Grossman and Helpman (1991) supposed that a host of products exist, and that each made respective stochastic improvements up a quality ladder. At this time, the terms of trade effects triggered by technical progress must be connected with the geographical position of products, with different levels of quality, and a general conclusion like that of the changes in terms of trade in South-North trade cannot be made. Besides above-mentioned literature that has touched upon the terms of trade effect, no other studies have included it in their analyses.

(3) Endogenous extension of productivity-growth-oriented technical progress. Debaere and Lee (2004) extended technical progress to research and development (R&D) and suggested that productivity improvement resulting from R&D and per capital GDP growth in the countries of trade partners had positive influences on the terms of trade.

(4) In-depth examination of the welfare effects of terms of trade changes. Analyses of the new trade theory have made breakthroughs in three aspects. The first is in the examination of the inherent differences between economies. Eaton and Panagariya (1982) extended the constant return to scale to the variable return to scale; Choi and Yu (1985) introduced external economy and extended it to the circumstances of large countries. The second is in the analysis of the international welfare effects caused by the changes in the terms of trade. Studies of Dedola and Leduc (2006) have showcased the latest progress. They pointed out, on the basis of systematically generalizing recent relevant analyses, that international spillover not only relies on the terms of trade changes, but also lies in the influence of accessible product variety changes on customers' welfare when the supply of product variety is endogenous. The third is the analysis of cross-time dynamic welfare effects, especially the welfare for representative manufacturers or families. Analyses of Grossman and Helpman (1989, 1991), Krugman (1990), and Aghion and Howitt (1990) all belong to this category.

Theoretical progress II— the introduction of the open macroeconomics framework and its integration and application with the theoretical framework of endogenous technical progress

Many scholars have used the open macroeconomics framework to prioritize the analysis of the implications of endogenous technical progress for the terms of trade changes and its welfare effects.

(1) The introduction of the theoretical framework of open macroeconomics. Among the wave of research on open macroeconomics, New Keynesianism has developed the Redux model from many aspects. Thanks to the efforts of many researchers, the framework of New Keynesianism has effectively extended technical progress analysis from static balance framework to dynamic optimized general equilibrium framework, which has also provided a dynamic basis of open economy for research on the terms of trade effects of technical progress.

(2) Integration of the two theoretical frameworks. The endogenous growth theory created by Romer (1990), Grossman, Helpman, and Krugman (1985,

1989), described the characteristics of a new product and its influence on economic development, which have effectively intensified the examinations of technical progress. Under their guidance, quite a number of experts in trade, including Feenstra (1999), have analyzed the implications of introducing new products and the variety of export products for trade performance. At the same time, some macroeconomics literature have also examined enterprise dynamics and endogenous product variety; that is, to take into consideration the businesses' internal marginal choice (choosing the production scale of a given product variety) and external marginal choice (introducing a new product) and its influence on the macro-economy. This has integrated the two theoretical frameworks: the endogenous technical progress theory, and the open macro-economics.

(3) The application of the theoretical framework of open macro-economics and its micro expansion. The research results of Corsetti, Dedola and Leduc (2006) have demonstrated the latest progress in the application of this theoretical framework. They introduced factors influencing domestic demand, such as capital market and consumption risk, and examined terms of trade effects triggered by technical progress and productivity changes from short-term and long-term perspectives. Meanwhile, some international economists have tried to make micro-expansion of open macro-economics. Ghironi and Melitz (2006) brought technical progress into the international business cycle model. Studies of Corsetti, Martin and Pesenti (2005) relaxed this assumption and proved that the difference in the response of the terms of trade to productivity rests with which cost is affected.

Improvement of mathematical model

(1) The theoretical basis for the mathematical model. As previously stated, the theoretical framework of New Keynesianism and the endogenous technical progress theory have become the basis for modelling at this time.

(2) Demand equation and supply equation. Demand equation is built on the basis of two different schools of thought. One is the indirect social effect. Joy and Yu (1985) used quasi concave social effect function to describe the demand of an economy. The other is the reduced form consumption effect of representative consumers (families). This thought originated from Dixit and Stiglitz (1977), which added influential variables like exchange rate in follow-up studies. The theoretical model of the influences of product quality improvement–oriented technical progress, in terms of trade, has been following the thought of quality gradient theory.

The supply equation can be described in two ways. The first is the production function under the model of two states, two products (industries) and two factors. Joy and Yu (1985) took the lead in building the production function of variable return to scale, analyzed by terms of trade. The second is the reduced-form supply equation, including product diversity and product quality.

(3) Analysis of changes in terms of trade and the effects triggered by technical progress. At this moment, relevant analyses have been divided into two. The first is the differential mode. Under the lead of Bhagwati, followers are represented by Eaton and Panagariya (1979, 1982), Joy and Yu (1985). The second is the regression equation mode. Its mathematic expression includes reduced form and structural form; the latter added some structural coefficients on the basis of the former.

Quantitative analysis

Definition and measurement of correlated variables

Under the endogenous technical progress and macro-economics framework, the uses of the terms of trade include the international influences of supply and demand factors, including technical progress, exchange rate fluctuations, and economic structure. Kouparitsas (1994) built an index that can be used to analyze the influences of trade structure and industrial structure; Baxter (2000) introduced commodity price and country price to examine the influence of import and export commodity structure; Broda and Weinstein (2004) came up with a new definition of welfare price index by leveraging on the introduction of Product Differentiation by Lrugman (1989) and the definition of the trade net welfare effect under the background of productivity growth by Feenstra (1994); Corsetti, Dedola and Leduc (2006) even directly took the actual exchange rate with export commutation factor being the cardinal number as a similar representative of the terms of trade.

One part of technical progress is still represented by the TFP, while the other part is represented by the changes in new product quantity. Krugman (1989) used the potential export growth of exporting countries to represent the changes of product variety in the demand for import. Gagnon (2004) used the actual output growth rate of exporting countries to represent the changes of product variety in the demand for import. However, technical progress for product quality improvement has never been involved.

Regarding the effects of the terms of trade changes, different methods are used to make the definition. One method is to take the actual GNP or actual GDP as a welfare index under the conditions of dynamic open economy. The international economics community tends to settle for the actual GNP, as the actual GDP has always been questioned. Another method is to take the changes in the consumption and investment demand of the entire nation and families or individuals as the indicator for the welfare effect of the terms of trade changes. Rogoff (1995) is one of the pioneer researchers in this aspect.

Quantitative analysis of the correlations between variables

The quantitative analysis at this time is to use the actual trade data of a nation to carry out statistical tests through measuring techniques, such as Cointegration Analysis, Unit Root Test and Granger Test. Gagnon (2004) testified to the conclusion of Krugman (1989) via empirical research on American export and import. Acemoglu and Ventura (2003), Debaere and Lee (2004) also successively proved in their empirical analyses the conclusion of Corsetti, Martin and Pesenti (2005).

Brief review

It is unquestionable that numerous trade theories have continuously enriched and developed the classical trade theory since its emergence, and have successfully built the theoretical model. Yet, so far there are no effective and in-depth theoretical examinations and quantitative studies on the terms of trade and its changes in the context of technical progress, and fewer still the precise measurement of the terms of trade changes and its welfare effects triggered by technical progress, especially dynamic welfare effects. The reason behind this is that all research on these correlated variables includes a series of important hypotheses and a whole set of theoretical systems. For instance, technical progress has always been a general concept, but in effect, it not only involves productivity improvement and new product creation, but also includes the vertical innovation that is quality improvement and technological innovation. Yet, the origin of the terms of trade theory is the comparative cost (interest) theory, or the supplement and improvement of the comparative cost theory. Its basic theoretical hypothesis is the precondition for the analysis of two states and two commodities. The "two countries, two products" hypothesis can by no means realize the measuring research on bilateral trade due to the so-called "trade theory testing" in the history of economic theory (Gu Kejian, 2006).

Because of these shortcomings, it is necessary for follow-up studies to focus on the following three aspects: to build a theoretical analysis framework including macro and micro aspects; to build the concept system that is capable of correctly describing terms of trade in modern economic and trade practices to clarify the identification of technique progress; to have an in-depth exploration of terms of trade effects of technical progress and its transmission mechanism, especially the dynamic effects of the terms of trade changes. For China, facing rising import price and falling export price, the top priority is to build a supply-demand equation that analyzes the terms of trade reflecting Chinese economic transition, economic openness, and economic and trade development, and carry out data tests. Beyond that, it is also important to illustrate the practical implication of institutional factors and other factors in this Chinese economic transition.

Research Method

This text attempts to use the equilibrium analysis framework to examine terms of the trade effects of technical progress and tries to encompass the profound implications of technical progress for the terms of trade changes and its welfare effects.

Basic assumption

As noted by Schumpeter and other scholars, innovation has increasingly become a conventional and predictable process, and both industries and enterprises have routinized innovation under the pressure of competition. In particular, many developing economies, including China, have built and will keep building the platform for product innovation and technological innovation. Classic cases include the software industry of India and the electronic and mechanical industry of China. This phenomenon will become more prominent amid economic globalization. Developing countries must no longer rely solely on introducing and absorbing foreign technologies. It is necessary for them to abandon the traditional "South-North Innovation–Expansion" assumption. Rather, they must presume that any country, any industry and even any product are undergoing technical progress. What is different is the speed and the type which the state, industry and product belong to. During the analysis process, as for one country, the whole world is composed of itself and foreign countries; but as for different nations, the entire world is comprised of large countries and small ones.[12] For the industries involved in trade, the world is divided into exportable and importable industries.[13] In addition, the three kinds of technical progress in certain industries are a relative result after being compared to not only trade partners but also to another industry, i.e. "relative-relative."[14]

Definition of correlated variables

Technical progress here no longer includes spread. Rather, it is classified in line with the ultimate influence of technical progress on production, including productivity growth–oriented technical progress, technological innovation–oriented technical progress, and product innovation–oriented technical progress. Product innovation–oriented technical progress further includes the emergence of new products and the improvement of product quality.

Although Net Barter Terms of Trade (NBTT) is not the most rational explanatory index, this text will continue using it.

Origin of theoretical framework

The economics theory of Prebisch's technology dependency represents the theoretical and practical source of research for this book. The two industries assumption and the theme of the influence of economic growth on the terms of trade in Prebisch's analytical framework of "trade immiserizing growth" have further provided theoretical support for this book.

Theoretical framework setting

Since we can break technical progress down into three levels in theory, we can write down technical progress as a product form between the three. That is:

$$TP = PG^{\alpha}PI^{\beta}PCI^{\delta} \qquad (1)$$
$$\alpha + \beta + \delta = 1$$
$$1 \geq \alpha$$
$$\beta, \delta \geq 0$$

In the formula, TP, PG, PI and PCI respectively refer to the overall technical progress, productivity growth–oriented technical progress, product innovation–oriented technical progress, and technological innovation–based technical progress of a state, an industry and even a product.

When examining the terms of trade effects of technical progress, we can simplify a reduced differential mode, that is, to use the terms of trade (represented by P) to derive technical progress. Then

$$\frac{\mathrm{d}p}{\mathrm{d}TP} = \frac{\mathrm{d}p}{\mathrm{d}PG} \cdot \alpha \cdot \frac{TP}{PG} + \frac{\mathrm{d}p}{\mathrm{d}PI} \cdot \beta \cdot \frac{TP}{PI} + \frac{\mathrm{d}p}{\mathrm{d}PCI} \cdot \delta \cdot \frac{TP}{PCI}$$

$$\frac{\mathrm{d}p}{\mathrm{d}TP} \cdot \frac{1}{TP} = \frac{\alpha}{PG} \cdot \frac{\mathrm{d}p}{\mathrm{d}PG} + \frac{\beta}{PI} \cdot \frac{\mathrm{d}p}{\mathrm{d}PI} + \frac{\delta}{PCI} \cdot \frac{\mathrm{d}p}{\mathrm{d}PCI} \qquad (2)$$

The above formula tells us that the elasticity of the influence of technical progress on the terms of trade equals the total sum of the elasticity of the influence of the three kinds of technical progress on the terms of trade.

In this sense, we can eventually analyze the implications of technical progress on the terms of trade and then analyze the overall influence on this basis. However, these three kinds of technical progress belong to the supply side, and although their ultimate effects on the terms of trade must be realized through the combined function of supply and demand together, the economics community has accepted that productivity growth is the economic result of technological innovation. This calls for the logical analysis of the two kinds of innovation-based technical progress, followed by the analysis of productivity growth–based technical progress. At the same time, this text also makes use of China's data to carry out empirical research.

Basic research method

The research method in this text is to combine the normative theoretical analysis and the empirical quantitative analysis, the static and dynamic equilibrium analysis, and incorporate macro-analysis into micro-analysis.

Analysis of the Terms of Trade Effects of Technological Advance

Analysis of the terms of trade effects of technological advance based on product innovation[15]

Baumol (2004) observed that product innovation refers to the innovation whereby the demand curve of end products moves to the right, and successful product innovation can increase the output of end products. Technical progress for product innovation refers to technological changes corresponding to successfully producing innovative products and introducing them into the market.

Theoretical examinations

General theoretical analysis

We can differentiate the circumstances between large countries and small ones and discuss the terms of trade effects in the light of the industries undergoing product innovation-based technical progress through defining product innovation and reviewing the basic law of supply and demand.

(1) Circumstances in small countries

First, the analysis of terms of trade effects of product innovation–based technical progress in export industries.

Assuming that, compared to the export destination, domestic export industries (manufacturers) have seen product innovation–based technical progress, or the speed of product innovation is faster, then the innovative product manufactured domestically can effectively meet and, to a certain extent, guide the demand of the export market at the introduction stage, and the curve of demand of the export destination for the product will move up within a certain time. However, due to the fact that the supply of domestic innovative products is rigid and the nation is a price taker, the quantity supplied would not increase rapidly since domestic products for export would hit both the international market and domestic market and the country would channel more of the product to its domestic market. It can be seen in Fig. 2.1 that domestic consumption and production are relatively close to export industries. The production capacity of exportable and importable products increases from OA and OQ to OB and OT respectively. The consumption demand for importable and exportable products rises from OP and OS to OP_1 and ON, respectively. Obviously, the consumption of exportable products at home grows faster than that of importable products, meaning domestic import quantity increases more than the export quantity. At this moment, line UR is evidently steeper than line HM. That means the domestic terms of trade improves, but relatively little.

Once the innovative product becomes mature, domestic export supply will grow faster compared with that at the introduction stage. At the same time, the domestic demand of exportable products will rise accordingly and grow more than that of importable products. That means the difference between the growth rate of the domestic import quantity and that of the export quantity is greater than that at the introduction stage. At this moment, line QS is steeper than line RM (see Fig. 2.2), the domestic terms of trade improves and it is larger than that at the introduction stage.

However when innovative products enter into the standard (recession) stage, the rigid supply of small countries makes the supply of exportable products continue to increase. But consumption happens more to import industries, and the

Fig. 2.1.   **Introduction stage of innovative product of export industries in small countries**

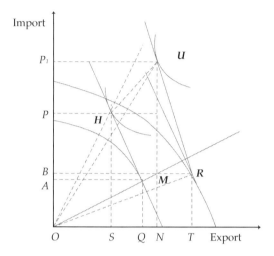

Fig. 2.2.   **Mature stage of innovative product of export industries in small countries**

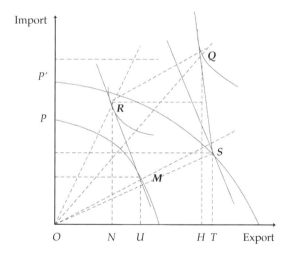

deterioration of the domestic terms of trade is reasonable. Fig. 2.3 shows that line *UH* becomes flatter than line *RM*. This indicates that product innovation–based technical progress in export industries of small countries must be the normality.

Second, analysis of the terms of trade effects of product innovation–based technical progress in export industries.

Assuming that, compared to the country for the source of import, domestic

Fig. 2.3. **Standard (recession) stage of innovative product of export industries in small countries**

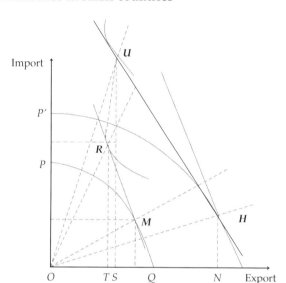

import industries (manufacturers) have seen product innovation–based technical progress (or the speed of product innovation is faster), then the innovative product can effectively meet and, to a certain extent, guide the domestic market demand at the introduction stage, and the curve of domestic demand for the product will move up. Butas the nation is the price taker in the international market and the price maker in domestic market, the supply of domestic innovative products enjoys elasticity in the domestic market and can gradually cut down the import demand. Therefore, as is shown in Fig. 2.4, the production quantity of domestic importable and exportable products rise respectively from OB and OQ to OA and OT; and the consumption quantity rises from OS and OP to ON and OP_1 respectively. That drives up the import prices compared with the export prices. Line UR is steeper than line HM, which means the domestic terms of trade improves, but relatively little in degree. The improvement depends on the capability of the innovative product to replace the former import product and the domestic market acceptance.

Once the innovative product becomes mature, the consumption quantity of domestic importable products at home will surge, while domestic import quantity will show a downward trend and the rate of decline will be faster than the growth rate of the export quantity. At the same time, the import prices will increase compared with the export price. Line UR is flatter than line HM (see Fig. 2.5), and domestic terms of trade will start to worsen.

Fig. 2.4.   Introduction stage of innovative product of import industries in small countries

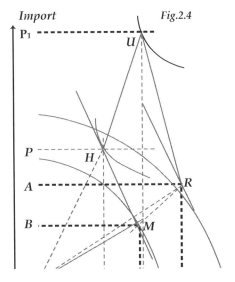

Fig. 2.5.   Mature stage of innovative product of import industries in small countries

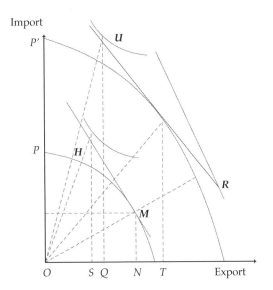

When the innovative product enters into the standard (recession) stage, the rigid supply of small countries will increase supply. But consumption has become saturated; that will result in further shrinking of the country's import quantity, while the export quantity will keep escalating and the terms of trade will keep worsening.

(2) Circumstances in large countries

First, analysis of terms of trade effects of product innovation–based technical progress in export industries.

Assuming that, compared to the export destination, domestic export industries (manufacturers) have seen product innovation–based technical progress or the speed of product innovation is faster than that in the export destination, then the innovative product manufactured domestically can effectively meet and, to a certain extent, guide the demand of the export market at the introduction stage, and the curve of demand from the export destination for the product will move up within a certain time. In addition, the supply of domestic innovative products is elastic (as the nation is a price maker); thus, domestic products for export hit the international market and domestic market at the same time. As a price maker, the nation will channel more of the product to the international market and its supply quantity will be determined by itself, according to the capacity of its followers in supply. As is shown in Fig. 2.6, domestic consumption and production are relatively close to export industries; the production capacity of exportable and importable products increases from OA and OQ to OB and OT respectively; the consumption demand for importable and exportable products rises from OP and OS to OP_1 and ON respectively. Obviously, the consumption of exportable products at home grows faster than that of importable products, which means domestic import quantity increases more than the export quantity. Line UR is steeper than line HM, which means domestic terms of trade improves, but relatively little in degree.

Once the innovative product becomes mature, then domestic export supply grows faster than that at the introduction stage. At the same time, the domestic demand for export goods also experiences a rising growth rate which increase far more rapidly than the growth rate of the demand for import goods. In essence, the difference in the growth rate between the import quantity and the export quantity is bigger than that at the introduction stage. At the same time, line QS is steeper than line RM (see Fig. 2.7), which means that the domestic terms of trade improves and the degree of improvement is larger than that at the introduction stage.

However, when the innovative product enters into the standard (recession) stage, the supply elasticity of large countries will decrease the supply of its exportable goods accordingly. Domestic consumption of this innovative product will go up and the export quantity will drop faster than the growth rate of the import quantity. In fact, the same situation appears in the introduction stage (see Fig. 2.8) and its terms of trade also improve, whereas the degree of improvement

Fig. 2.6. **Introduction stage of innovative product of export industries in large countries**

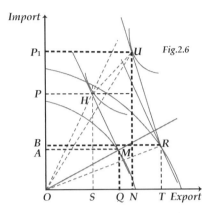

Fig. 2.7. **Mature stage of innovative product of export industries in large countries**

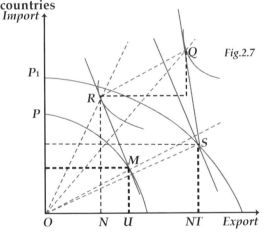

relies on the effective market demand at home. That implicates that technical progress for product innovation shall not just focus on the export market; rather, it shall attach more importance to the cultivation of domestic demand.

Second, analysis of the terms of trade effects of product innovation–technical progress in export industries.

Assuming that, relative to the country of the source of import, domestic import industries (manufacturers) have seen product innovation–based technical progress (or the speed of product innovation is faster), then the innovative product can effectively meet and, to a certain extent, guide domestic

Fig. 2.8. **Standard (recession) stage of innovative product of export industries in large countries**

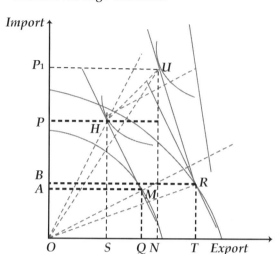

market demand at the introduction stage, and the curve of domestic demand for the product will increase. But as the nation is the price marker in both the international market and the domestic market, the supply of domestic innovative products is elastic. Therefore, as is shown in Fig. 2.9, the emergence of the innovative product in import industries can greatly influence its demand for importable products and the import demand fluctuates dramatically. The production quantity of domestic importable and exportable products rises respectively from OB and OQ to OA and OT and the consumption quantity rises from OS and OP to ON and OP_1 respectively. The import price drops compared with the export prices., This means that the domestic terms of trade improves and the degree of improvement is larger compared with that of small countries; after all, the influence of the domestic demand of large nations on the world demand should not be underestimated.

Once the innovative product matures, the difference between the production and consumption of domestic importable goods will narrow significantly and the domestic import quantity will contract remarkably. That will lead directly to a sharp decline of the import price and the rate of decline will be faster than the growth rate of the export quantity. Line UR is steeper than line HM (see Fig. 2.10), which means the domestic terms of trade further improves.

When the innovative product enters into the standard (recession) stage, the supply elasticity of large countries will decrease the supply and even change it into an exportable product. This can maintain the low import price and even create a high export price, which keeps improving the terms of trade.

Fig. 2.9. Introduction stage of innovative product of import industries in large countries

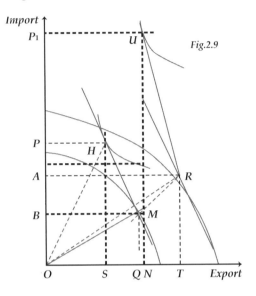

Fig. 2.10. Mature stage of innovative product of import industries in large countries

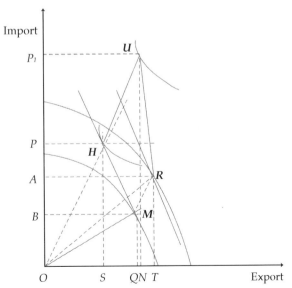

It is easy to see that the above analyses are all made under the conditions of a normal product. But technical progress caused by product innovation has strongly changed the inherent attribute of the goods mutually exchanged

between countries, especially between developed and developing countries. More new "inferior goods" and corresponding "normal products" have emerged. As a result, it is necessary at this point to introduce the inferior products.

Extension of the basic analytical framework of terms of trade effects: inferior products

(1) Further interpretation and introduction of inferior commodities

Most economists are satisfied with the explanation of inferior goods by Hicks (1946), but this explanation fits well to small-scale production (Gould, 1981). Some scholars have made regressions of the consumer theory and made further explanations. Gould believes that, during the actual consumption process, consumers also face some other restrictions that cannot be reflected by standard effects, such as consumer's preference and non-budget restrictions, as well as the non-cooperation between two kinds of products consumed, i.e. the increase of one kind of product in consumption will disrupt the consumption of another product. Garratt (1997) introduced the inseparability and separability of commodities and observed that separability can easily give rise to inferior goods while inseparability might result in discrete purchases. An important implication of technical progress resulting from product innovation for the economy is that it has added to the diversity and difference of commodities. It can directly promote and add to non-cooperativeness, inseparability and non-budgetary limitations of products.[16] For that matter, we must introduce inferior commodities when analyzing terms of trade effects triggered of product innovation–based technical progress.

(2) Equilibrium analysis and graphic analysis based on the Rybzinski theory

The following analyses are mainly to examine the influential mechanism among technical progress, commodity attribute, relative demand, and terms of trade, based on the above analytical framework.

Circumstance: "Relative-relative" product innovation–based technical progress happens domestically.

If this kind of "relative-relative" technical progress happens to domestic export industries, both domestic and foreign demand for new products become stronger and these types of new products become inseparable, such as a finished automobile and its complete equipment. The original competing products of the same kind, such as automobile parts, equipment parts and components will then become inferior commodities. At this moment, the market demand for export industries continues to expand, but the supply for the international market is much lower than the demand, which drives up the price of export products; meanwhile, domestic demand for export products keeps falling, decreasing the import price and improving the terms of trade. In contrast, if this kind of technical progress happens to the export industries of a nation under the same conditions, then it will directly increase domestic demand for the export products manufactured at home, while domestic demand for foreign commodities will drop in relation. If the country is a large, international, economic power, the import price will decrease; if it is a small nation, the import price will basically remain unchanged or drop moderately. At the same time, domestic demand for exportable products will decline while domestic export supply will increase in relation. If it is a small economy, the export price will decrease dramatically; if it is a large nation, then the export price will basically remain unchanged. Accordingly, the terms of trade will improve if it is a large country or worsen if it is a small economy.

Fig. 2.11 demonstrates the circumstances of a small nation: the horizontal axis represents the exportable produces of a nation, the vertical axis refers to the importable products of a nation, PP is the possible production curve of the nation, and R refers to its consumption indifference curve. After export industries witness the creation of a new product and product quality improvement–based technical progress, the new possible production curve changes into $P'P'$; this technical progress has brought about "inferior commodities" in exportable products and original importable goods. Yet, under the circumstances of small countries, the export quantity increases faster than the shrinking speed of the import quantity. That means the new consumption indifference curve is on the upper left of the URT interval, the consumption point is Q, the new possible production point is H, the new terms of trade line is QH, and the original terms of trade line is RM. Fig. 2.11 tells us that line RM is steeper than line QH, which means the terms of trade in small countries will worsen.

Fig. 2.12 indicates the circumstances in large countries: the horizontal axis represents the exportable products of a country, the vertical axis refers to the importable goods of a country, PP refers to the possible production curve of the country, and R refers to the consumption indifference curve. After import industries witness the creation of a new product and product quality improvement–based technical progress, the new possible production curve changes into $P'P'$; this technical progress has brought about "inferior commodities" in exportable products and original importable goods. Yet, under the circumstances in large countries, the export quantity increases at a slower rate than the shrinking speed of the import quantity. Therefore, the new consumption indifference curve is on the upper left of the URT interval, the

Fig. 2.11. *R* theorem expansion—technical progress in import industries of small countries

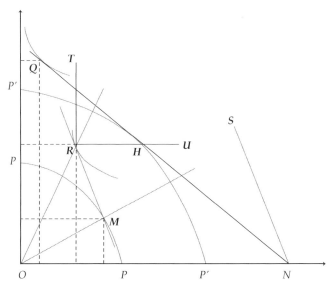

Fig. 2.12. *R* theorem expansion—technical progress in import industries in large countries

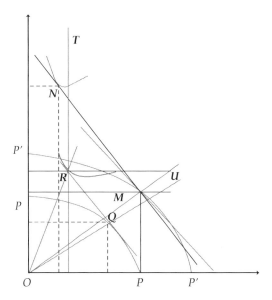

consumption point is *N*, new possible production point is *M*, the new terms of trade line is *MN*, and the original terms of trade line is *RQ*. Fig. 2.12 tells us that line *MN* is steeper than line *RQ*, which means the terms of trade in large countries will improve.

To be sure, this analysis framework also applies to the circumstances that trade partners witness "relative-relative" product innovation-based technical progress and the results will be vice-versa.

The practical examination of China: empirical test based on the panel data of relevant industries

Empirical model and data

Studies in this section adopt the fixed effect model in balanced panel data to make analyses from different industries and use a practical, generalized least square method to estimate with the purpose of reducing the heteroskedasticity influenced by sectional data.

(1) Empirical model

In order to simply analyze the influences of product innovation–based technical progress in different industries on export price and terms of trade, we only use the varying-coefficient model for the certain effects of a single variable,[17] and we will not carry out F test of the sample data, considering the differences in the export price and the terms of trade between different industries. Rather, we directly use the variable intercept model , built as:

$$TOT_{ids} = a_{ids}\, PR_{ids} + u_{ids}$$
$$EP_{ids} = a_{ids}\, PR_{ids} + u_{ids} \qquad (3)$$

In the formulas, *TOT*, *EP* and *PR* respectively represent terms of trade index, export price index and the actually approved patent mathematical index; subscript *ids* refers to different industries. Research in this section includes the entire nation, labor and resource intensive industry, low-tech industry, middle-tech industry, high-tech industry, the textile and garment industry and the electronic and mechanical industry.[18]

(2) Data specification

Data in this section are all from *China Customs Statistics Yearbook 1995–2005* and *China Science and Technology Statistical Yearbook 1995–2005*. The data frequency is one year and they are all in the form of index. The time interval is between 1995 and 2005 and the year of 1995 is taken as the base year.

Patent grant number index: the patent grand number in this section is from the annual data published by China Statistical Yearbook on Science and Technology 1995–2005. It includes the invention and utility model patent, and its classification adopts the internationally agreed standard of eight categories. After obtaining the index for the number for different industries, the year of 1995 is defined as the base year with the index at 100, and other years are presented correspondingly.

Export price index: the export price index in this section is from *China Customs Statistics Yearbook 1995–2005*. Based on the calculation method of the standard import price index that eliminates all price fluctuation data between the years, except the data between 0.5–3.0, the export price indices are all represented in the form of Fisher Index.[19] The base year is also defined as 1995 and other years correspond in order.

Terms of trade index: the basic calculation method of the terms of trade index and data source in this section includes the calculation of the import price index as with the export price index. It ultimately takes the ratio of the export price index to the import price index as the terms of trade index.

Empirical results

(1) Effects model of product innovation–based technical progress on export price

After the processing of EVIEWS5.0 software, we can divide the basic model results into seven parts for demonstration:

Effects model on labor and resource-intensive industries:

$$EP\_LIP = 7.619\ 966 + 0.047\ 014 \cdot PR\_LIP \tag{4}$$

Effects model on low-tech industries:

$$EP\_LTP = -13.545\ 10 + 0.179\ 709 \cdot PR\_LTP \tag{5}$$

Effects model on middle-tech industries:

$$EP\_MTP = -1.637\ 304 + 0.130\ 090 \cdot PR\_MTP \tag{6}$$

Effects model on high-tech industries:

$$EP\_HTP = -5.442\ 440 + 0.111\ 662\ 2 \cdot PR\_HTP \tag{7}$$

Effects model on the electronic and mechanical industry:

$$EP\_MEP = 4.273\ 630 + 0.111\ 611\ 8 \cdot PR\_MEP \tag{8}$$

Effects model on the textile and garment industry:

$$EP\_TP = 10.456\ 15 + 0.036\ 885 \cdot PR\_TP \tag{9}$$

Effects model on the entire nation:

$$EP\_AP = -1.724\ 811 + 0.095\ 178 \cdot PR\_AP \tag{10}$$

The measuring result of the overall model signals that the adjusted coefficient of determination reaches 0.992 353, indicating that the model's goodness of fit is quite high and the *D.W* test value is 2.134 606. This proves that the residual error is absence of serial autocorrelation. It is obvious that the model effect is sound.

When it comes to the classification model of different industries, the measuring result signifies that the T test values of the effects model on the labor and resource intensive industries and the textile and garment industry are 0.755 111 and 0.469 357 respectively, and the concomitant probabilities are 0.453 5 and 0.640 7 respectively. The fitting effect of these two models is relatively poor; the T test values of the effects model on the low-tech industry, middle-tech industry, high-tech industry, electronic and mechanical industry, and on the entire nation are 2.385 704, 2.852 797, 4.231 307, 3.279 439, and 2.469 989; the concomitant probabilities are 0.020 7, 0.006 2, 0.000 1, 0.001 8, 0.016 8. The fitting effect of these models is rather strong and relevant coefficients are outstanding at the levels of 98%, 99%, 100%, 100%, and 99%.

(2)The effects model of product innovation–based technical progress on terms of trade

After the processing of EVIEWS5.0 software, we can divide the basic model results into seven parts for demonstration:

Effects model on labor and resource intensive industries:

$$TOT\_LIP = 7.952\ 766 + 0.105\ 660 \cdot PR\_LIP \tag{11}$$

Effects model on low-tech industries:

$$TOT\_LTP = -8.456\ 163 + 0.175\ 657 \cdot PR\_LTP \tag{12}$$

Effects model on middle-tech industries:

$$TOT\_MTP = 2.134\ 052 + 0.103\ 185 \cdot PR\_MTP \tag{13}$$

Effects model on high-tech industries:

$$TOT\_HTP = -8.008\ 325 + 0.088\ 974 \cdot PR\_HTP \tag{14}$$

Effects model on the electronic and mechanical industry:

$$TOT\_MEP = 7.586\ 220 + 0.069\ 706 \cdot PR\_MEP \tag{15}$$

Effects model on the textile and garment industry:

$$TOT\_TP = -0.143\ 274 + 0.161\ 444 \cdot PR\_TP \tag{16}$$

Effects model on the entire nation:

$$TOT\_AP = -1.065\ 275 + 0.096\ 989 \cdot PR\_AP \tag{17}$$

With regard to the overall model effects, the measuring result signals that the adjusted coefficient of determination reaches 0.994 057, indicating that the model goodness of fit is quite high and the $D \cdot W$ test value is 2.113 002.

This proves that the residual error is the absence of serial autocorrelation. It is obvious that the model effect is good.

When it comes to the classification model of different industries, the measuring result signifies that the T test values of the effects model on the labor and resource intensive industry, the electronic and mechanical industry, and the textile and garment industry are 1.475 223, 1.839 630 and 1.837 278 respectively, the concomitant probabilities are 0.146 1, 0.071 4 and 0.071 8 respectively. The fitting effect of these three models is relatively poor; the T test values of the effects model on the low-tech industry, middle-tech industry, high-tech industry, and on the entire nation are 2.222 097, 2.061 621, 2.832 187 and 2.290 174, and the concomitant probabilities are 0.030 6, 0.044 2, 0.006 5 and 0.026 0. The fitting effect of these models is rather strong and relevant coefficients are outstanding at the levels of 97%, 96%, 99%, and 97%.

Brief summary

(1) From the perspective of either the overall landscape or of the six different industries, product innovation–based technical progress has a positive correlation with export price and terms of trade, and related coefficients of the whole model are all at about 10%. Product innovation–based technical progress should become a major channel for the improvement of China's terms of trade.

(2) When it comes to China's leading export industries, such as the labor and resource industry, the electronic and mechanical industry, and the textile and garment industry, product innovation–based technical progress does not have an outstanding influence on export price and terms of trade. This partially offers us two inspirations: first, the source of the comparative advantage of these industries remains to be the cheap labor force due to the effect of Factor Price Insensitivity Theorem; second, China needs to work harder to promote the product innovation–based technical progress in these industries, especially to speed up the building of an immediate feedback mechanism between domestic product innovation and international market demand.

(3) When we look at China's leading technology-based industries, we can find that the product innovation–based technical progress in high-tech industries, middle-tech industries and low-tech industries can make a great difference in export price and terms of trade. Yet, one phenomenon requiring attention is that related coefficients between the two types of the model in low-tech industries is the highest, and the two are basically the same, followed by high-tech industries and then middle-tech industries. This adds to the evidence of the standpoint of Gu Kejian (2000) that China's low-tech industries

and high-tech industries are more competitive than its middle-tech industries in export. This also reminds us that the terms of trade in low-tech industries and high-tech industries must rely on proprietary product innovation for further improvement and can give play to the advantages of a latecomer and the technological leapfrog effect, while middle-tech industries must enhance their competitiveness in export through integrating the bring-in strategy with independent innovation.

Analysis of the terms of trade effects of technological advance based on process innovation

Baumol (2004) observed that process innovation is the innovation to push the relevant cost curve down.Successful process innovation can add to output, slash product price and increase welfare. Technological advance of process innovation refers to technological changes that happened during the process innovation; process innovation is the outcome of technological advance of process innovation. Following this definition, we introduced the cost-plus pricing theory, based on the basic framework of technological advance of process innovation–cost–price–terms of trade and the circumstances in large and small countries, and will discuss the terms of trade effects in line with different industries.

General theoretical analysis

Circumstances in small countries

(1) Analysis of terms of the trade effects of the technological advance based on process innovation in export industries

As is shown in Fig. 2.13, if we presume that "relative-relative" technological advance of process innovation happens to a domestic export industry (manufacturer) of a nation compared with the export destination, but the country remains to be a price taker, then it is impossible for the industry to increase the cost-plus proportion. Itmay even take the initiative to decrease the proportion in pricing regardless of the stage: just starting to introduce the product into the international market, or the stage of exporting the product in a large-scale, or the scale stable and recession stage. This makes the export unit price of the industry basically stable and even decrease within a certain range. Against this background, domestic terms of trade remain basically stable and even deteriorate to a certain extent.

Fig. 2.13.   Analysis of the terms of trade effects of technological advance of process innovation in export industries in small countries

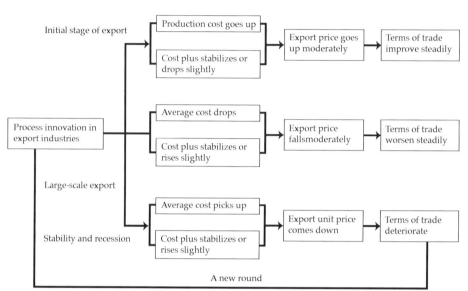

(2) Analysis of terms of trade effects of technological advance based on process innovation in import industries

As is shown in Fig. 2.14, if we presume that "relative-relative" technological advance of process innovation happens to a domestic import industry (manufacturer) of a nation compared with the source of import, but the country remains to be a price taker, then it is impossible for the industry to increase the cost-plus proportion. It may even take the initiative to decrease the proportion in pricing regardless of the stage. This makes the unit price of the product and even of the industry basically stable or even decrease within a certain range, which will attract domestic consumers from the international market to domestic market. It might enter a virtuous cycle of price fall–quantity demand rise–process innovation–price fall, which will make the shrinking demand for import and the falling import price reasonable. At this moment, domestic terms of trade will improve gradually.

Circumstances in large countries

(1) Analysis of terms of trade effects of the technological advance based on process innovation in export industries

As is shown in Fig. 2.15, if we presume that "relative-relative" technological advance of process–innovation happens to a domestic export industry (manufacturer) of a nation compared with export destination, and the country

Fig. 2.14. **Analysis of the terms of trade effects of technological advance of process innovation in import industries in small countries**

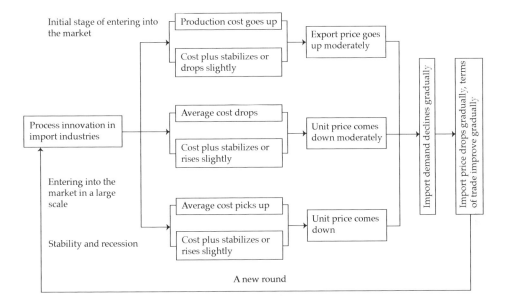

Fig. 2.15. **Analysis of the terms of trade effects of technological advance of process innovation in export industries in large countries**

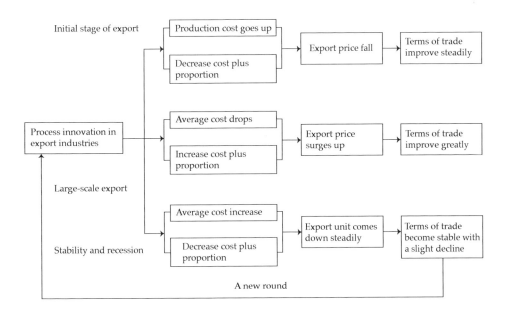

remains to be a price maker, then regardless of which stage it is, it can slightly decrease or dramatically increase the cost-plus proportion and make it bigger than the increase rate or decrease rate of costs. This makes the export unit price of the industry keep growing steadily. To make that happen, within this context, domestic terms of trade should keep improving. Even if temporary deterioration happens, terms of trade are expected to reverse the trend rapidly, driven by a new round of process innovation.

(2) Analysis of terms of trade effects of technological advance based on process innovation in import industries

As is shown in Fig. 2.16, if we presume that "relative-relative" technological advance of process innovation happens to a domestic import industry (manufacturer) of a nation compared with the source of import, and the country remains to be a price maker, then regardless of its stage, it can slightly decrease or dramatically increase the cost-plus proportion in pricing so that it can lower the import price to become monopolistic. At this moment, the import unit price will keep dropping. Once the product hits the market in a large scale, it can gradually increase the cost-plus proportion and make the proportion bigger than that of the decrease of cost. This makes the unit price of the product and the industry rise relatively and makes it higher than that of the foreign imported

Fig. 2.16.   Analysis of the terms of trade effects of technological advance of process innovation in import industries in large countries

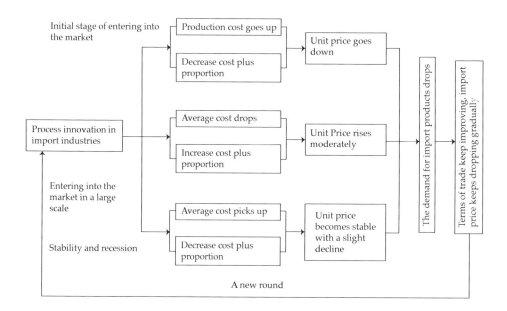

product. As foreign imported products have been driven out of domestic market by domestic innovation products over time, this demonstration effect can be used to guide other manufacturers and industries to step up efforts in process innovation, which will keep pressurizing the import price into going down. In this case, domestic terms of trade will keep improving during this process.

China's practical examination: empirical test based on the panel data of China's industries

Empirical model and data

(1) Balanced panel model based on data of listed companies

This study still adopts the fixed effect model of the balanced panel data to make an analysis of 12 different listed foreign trade companies. At the same time, the study also employs a workable broad sense least square method to make its estimations.

In order to simply analyze the influences of the technological advance of process innovation in different public companies on the cost and price of products, we respectively use the varying coefficient model and varying intercept model to accomplish a certaineffect of a single variable.[20] Taking into consideration the differences in export price, cost and price between different companies, we will not carry out F test of sample data. Rather, we will directly use the varying intercept model. The model is built as:

$$ROCP_{ids} = a_{ids}\,RONM_{ids} + u_{ids} \tag{18}$$

$$ROCP = a\,RONM + u$$

In the formula, $ROCP$ and $RONM$ represent the ratio of cost to price and the proportion of newly added equipment in the company respectively,[21] and subscript ids refers to different listed companies. Research objectives of this section include G Sinochem (600500–ZH), Orient International Enterprise (600278–DF), G Zongda (600704–ZD), Ruyi Group (000626_RY), CNTIC Trading Co., Ltd. (600056_ZJ), Suntime International Economic-Trading Co., Ltd. (600084_XT), G Lansheng (600826_LS), Nanfang Textile (600250_NF), Shanghai Commodity Trade Building (600822_WM), Xiamen Xindeco Ltd. (000701_XM), G Jilin Wuhua Group Co., Ltd. (600247_WH) and Orient Group (600811_DG).

(2) Data explanation

Data in this section are all taken from the 2002–2004 annual report of the 12 listed companies and the Genius Database. The data frequency is year and all adopt relative indices, i.e. in the form of proportion.

Ratio of cost to price: the ratio of cost to price in this section is taken from data of the gross profit margin of listed companies provided by the Genius Database. After obtaining the indices, we use 1 to reduce it and obtain the ratio.

Proportion of newly added equipment: as it is difficult to access the data of R&D input of Chinese listed companies, we adopt the data of newly added equipment. The newly added equipment data in this text are all learned from the 2002–2004 annual report published by the above-mentioned 12 listed companies. They are calculated by using the total of the newly added equipment published in the annual reports to divide the amount of equipment at the end of the previous year. The equipment hereby includes mechanical equipment and electronic equipment.

Empirical results

(1) Varying-coefficient model of certain effect

After the processing of the EVIEWS5.0 software, we can divide the results of the basic model into 12 sections for presentation.

$$ROCP\_WM = -0.041\,351\,RONM\_WM + 8.026\,045 \tag{19}$$
$$ROCP\_LS = 0.075\,709\,RONM\_LS + 3.128\,189 \tag{20}$$
$$ROCP\_NF = -0.000\,815\,RONM\_NF + 3.795\,187 \tag{21}$$
$$ROCP\_DF = -0.244\,923\,RONM\_DF + 0.829\,270 \tag{22}$$
$$ROCP\_RY = 0.180\,662\,RONM\_RY + 3.545\,954 \tag{23}$$
$$ROCP\_XM = -0.059\,966\,RONM\_XM + 7.556\,283 \tag{24}$$
$$ROCP\_WH = -0.196\,488\,RONM\_WH - 31.980\,12 \tag{25}$$
$$ROCP\_XT = -0.173\,535\,RONM\_XT + 0.767\,199 \tag{26}$$
$$ROCP\_ZJ = -0.000\,826\,RONM\_ZJ + 7.948\,601 \tag{27}$$
$$ROCP\_DG = -0.221\,284\,RONM\_DG + 1.184\,481 \tag{28}$$
$$ROCP\_ZD = 0.470\,867\,RONM\_ZD - 3.457\,250 \tag{29}$$
$$ROCP\_ZH = 0.139\,151\,RONM\_ZH - 1.343\,840 \tag{30}$$

(2) Varying-intercept model of certain effect

After the processing of the EVIEWS5.0 software, we can divide the results of the basic model into 12 sections for presentation.

$$ROCP\_WM = -0.001\,138\,RONM\_WM + 8.505\,403 \tag{31}$$

$$ROCP\_LS = -0.001\ 138\ RONM\_LS + 4.894\ 712 \tag{32}$$

$$ROCP\_NF = -0.001\ 138\ RONM\_NF + 4.846\ 58 \tag{33}$$

$$ROCP\_DF = -0.001\ 138\ RONM\_DF + 0.363\ 036 \tag{34}$$

$$ROCP\_RY = -0.001\ 138\ RONM\_RY + 5.704\ 689 \tag{35}$$

$$ROCP\_XM = -0.001\ 138\ RONM\_XM + 5.953\ 918 \tag{36}$$

$$ROCP\_WH = -0.001\ 138\ RONM\_WH - 35.986\ 33 \tag{37}$$

$$ROCP\_XT = -0.001\ 138\ RONM\_XT - 5.273\ 472 \tag{38}$$

$$ROCP\_ZJ = -0.001\ 138\ RONM\_ZJ + 9.120\ 854 \tag{39}$$

$$ROCP\_DG = -0.001\ 138\ RONM\_DG - 3.093\ 488 \tag{40}$$

$$ROCP\_ZD = -0.001\ 138\ RONM\_ZD + 2.951\ 219 \tag{41}$$

$$ROCP\_ZH = -0.001\ 138\ RONM\_ZH + 2.012\ 878 \tag{42}$$

Simply looking at the overall effect of the model, we can find from the measuring results that the adjusted coefficients of determination have reached 0.999 957 and 0.999 873, signifying that the model goodness of fit is rather high. The $D.W$ test values are 2.375 633 and 3.190 358 respectively, testifying the absence of series autocorrelation of residual error. It is clear that the model effect is good.

When we look at the varying-intercept model of all listed companies, the measuring results indicate that the regression coefficient is outstanding at the level of 100% and the related coefficient is −0.001 138. That means one percentage point of growth in the proportion of the newly added equipment of these companies means that the ratio of cost to price will fall 0.1 percentage point accordingly; the price rises compared to the cost.

When it comes to the variable coefficient model of listed companies, the measuring results demonstrate that the T test values of the influence model of Orient International Enterprise (600278_DF), Xiamen Xindeco Ltd. (000701_XM), G Jilin Wuhua Group Co., Ltd. (600247_WH), Suntime International Economic-Trading Co., Ltd., (600084_XT), CNTIC Trading Co., Ltd (600056_ZJ), Orient Group (600811_DG) are −4.676 640, −4.957 474, −2.924 601, −2.492 823, −4.264 548 and −3.527 376 respectively and the concomitant probabilities are 0.000 9, 0.000 6, 0.015 2, 0.031 8, 0.001 7 and 0.005 5. The goodness of fit of these six models is comparatively good; related coefficients are outstanding at the levels of 100%, 100%, 98%, 97%, 100% and 99%, and related coefficients are all negative values. The goodness of fit of the other six companies is poor and the concomitant probabilities are generally more than 5%, with some even more than 76.49%, and most of related coefficients (2/3) are positive values.

Brief summary

(1) From either the varying-intercept model or the variable-coefficient model, we can see that technological advance of process innovation has a prominent and negative correlation with export price. That means process innovation can become an important channel to cut down product cost, increase the price of export products, and further improve terms of trade.

(2) The varying-intercept model of the 12 listed companies shows that the related coefficients of Orient International Enterprise and Oriental Group are as high as –25% and –22% respectively, and the concomitant probability is 0. This indicates that the technological advance of process innovation in high-tech enterprises has more remarkable influence on cost price. A rational explanation for this phenomenon is that process innovation requires enterprises to have stronger strength in technology and it cannot exert an influence on the business's economic performance without the support of a relevant technology platform. To make that happen, accelerating the building of the platform for enterprise process innovation is the priority for business technological innovation, otherwise, it will lose its economic significance.

Analysis of the terms of trade effects of technological advance based on productivity growth

General theoretical examination: HBS deduction and the exchange rate transmission mechanism

At this point, a corresponding improvement and expansion are made based on HBS theory and the theoretical framework of exchange rate transmission.

Definition of correlated variables

(1) Definition of two industries or sectors

The basic assumption of this section is that a country has two industries: the export industry and the import industry; the demarcation between these two industries is still 10% of the export volume, and we can say for certain that the growth rate of labor productivity in these export industries is faster than that in import industries.

(2) Definition and measurement of technical progress caused by productivity growth

This analysis follows the trains of thought of Gu Kejian (1998) and adopts the all-personnel labor productivity. The general computational formula is:

All-personnel labor productivity = Industrial added value/Average number of all employees

(3) Measurement and calculation of the exchange rate in real terms

The means of calculation adopted for the actual exchange rate is:

Exchange rate = Nominal exchange rate × Foreign price index/Domestic price index

Basic theoretical model

We assume that the economy of a nation is composed of the export industry and the import industry and the two sectors meet the following Cobb-Douglas production function form:

$$Y_E = A_E F (K_E, L_E) = A_E K^a_T L^{1-a}_T$$
$$Y_I = A_I F (K_I, L_I) = A_I K^b_I L^{1-b}_I \qquad (43)$$

In the formula, subscript E and I represent the export industry and import industry; Y, A, K and L represent output, labor productivity, capital and labor respectively; $0 < b < a < 1$, means that both the export industry and import industry have high capital density.

Representative manufacturers of these two industries face the following problems of maximum:

$$max \ \sum_{S=1}^{\infty} \left(\frac{1}{1+r} \right)^{s-1} [P_E A_{E,S} F (K_{E,S}, L_{E,S}) - \omega_S L_{E,S} - VK_{E, S+1}]$$
$$max \ \sum_{S=1}^{\infty} \left(\frac{1}{1+r} \right)^{s-1} [P_I A_{I,S} F (K_{I,S}, L_{I,S}) \ \omega_S L_{I,S} - VK_{I, S+1}] \qquad (44)$$

In the formula, $VK_{C,S+1} = K_{C,S+1} - K_{C,S}$, $C = E, I$ (no depreciation); the price of exportable products and importable products are P_E and P_I. If $k_E = K_E/L_E$, $L_I = K_I/L_I$, the capital and labor marginal conditions meeting the aforesaid conditions of maximum include:

$$r = P_E \ A_E \ a \ k^{a-1}_E$$
$$r = P_I \ A_I \ a \ k^{a-1}_I$$
$$w = P_E \ A_E \ (1-a) \ k^a_E$$
$$w = P_I \ A_I \ (1-a) \ k^a_I \qquad (45)$$

In the formula, r, w, k_E, k_I respectively refer to marginal return on capital, wage, capital-labor ratio of the export sector and the capital-labor ratio of the import sector.

If we transform the above four formulas, then we can conclude:

$$\omega / r = \left(\frac{1-a}{a}\right) k_E = \left(\frac{1-b}{b}\right) k_I$$

$$k_I = \left(\frac{1-a}{a} \Big/ \frac{1-b}{b}\right) k_E \tag{46}$$

If we combine the transformation, we can conclude:

$$k_E = k_E \left(\frac{P_I}{P_E}, \frac{A_E}{A_I}, a, b\right)$$

$$k_I = k_I \left(\frac{P_I}{P_E}, \frac{A_E}{A_I}, a, b\right)$$

$$r = r \left(\frac{P_I}{P_E}, \frac{A_E}{A_I}, a, b\right)$$

$$\omega = \omega \left(\frac{P_I}{P_E}, \frac{A_E}{A_I}, a, b\right) \tag{47}$$

We can also conclude:

$$\partial \omega / \partial \left(\frac{A_E}{A_I}\right) < 0 \tag{48}$$

The economics meaning of Formula (48) is that when other conditions remain unchanged, if the growth rate of export sector productivity to that of import sector of a country grows faster than that of foreign countries, then the domestic wage level decreases. Dropping wage level signifies the falling of the price level.

The actual exchange rate of a country shall be affected by the nominal exchange rate, foreign price index and domestic price index, and the expression is:

$$ERR = NRR \cdot \frac{FPI}{DPI} \tag{49}$$

In the formula, ERR, NRR, FPI and DPI respectively refer to exchange rate in real terms, nominal exchange rate, foreign price index and domestic price index.

If we take the logarithm of both the left and the right of the above formula, then

$$\ln ERR = \ln NRR + \ln FPI - \ln DPI \tag{50}$$

In order to have the rate of change, we calculate the derivatives of Time T in both the left and the right of above formula, then get:

$$d\frac{\text{In}ERR}{ERR}/dT \; - \; d\frac{\text{In}NRR}{NRR}/dT + d\frac{\text{In}FPI}{FPI}/dT \; - \; d\frac{\text{In}DPI}{DPI}/dT \qquad (51)$$

Formula (51) tells us that the rate of change of a country's actual exchange rate is determined by the volatility of nominal exchange rate, the volatility of foreign price level, and the volatility of domestic price level together. But during the analysis process, we usually presume that foreign price index remains basically stable or fluctuates moderately. In this case, the fluctuation of the actual exchange rate will be determined by the volatility of nominal exchange rate; the volatility of the domestic price level and has a positive correlation with the former and a negative correlation with the latter. In particular, when the volatility of nominal exchange rate is not taken into consideration and the "relative-relative" productivity growthoriented technical progress happens to the export sector of a nation, the domestic price index is bound to drop, while domestic exchange rate in real terms will increase due to the decrease of the wage level at home. Yet, once the volatility of the nominal exchange rate is taken into account, we need to differentiate between different countries.

What is stated above is only the first part of the basic analytical model. The second part of the model focuses on the influence of the volatility of the actual exchange rate on the terms of trade.

According to the definition of basic terms of trade, we can use a formula to express this as:

$$TOT = \left(\frac{P_E}{P_I}\right) \qquad\qquad (52)$$

If we take the logarithm of both the left and the right of the above formula, then

$$\ln TOT = \ln P_E - \ln P_I \qquad\qquad (53)$$

In order to have the rate of change, we calculate the derivatives of Time T in both the left and the right of the above formula, then get:

$$d\frac{\text{In}TOT}{TOT}/dT \; = \; d\frac{\text{In}P_E}{P_E}/dT + d\frac{\text{In}P_I}{P_I}/dT \qquad (54)$$

Formula (54) tells us that the volatility of the terms of trade is determined by the volatility of the export price and import price. Both the import price and export price are affected by the volatility of the actual exchange rate. In this connection, the ultimate direction of change of the terms of trade will be determined by the difference between the transmission of the actual exchange rate to the export price, and by the transmission of the actual exchange rate to the import price. The expression formula is:

$$\left(d\frac{\ln P_E}{P_E}/dT\right)\bigg/\left(d\frac{\ln TOT}{TOT}/dt\right) = pte$$

$$\left(d\frac{\ln P_I}{P_I}/dT\right)\bigg/\left(d\frac{\ln TOT}{TOT}/dt\right) = pti \tag{55}$$

Application of basic theoretical model I—circumstances in small countries

For an economically small country, if the speed of the technological progress resulting from productivity growth in its export sector to that in its import sector is faster than that in foreign countries, then the domestic wage level will decreaseand the price index will drop. In addition, the nominal exchange rate of small countries will remain basically stable,[22] the volatility will be small or basically absence in the short run, and we can conclude the following formula, i.e. actual exchange rate appreciation:

$$d\frac{\ln ERR}{ERR}/dT = d\frac{\ln NRR}{NRR}/dT + d\frac{\ln FPI}{FPI}/dT - d\frac{\ln DPI}{DPI}/dT > 0 \tag{56}$$

Under the assumption that the actual exchange rate of a country rises, if according to the exchange rate transmission theory the country is small in the international economics sense, negative fluctuation of the export price triggered by the rising exchange rate is drastic at this moment, and the transmission elasticity of the export price is larger than that of the import price, then $-1 < pte < 0$, $1 > pti > 0$, and $|pte| < |pti|$.

But, as is shown in the following formula,[23] the difference between the new and old terms of trade index lies in the difference in the transmission elasticity of the exchange rate to the export price and the import price.

$$\ln TOT = \ln\frac{P_E}{P_I} = \ln P_E - P_I$$

$$\ln TOT^* = \ln\frac{P_E^*}{P_I} = \ln P_E(1 + pte) - \ln P_I(1 + pti)$$

$$\ln TOT^* - \ln TOT = \ln(1 + pte) - \ln(1 + pti) \tag{57}$$

Within this context, $\ln(1 + pte) - \ln(1+pti) < 0$, meaning that the new terms of trade index is smaller than that before the changes in the exchange rate, the Net Barter Terms of Trade (NBTT) of the country deteriorate and the deterioration is rather serious.

On the contrary, if the "relative-relative" productivity growth oriented technical progress happens to the import sector, then the actual exchange rate of the country will depreciate. If the actual exchange rate of the country devalues, according to the exchange rate transmission theory, the negative fluctuation of the export price triggered by the depreciation of the exchange rate at this moment is drastic, and the resulting export price transmission elasticity is larger than the import price transmission elasticity; that is $0 < pte > 1$, $0 > pti > -1$, and $|pte| > |pti|$, then $\ln(1 + pte) + \ln(1 + pti) > 0$; that means new terms of trade index is larger than that before the fluctuation of the exchange rate, and the NBTT of the country will improve.

Application of basic theoretical model II—circumstances in large countries

As for an economically large country, if the speed of technical progress resulting from productivity growth in its export sector to that in its import sector is faster than that of foreign countries, then domestic wage level decreases and the price index drops; but the nominal exchange rate of large countries is floating and these large countries have the capability to implement strong home currency. The United States is a typical example, as it has been implementing a strong U.S. dollar policy. The terms of trade of the United States has not seen deterioration under the circumstance that its home currency keeps appreciating. At this moment, the nominal exchange rate of large countries will response reversely, thus it can slow down the rising of the actual exchange rate and even depreciate the exchange rate so as to improve the terms of trade.

At the same time, what needs to be pointed out is that even if moderate appreciation happens, large countries can smooth the negative fluctuation of the export price sparked by exchange rate appreciation by leveraging on its position as large countries, and make the resulting export price transmission elasticity smaller than the import price transmission elasticity: $-1 < pte < 0$, $1 > pti > 0$, and $|pte| < |pti|$, then $\ln(1 + pte) - \ln(1 + pti) < 0$. That means the new price terms of trade index is still larger than the original price terms of trade index. At this moment, terms of trade will deteriorate slightly but the duration is short.

On the contrary, if the "relative-relative" productivity growth–based technical progress happens to the import sector of the country, then according to the exchange rate transmission theory, the exchange rate depreciates at this moment, the resulted negative fluctuation of export price is small, and the export price transmission elasticity is smaller than the import price transmission elasticity. That is: $0 < pte < 1, 0 > pti > -1$, and $|pte| < |pti|$, then $\ln(1 + pte) - \ln(1 + pti) > 0$, which means the new terms of trade index of the new price is still larger than the terms of trade index of the original price, and the terms of trade will improve.

The examination and quantitative analysis of China's practice

Does HBS effect exist in China?

In recent years, numerous foreign and Chinese scholars have used Chinese data to test the HBS theory. The academic community has come up with two contrary views: the first one has recognized the significant force of the theory in the explanation of the Renminbi exchange rate; the other holds that this theoretical hypothesis is not appropriate to explain Chinese economic growth (Lu Feng, 2006a).

The Chinese economy is undergoing rapid development and structural changes and it keeps integrating into the global economy; thus preconditions for the hypothesis of the expanded HBS theory in the text exist in reality.

(1) Changes in productivity of the export sector and the import sector

According to the calculation of Lu Feng (2006), between 1978 and 2004, the estimated value of the average annual growth rate of the labor productivity in China's manufacturing industry changed remarkably from low to high. Between 1978 and 1990, it was only 1.85%, much lower than the per capita GDP growth rate of 7.49% in the same period. Upon entering late 1990s, the labor productivity of Chinese manufacturing industry picked up dramatically–the average annual growth rate was 13.1% between 1991 and 2004, more than the 8.18% per capita GDP growth rate during the same period. The average annual growth rate reached as high as 15.5% between 1994 and 2004.

If we make a comparison of the productivity between Chinese manufacturing industry and the U.S. manufacturing industry, we can find that the labor productivity of the Chinese manufacturing industry was 100 in 1978 and the cumulative relative growth index dropped to 88 in 1990; however, it started to grow after entering the 1990s and rose to 100 again between 1992 and 1993. During the middle and late 1990s, this relative index elevated rapidly and the cumulative growth index rose to 241 in 2004 from 98 in 1994, representing an annual growth rate of 9.41%.

If we examine the relative growth rate of the labor productivity in the Chinese manufacturing industry compared with 13 OECD countries, we can find that the labor productivity of the Chinese manufacturing industry dropped during the 10 years before and after the 1980s, and the relative growth index dropped from 100 in 1978 to 88 in 1990. Since the 1990s, the labor productivity has increased relatively and the growth rate has been faster than that of the United States; and the accumulated relative growth index rose from 106 in 1994 to 308 in 2004, recording 11.26% in the annual growth rate.

However, once we further examine the relative difference in the growth rate of the internal labor productivity in the Chinese manufacturing industry, we will find that between 1978 and 2004 the average annual growth rate of the total productivity in the manufacturing industry was 8%. The sectors with the fastest growth rate include electricity and electronics, and mechanical equipment which recorded an average annual growth rate of 10%; the sectors including garment and fibre, tobacco processing and metal products grew relatively faster, about 7%; petroleum and coking grew the slowest, at only 0.2%. In fact, these sectors with relatively faster growth rate are all large export sectors while those with slower growth rate are typical import sectors. This indicates that, to a certain extent, the labor productivity of Chinese export sectors grew faster than that of import sectors within the manufacturing industry. If we include the service industry, which is a typical import industry, we can find that this tendency of having a relatively fast growth rate will be even more obvious. According to the calculation of Lu Feng (2006), the average annual growth rate of the labor productivity in the Chinese service industry during this period was relatively low, at only 2.6%.

(2) "Relative-relative" labor productivity in the two sectors and the RMB real exchange rate–taking the U.S. and China for example

Lu Feng (2006) took as examples the real exchange rate of Renminbi against the USD and the "relative-relative" labor productivity between the manufacturing industry and the service industry to carry out empirical research. After making the graphic description and regression of the two, it was concluded that the regression coefficient of the former to the later reached –0.61.

Yet, the analysis is only limited to the manufacturing and the service industry. According to the classification standard of "export sector" and "import sector" designed in this chapter, we can be certain of two points: first, the rising trend of "relative-relative" productivity will be more prominent; second, the correlation coefficient between "relative-relative" productivity and the real exchange rate of Renminbi against the U.S. dollar will increase.

The above two typical facts give us the reason to believe that the improved HBS model is still applicable in China. That means Chinese export sectors' "relative-relative" productivity growth will result in the appreciation of the Renminbi exchange rate in real terms.

Does stable exchange rate transmission exist in China?

(1) Empirical model and data

The study in this section adopts the *VAR* model for analysis. Compared with the least square method, VAR can solve the uncertainties rising from unbalanced data and the mutual influence between endogenous variables. Therefore, the VAR model can better estimate the real influence of real exchange rate fluctuation on trade.

a. Empirical model

We follow the model specification in the study of Parsley (1993) on the exchange rate transmission to the entire export of Japan. The import and export price of a nation are: the function of domestic and foreign demand, domestic production cost, and exchange rate. The four control variables are: domestic real income, the world total real income (excluding China), domestic unit labor cost, and real exchange rate. But for the convenience of studies, the linear form under the natural logarithm is usually adopted to estimate this kind of export equation. The initial general estimating equation form is:

$$\text{Log}ep = a \log err + b \log lc + c \log nc + d \log nw + \varepsilon$$

$$\text{Log}ip = a \log err + b \log lc + c \log nc + d \log nw + \varepsilon \qquad (58)$$

In the formula, *ep* is export price index; *ip* is import price index; *err* is real effective exchange rate index; *lc* is unit labor cost index; *nc* is Chinese domestic real income index (replaced by Gross National Income index); *nw* is the world total real income (replaced by Gross National Income index). We take the logarithm value of these series, which can reduce dramatic fluctuation and heteroscedasticity of real values on the one hand and demonstrate the elasticity of export price to real exchange rate on the other.

But, the time series of these variables are not always balanced, and the introduction of some variables might result in the failure to conclude an estimating equation. We can find through testing two facts: first, all variables are unbalanced; second, the variable of Chinese domestic real income in the export price equation and the variable of the world total real income in the import price equation are all redundant variables, because the above variables are, in effect, an overall index reflecting China's labor productivity. China's real exchange rate is also affected by the "relative-relative" fluctuation of China's labor productivity, and there is an autocorrelation between the two. For that matter, we need to get rid of these variables.

After getting rid of these variables, we build a VAR model of the remaining variables. Then we make a Cointegration Test of the four variables and build a VAR extended VEC (Vector Error Correction) to estimate the influence of exchange rate on export price fluctuation, which is the transmission of exchange rate.

On the basis of the Cointegration Test result, we write the cointegration relationship between the export equation and import equation as the mathematical relationship formula and make it equal *VECM*:

$$VECM = lep + 1.55lerr + 0.83llc + 0.04\ln w - 15.75$$

$$VECM = lip - 0.33lerr + 0.73\ln c + 0.49llc - 8.66 \tag{59}$$

After conducting the Unit Root Test of the two series of *VECM*, we found they are a balanced series that has proved the correctness of the cointegration relationship. As a matter of fact, the above series are error correction terms that have reflected the long-term trend of export price changes. The long-term trend of export price changes has demonstrated a long-term stable relationship between import and export price and national income, unit labor cost and actual effective exchange rate, and this long-term relationship will play an inhibitory role in short-term price fluctuations.

After building this cointegration equation, we can carry out the Granger Causality Test of various variables, and the real exchange rate in the Vector Autoregression Model, and then find out whether Spurious Regression exists.

The measurement result indicates that, in terms of China's export price equation, it can be believed, with at least 95% of the confidence level, that the real exchange rate, China's unit labor cost, and the world total real income are all Granger Cause. When it comes to the import price equation, it can be believed with at least 90% of the confidence level that the real exchange rate, China's actual income, and the world unit labor cost are Granger Cause. That is to say, Spurious Regression does not exist in our model.

b. Data explanation

Real exchange rate: the real exchange rate index in this text is learned from the annual real effective exchange rate index published by the *International Financial Statistics Yearbook*. This data is calculated based on unit labor cost of domestic and the representative 16 trade partners,[24] which basically represents the changes of domestic real exchange rate and the international competitiveness, and thus it is a reliable index (Li Yaxin, and Yu Ming, 2002).

Import price: due to the lack of the data before 1990 in *China Customs Statistics Yearbook* and difficulties in finding *China Customs Statistics Yearbook*

1994 in domestic libraries, all data in this section is taken from the annual import and export unit values in domestic currency and published by the United Nations Conference on Trade and Development (UNCTAD). The original data take the year of 2000 as the base year which has been changed in this analysis to the year of 1995.

Unit labor cost: on the basis of the studies of Wang Huimin, Ren Ruo'en, and Wang Huiwen (2004), this section continues to use the same method to calculate the data between 2001 and 2005 and the data of China's labor cost. , Another estimation method is used to correspond the unit labor cost data to import products Utilize the data published by the International Labor Statistics, select four different countries and regions, including the United States, Singapore, South Korea, and Hong Kong,[25] and calculate the weighted average of the corresponding cost based on the weighting used by IMF in calculating the real exchange rate.

The world total real income and China's total national income: the data are from the total national income denominated in the U.S. dollar and published in the database of the United Nations Statistics Division (UNSD). The world total real income is the total income of all other countries except China. The year of 1995 is the base year to calculate the index corresponding to each year.

(2) Empirical results

First, VEC model of the transmission of real exchange rate to export price

According to the analysis of the VAR model, the lag order is determined by AIC and SC indices and generally the smaller, the better. Therefore, the lag order shall be 3, and the number of the cointegration relationship of the entire model is 2. We can directly choose 1 or 2 in the lag phase of VEC model, and the final mathematical expression of the import price VEC model is:

$$
\begin{aligned}
D\ (LEP) = \quad & -0.69D\ (lep(-1)) - 0.21D\ (lep(-2)) \\
& + 0.33\ (D\ (lerr\ (-1)) + 0.19D\ (lerr(-2)) \\
& + 0.25D\ (\ln\omega(-1)) + 0.32D\ (\ln\omega(-2)) \\
& + 0.04D\ (llc\ (-1)) + 0.05D\ (llc\ (-2)) \\
& + 0.08VECM - 0.01
\end{aligned}
\tag{60}
$$

The measurement result indicates that the fitting effect is good regardless whether it is the whole model or a single model focusing on the research on the transmission of real exchange rate to export price. The coefficient of the determination of a single model reaches 90.15%, the overall AIC and SC value is small, and the absolute value of the T test statistics value of the parameter test of one-period lag and two-period lag of the real effective exchange rate, unit labor cost and the world real total income, in most cases, is bigger than 1.64 or 2.35–significant at the level of 10% and 5%. The coefficient of the cointegration item is basically a negative number, signifying the existence of the negative error auto-correction mechanism.

Second, VEC model of the transmission of real exchange rate to import price

Similar to theexport price model, we can conclude that lag order should be set to 3, and the number of the cointegration relationship in the whole model is 1. We can directly choose 1 or 2 in the lagging phase of the VEC model, and the mathematical expression of the VEC model of the final import price is:

$$
\begin{aligned}
D\,(lip) = \quad & -0.13D\,(lip(-1)) - 0.10D\,(lip(-2)) \\
& -0.07\,(D\,(lerr\,(-1)) - 0.09D\,(lerr(-2)) \\
& -0.11D\,(\ln c(-1)) - 0.08D\,(\ln\omega(-2)) \\
& +0.23D\,(llc\,(-1)) - 0.18D\,(llc\,(-2)) \\
& +0.32VECM - 0.12
\end{aligned}
\tag{61}
$$

The measurement result indicates that the fitting effect is good regardless of whether it is the whole model or a single model focusing on the research on the transmission of real exchange rate to import price. The coefficient of the determination of a single model reaches 53.84%, and the overall AIC and SC value is small. The absolute value of the T test statistics value of the parameter test of one-period lag and two-period lag of the real effective exchange rate, unit labor cost and the world total real income, in most cases, is bigger than 0.76–significant at the level of 25%. The coefficient of the cointegration item is basically a negative number, signifying the existence of the negative error auto-correction mechanism.

Brief summary

The final VEC model has effectively analyzed the causes for Chinese import and export price fluctuation.

(1) Export price model

First, the transmission of real exchange rate to export price. The model demonstrates that the fluctuation of the exchange rate in real terms has a significant influence on Chinese export price. The one-period lag and two-period lag of exchange rate transmission elasticity reached 0.328 2 and 0.192 1 respectively, and statistics of the former is higher than that of the latter.

Second, the influence of unit labor cost changes on export price fluctuation. The result of the model demonstrates that changes of unit labor cost have little impact on export price fluctuation, only about 0.04–0.05. This further indicates that China's advantages in labor cost will continue to play a role in the long run. The underlying factor for this phenomenon is that the increase of productivity has provided possible space for the rise of factor price, but the Factor Price Insensitivity Theorem has made it difficult for some production factors, especially those factors demonstrating fixed claims to share the interests bought about by productivity increase. This in turn has made the productivity increase as the source of competitiveness used by enterprise, and has driven the expansion of export.

Third, the influence of the world total income in real terms on the fluctuation of the export price. The income increase of other countries as the importer of Chinese products has showcased the consistency with the fluctuation of Chinese export price. The elasticity of one-period lag and two-period lag has reached 0.25 and 0.32 respectively. This indicates, to a large extent, that China's export products have had certain competitiveness in export in the long run and China has become one of the leading suppliers of the majority of products worldwide. But China's pricing power in export products needs to be strengthened.

(2) Import price model

First, the transmission of real exchange rate to import price. The model demonstrates that the fluctuation of real effective exchange rate has little influence on Chinese import price, the one-period lag of exchange rate transmission elasticity stands at –0.072 7, two-period lag of exchange rate transmission elasticity stands at –0.090 3, and the transmission of two-period lag is more significant than that of the one-period lag. This demonstrates China is basically still a price taker in export, which is a leading factor behind the much lower elasticity of Chinese import price transmission than that of developed countries.

Second, the influence of unit labor cost changes on import price fluctuation. The result of the model demonstrates a phenomenon that merits our attention: The changes of unit labor cost have great impact on import price fluctuation; one-period lag elasticity reached 0.233 3 and two-period lag elasticity reached –0.176 4. This has offered two inspirations: first, the source countries of Chinese import leverage on their role as the price maker to rapidly transfer the negative impact of increased labor cost; second, the source countries of Chinese import could rapidly leverage on all kinds of technologies to assimilate the increasing labor cost within a short time period (two years at most), and rapidly create a new competitive edge with new products and lower prices.

Third, the influence of the changes of Chinese national income on the fluctuation of export price. Chinese national income growth also showcases the consistency with the fluctuation of Chinese import prices. One-period lag and two-period lag elasticity reached –0.11 and –0.08 respectively. A possible explanation for this phenomenon is that Chinese national income is not only a defining factor for consumption but also closely linked with Chinese real output. The increase of Chinese national income is also coupled with the enhanced capacity of China to supply all kinds of goods. In particular, China as a large developing economy has a strong industrial base and huge consumption capacity. Therefore, China could build a domestic supply platform for original imported products by means of reform, introduction, and cooperation within a relatively short time. However, due to unforseen circumstances, China might also experience over-capacity rapidly and turn to the international market, which has reduced its reversed impact.

Conclusions

This project can be extended in the following research conclusions and policy implications.

Research conclusions

Technical progress is an effective way to improve the terms of trade of a country, but different kinds of technological advance in different countries and different industries have different action mechanisms and ultimate influences on the terms of trade.

Circumstances in small countries

Under the hypothesis which takes no account of inferior goods, if the "relative-relative" technological advance of product innovation happens to domestic export industries, its different influences on domestic import and export volume at the product introduction stage, mature stage, standard, and recession stages will lead to deterioration of domestic terms of trade in the end; likewise, if the "relative-relative" product innovation happens to import industries, then domestic terms of trade will only improve slightly at the introduction stage but will start to deteriorate at the next two stages. Once taking into account the inferior goods hypothesis, then the "relative-relative" product innovation will improve the terms of trade if it happens to export industries, and will deteriorate the terms of trade if it happens to import industries.

If the "relative-relative" technological advance of process innovation happens to export industries, its different influences on the cost and pricing of the product at the introduction stage, large-scale export stage, stable, and recession stages will make domestic terms of trade basically stable and even deteriorated to a certain extent. On the contrary, if the "relative-relative" technological advance of process innovation happens to import industries, domestic terms of trade will gradually improve.

If the "relative-relative" technological advance of productivity growth happens to domestic export industries, then the domestic real exchange rate will appreciate. Terms of trade will deteriorate due to the differences of the real exchange rate in transmission elasticity to the import and export. Likewise, if the "relative-relative" technological advance of productivity growth happens to domestic import industries, then the domestic real exchange rate will depreciate, and the terms of trade will improve accordingly.

Circumstances in large countries

Under the hypothesis which takes no account of inferior goods, if the "relative-relative" technological advance of product innovation happens to domestic export industries, its different influences on domestic import and export volume at the product introduction stage, mature stage, standard, and recession stages will be under control as the country is a price maker, and domestic terms of trade will only improve. Likewise, if the "relative-relative" product innovation happens to import industries, then domestic terms of trade will also deteriorate. Taking into account the inferior goods hypothesis, then the "relative-relative" product innovation in export industries will improve the terms of trade. The "relative-relative" product innovation in import industries will also improve the terms of trade under the strong influence of the "price maker" on demand and supply.

If the "relative-relative" technological advance of process innovation happens to domestic export industries, its different influences on the cost and pricing of the product at the introduction stage, large-scale export stage, stable, and recession stages will make domestic terms of trade basically stable or even deteriorate slightly in the end. On the contrary, if the "relative-relative" technological advance of process innovation happens to import industries, domestic terms of trade will keep improving.

If the "relative-relative" technological advance of productivity growth happens to domestic export industries, then domestic real exchange rate will appreciate. Terms of trade will deteriorate for sure due to the differences of the real exchange rate in transmission elasticity to the import and export; but, the strong monetary policy supported by the floating exchange rate system in large countries will make the real exchange rate basically stable or even depreciate, which will result in the stability and even improvement of the terms of trade. Likewise, if the "relative-relative" technological advance of productivity growth happens to domestic import industries, then the domestic real exchange rate will depreciate and the terms of trade will improve accordingly.

Policy implications

Given that terms of trade are the decision-making elements and incentive and constraint conditions for building the international environment for enterprise-independent innovation, the policy choice for enterprise independent innovation should pay close attention to the international environment and its transmission carrier–trade flow and trade policy–and determine the industrial policy, technical policy and trade policy. This could encourage the technological innovation according to the growth rate of demand and the growth rate of productivity.

This project continues to employ the Rybczynski Theorem in international trade theory to study the influence effect of different types of technological innovations under the conditions of inferior goods and normal products, in the economics sense, on terms of trade. It indicates that accelerating technical progress, carrying out technological innovation, and ensuring enterprises often remain at the new product introduction stage in the rapid development of world scientific and technological pattern can help the country to leverage on favourable terms of trade changes and seek interest maximization in the international division of labor. In particular, China has become extensively involved in the global value chains specialization, and the positioning in the product growth cycle built by the independent innovation level has determined the long-term strategic interests of Chinese enterprises in both the world market and the international environment. Amid fierce international competition, there is an urgent need for Chinese enterprises to change the mode of expanding to the international market with low-end and low price products. They must participate in international competition through technological innovation and innovative ideas, and with the aid of high quality and high-tech products.

The studies show that as the unified domestic factor market and commercial market may promote the price signal to form a unified factor cost, and even the commodity price, according to the so-called "One Price Law" under the market condition, it is very likely to create positive or negative changes in terms of the trade of export sectors and import-competing sectors. In this sense, we need to adopt a means of an effective exchange rate, except the nominal exchange rate, to build a different profit level or nominal effective exchange rate level of the export sectors and import-competing sectors to ensure that technological innovation and technical progress can generate widespread and similar interests through the market.

Our studies further indicate that, apart form ensuring domestic technological innovation and technical progress to get universal incentive, the country needs to expand the implementation of trade policy to competition policy, realize international coordination in competition policy, and work hard to avoid the distortion of international environment and the losses of the effects of enterprise technological innovation caused by excess competition in domestic markets and the circumstances in small countries.

3

Chapter

The Financing Environment for Independent Innovation

Guo Yan, Yi Zhihong, and Li Yan

Introduction

Innovation has always been the driving force for social development, and enterprise independent innovation is the basic pillar for promoting technical progress and economic development. With the further evolution of Chinese economic structure, independent innovation, including technological, product, production organization (including outsourcing, original equipment manufacturer (OEM), franchising and so on), and management innovation has developed in a growing number of enterprises and has become the core of transforming economic development patterns and realizing sustainable development. It also represents the new direction of economic development. Both technological and product innovation are top priorities among the types of business innovations. Technological and product innovation differ from organizational and management innovation in that they need more investment of social resources. Therefore, technological and product innovation are equivalent to the investment of enterprises and thus need financing. Yet, technological innovation and product innovation for one part and financing for another are contradictory. From the perspective of the Chinese experiences and the world at large, most enterprises, especially small and medium-sized enterprises (SMEs) which are the important carrier of independent innovation, will encounter difficulties in financing during the process of independent innovation. For SMEs with innovation characteristics, the capital they obtained from financial institutions can hardly match their contributions to economic growth. What is the reason behind this contradiction? How can this contradiction be settled? An in-depth discussion of these questions is of great importance for a government to roll out policies to effectively boost the development of enterprise independent innovation. We explain this phenomenon from the perspectives of information asymmetry, business lifecycle and formal and informal finance, based on the research results in relevant disciplines, such as economics, finance and enterprise theory. We also try to find a favorable financing environment for business innovation in line with the historical development of the financing system, supportingenterprise independent innovation in China.

According to the above analyses, the objectives for our analyses in this research are concentrated on the financing environment for small and medium-sized innovative enterprises facing larger constraints in financing during the process of technological innovation and product innovation; we call them "innovative enterprises" in short. We will not take into consideration the technological innovation of large enterprises and the non-product and technological business innovation.

Literature Review

Economists have explained the causes for the financing difficulties faced by enterprises from the perspective of information theory. Stiglitz (1981) believed that information asymmetry and the adverse selection resulting from information asymmetry would make commercial banks adopt credit rationing behavior. That means banks would take rationing measures for credit borrowers at an interest rate level that is lower than the competitive equilibrium interest rate. It could also make them maximize the expected profits when they face an excess demand for loans but could not distinguish risks of individual borrowers. Rejected applicants could not be granted loans because they might choose high risk projects, even if they are willing to pay a higher interest rate. The credit rationing theory of Stiglitz and Weiss is now an important cornerstone for studies on the information asymmetry between lenders and borrowers. This theory appears imperfect, as it regards the interest rate as the only endogenous decision-making variable during the process of credit rationing. After all, the requirement for loan collateral and the costs for examination and supervision of the size of loans might also become endogenous decision-making variables.

When it comes to how to work out issues about the credit rationing, there are several research results, as follows:

Information screening mechanism

Bester (1985) demonstrated the effect of mortgage on information asymmetry from the information screening perspective. He argued that mortgage and interest rate can simultaneously serve as the screening mechanism of banks to distinguish the different types of risk of loan projects, and banks can differentiate between high-risk and low-risk loan projects through the sensitivity of enterprises to the changes in quantity of mortgage. The essence of Bester's suggestion is to make low-risk enterprises promise a higher level of mortgage but enjoy a lower interest rate and make high-risk enterprises promise a lower level of mortgage but pay a higher interest rate.

Advantages of small and medium-sized financial institutions in "soft information"

The differentiation of "soft information" and "hard information" was first proposed by Stein (2002) to employ the incentive from business asset control right in the studies on the organizational structure of banks. He called the objective and easily-transmitted and testified information, like a financial statement, as "hard information," and named the intangible information, like the operation capacity and personal quality of entrepreneurs and market factors, as "soft information." Large enterprises usually have complete financial statements and credit records, but SMEs generally lack "hard information" and may only rely on "soft information" to make loan decisions. Given this situation, small banks have more incentives to gather this kind of information due to the short information chain for transmission, and therefore have an advantage providing loans to SMEs.

The follow-up studies of Berger and Udell (2002) concluded that relationship lending can help banks access "soft information" through building a close relationship with enterprises so as to overcome issues brought about by information asymmetry. But, large banks are at a disadvantage in terms of handling relationship lending as a result of the complex bureaucratic structure and the long agent chain. Research of Banerjee (1994) on this issue has narrowd down two hypotheses: the first is the long-term interaction hypothesis, believing that a long-term partnership can help address the issue of information asymmetry between small and medium-sized financial institutions and SMEs; the second is the peer monitoring hypothesis, believing that cooperative small and medium-sized financial institutions will carry out mutual supervisions which are more effective than the supervision of financial institutions.

Chinese scholars such as Zhang Jie (2002) indicated, through a model of organizational theory judging and weighing both information and agent cost in order make sound lending decisions, that large banks prefer concentration of powers in the allocation of decision-making rights in lending, while small banks tend to separate decision-marking powers in lending. However, relationship lending relies on whether the specific enterprise knowledge of frontline managers and loan officers matches their decision-making power in lending. This model has proved the strong points of small banks in relationship lending from the perspective of organization theory. Li Zhiyun (2002) emphasized that the heterogeneity, mortgage, and transaction costs are factors influencing the access for SMEs to loans. He observed that bringing in small and medium-sized financial institutions should increase the credit for SMEs and enhance the overall economic efficiency under the conditions that "there are relatively abundant capitals," or that the marginal productivity of capital of SMEs is larger than that of large enterprises.

Discussions about formal finance and informal finance

Discussions about the evolutionary game between formal finance and informal finance were mainly accomplished by domestic scholars. The informal finance is a relative concept to formal finance and mainly refers to the financial transactions that happen beyond the functional scope of all kinds of existing formal institutions, free from supervisions of the regulatory authority. Informal finance, more often than not, emerges with market deficiency, such as free borrowing in the private sector, business financing from the public, and borrowing from private lenders. A crucial part of these studies is to investigate and inspect regions with developed private finance during the process of reform. Shi Jinchuan and Ye Min (2001) pointed out, through case studies on private finance in Wenzhou, Zhejiang province, that non-state economic sectors have realized preliminary internal accumulation by relying on the abundant original capital from informal financial sectors, which has made them qualified to get loans from formal financial institutions. Guo Bin and Liu Manlu (2002) believed in the need to put in place a diversified financing service system for SMEs and guide private financial organizations to gradually develop into private financial institutions with standardized operation and services targeting SME financing. While studying the cases of private enterprises in Ningbo city, Zhejiang province, Luo Danyan and Yin Xingshan (2006), with the help of the concept of "social capital," put forward that the informal financing relationship between enterprises, individuals, and different enterprises as an important social capital, can help overcome the weakness of SMEs and strengthen the mutual trust and interdependency. This can make them form into an interest community, which can help SMEs to overcome not only the financial constraint of the supply side, but also that of the demand side.

In discussions about informal finance and formal finance, relevant policy suggestions prefer to set up and improve small and medium-sized financial service systems. Lin Yifu and Li Yongjun (2001) believed that the cost and efficiency of financial services provided by different financial institutions to enterprises of different sizes are diverse, and large banks should not bear the "policy burden" for providing loans to SMEs. There are still shortcomings in developing the capital market focusing on SME financing and establishing a guarantee and credit system at the current stage. The way to address the financing difficulties faced by SMEs in China is to establish and improve small and medium-sized financial institutions, consisting of small and medium-sized local commercial banks and local cooperative lending financial institutions. Lin Yifu and Sun Xifang (2005) further noted that it's difficult for formal finance to overcome adverse selections caused by information asymmetry. But, informal finance has strong points in gathering the "soft information" of SMEs and thus the existence of informal finance can effectively make a difference in the efficiency of the entire credit market in capital allocation. Studies of Guo Bin and Liu Manlu (2001) stressed that the small and medium-sized financial service system needs the involvement of both government capital and private capital, and relies on the establishment of infrastructure like SME credit and SME credit guarantee system. It's unrealistic to simply depend on the commercial banking system to solve the financing difficulties faced by SMEs. There is a need to guide private financial organizations to gradually develop into private financial institutions that are under standard operating and target SMEs for financing services.

Zhang Jie (2000) believed that financial difficulties faced by the private economy, after relying on internal financing and then pursuing technical progress and intensive capital, are rooted in the financial support of the state-owned financial system for state-owned enterprises, the solid dependence of state-run enterprises on this kind of support, and the credit capitalization resulting from this dependence. Relying on changing the credit behaviors of state-owned banks and the directions of their capital investment, developing exogenous small and medium-sized financial institutions, encouraging the entry into the stock market, and other means to solve the financing difficulties faced by private enterprises can produce limited effects. Therefore, there is a need to focus on building the external environment for the growth of the endogenous financial system and relax the limit on the innovation of it of the private economy. The exclusion of state-owned exogenous financial institutions was rooted in their emphasis of the "two-fold structure" of Chinese financial system (1998). The superstructure has a controlling position, which is the deep-seated reasons for the failure of many financial forms emerged in Chinese history to develop into modern financial system.

Research Objectives and Method

From the above analysis, we see that relevant studies at home and abroad are mainly conducted by centering on the information asymmetry of SMEs and have extended to the game between formal finance and informal finance. However, most studies treat innovative enterprises in different industries and stages in the same way, and remove the characteristics of different development stages and industries they belong to. We find from our investigations and survey on Zhongguancun Science and Technology Park and Shanghai Pudong Development Area that innovative enterprises in different industries have different financing demands, and the degree of information asymmetry is also different, especially in the selection of formal and informal finance. Therefore, we will discuss the financing environment for the independent innovation of enterprises from the following aspects:

(1) From the perspective of information economics, we take the lifecycle of independently innovative SMEs to study the characteristics of the financing demand of enterprises at different development stages for independent innovation and the characteristics of information asymmetry.

(2) Through comparative analyses, we study the applicability and efficiency of different ways of financing (endogenous financing, government aid plan, Angel Investment, risk investment, loans, and direct issuance in the capital market) in meeting the financing demand of independently innovative enterprises at different development stages, and research the internal and external factors influencing the SME financing performance.

(3) Under the premise of recognizing the path dependency, and in line with the national condition of China, we analyze the status quo of China's institutional supply for the financing for business independent innovation, study the challenges emerged during this process, and try to offer our policy suggestions.

Applied Research on the Financing Environment for Innovative Enterprises

Lifecycle of independently innovative enterprises: demand for financing and its satisfaction

Independent innovation is a kind of economic activity and the application of new technology, systems and management in production and marketing. Independent innovation has four evolution stages: the conception of innovation, research and development, trial production and trial marketing, and large-scale industrialization. This is valid regardless of whether it is driven by technology, demand, the interaction between technology and demand, or other factors. The four stages and the characteristics of each of the stage in this process is the lifecycle of independently innovative enterprises we stressed.[1] At different stages of the lifecycle, the information asymmetry between enterprises and investors is different in degree and nature, so business financing has different characteristics. Only when this kind of financing characteristics and the selected financing contract are closely matched up, will transactions be effective.

Stage features of the lifecycle of innovative enterprises

The stage of conceiving innovation ideas

At this stage, innovators just come up with the ideas and form the preliminary concept of innovation. In order to move forward with independent innovation and achieve success, innovators need to develop new plans, set up an organizational structure, and raise the funds needed. The conceptual stage of innovation is featured by: (1) high uncertainty—innovative enterprises might keep in mind or give up the ideas of innovation; there are many uncertainties in whether innovation will succeed. In addition, a successful understanding of this kind of innovation is rather opaque in information transmission. (2) Little demand for capital—innovative enterprises at the conceptual stage of innovation only need a small amount of funds to accomplish their plan on technological innovation. In this case, capital needed at this stage is deemed as "seed money" or "start-up capital." (3) The source of capital is mainly from internal financing and supplemented by external financing. Generally speaking, as a result of the high uncertainties, serious information asymmetry and little demand for capital, most innovative enterprises finance at the conceptual stage of innovation using their own capital or seek financing from their relatives and friends; only a small number of innovators seek external financing at this stage.

Research and development stage

At this stage, innovative enterprises need to organize a research and development (R&D) team, purchase equipment and materials for R&D, design the technological path for R&D, and deliver a technical R&D plan. The target at the R&D stage is the emergence of innovative products. As with the conceptual stage, the R&D stage is still a stage of pure investment. At the same time, the capital needed at the R&D stage is much more than at the conceptual stage of innovation. In this connection, it's difficult for an innovative enterprise to meet the financial demand on themselves. Generally speaking, external financing is a must at the R&D stage.

At the R&D stage, the main assets of an enterprise are the intangible assets brought by R&D and the largest risk is technical risk. Whether the technical path is correct and whether R&D of the product can succeed smoothly have a direct bearing on whether technological innovation can move into the next stage. Innovative enterprises and financial institutions often witness adverse selection and moral risks due to the confirmation, transmission and examination of information.

The stage of trial production and trial marketing

At this stage, innovative enterprises need to accomplish the trial production and trial marketing with its main objective being market testing. Innovators need to make necessary adjustments of the innovation plan and the product design according to market response. At this stage, the principal risk faced by innovators is market risk—whether the innovative product could effectively meet the market demand.

Innovative enterprises at this stage can generate a small amount of sales revenue through trial marketing, but generally they are still in financial losses and will need external financing. However, the assets of enterprises at this stage are still intangible assets; both the quantity and quality of collaterals that they are able to provide still cannot meet the requirements of large and medium-sized financial enterprises. Fortunately, the production capacity and sales volume at this stage are not large, thus the financing amount is limited.

The stage of large-scale industrialization

If the product is acceptable in the market, the innovative enterprise needs to carry out mass production and marketing. Substantial funds are needed to purchase a production facility and plant, establish marketing channels and after-sale service networks, and conduct appropriate advertising. The major risks faced by innovative enterprises at this stage include market risk, operational risk and technical risk. Market risk is mainly reflected by competition risk; operational risk is mainly demonstrated by the risk in business management; technical risk is typically embodied by the risk that the innovative technology might be replaced by more advanced or cheaper technology. In order to manage these risks, the innovative enterprise needs to set up effective market barriers to try to maintain the time for them to generate excess profit. The innovative enterprise needs to build a professional management team, follow up the trend of technology development at all times, and improve their innovative product when it is necessary to keep from losing the market due to the technology upgrading too fast.

At the initial stage of industrialization, the total capital flow is a negative value due to the large demand of the innovative enterprise for funding. The enterprise will obtain normal capital inflow and reach a balance in profit and loss as it starts the production and sales. Intangible assets are still one of the major assets,but as the enterprise starts the regular production and operating activities, the proportion of tangible assets will increase gradually while the proportion of intangible assets will decrease accordingly. In the end, tangible assets will exceed intangible assets to become the major assets of the enterprise at this stage.

Financing demand at different stages of the lifecycle

Table 3.1 has summarized the features of the risk, fund demand and asset structure of innovators (innovative enterprises) at the four stages of independent innovation.

Table 3.1.   Characteristics of different stages of independent innovation

Technological Innovation Stage	Principal Risks	Features of Fund Demand	Features of Asset Structure
Innovation conception	Technological risk	Pure fund investment; a small amount of capital demand	Basic intangible assets
R&D	Technological risk	Pure fund investment; more capital demand	Mainly intangible assets
Trial-production and trial-marketing	Market risk	A small amount of sales revenue; more capital demand	Mainly intangible sssets with a small amount of tangible assets
Large-scale industrialization	Technological risk, market risk, and management risk	A large amount of capital demand, and reach fund balancing at the middle and late stages	The proportion of tangible assets gradually exceeds that of intangible assets and eventually become the main assets

Fig. 3.1 has summarized the cash flow of innovators (innovative enterprises) at the four stages of independent innovation.

According to the previous analyses, we can conclude the features of the financing demand of enterprises during the process of independent innovation:

First, the financing demand of independently innovative enterprises also showcases the stage features resulting from the features of their lifecycle.

Second, the asset structure of innovative enterprises is mainly intangible assets. After directly entering into the stage of large-scale industrialization, the asset structure of innovative enterprises is mainly intangible assets that might include technology, patent, trademark and so on. Innovative enterprises lack these intangible assets as their products have yet to become famous brands compared with businesses with mature products. Therefore, financing capacity with intangible assets as the collateral is weaker than that with tangible assets as the collateral.

Third, most innovative enterprises are also start-up enterprises and they lack the necessary credit record for financing. On top of that, the extensive management at the early stage of innovation causes them to lack strict and standard financial statements and the "hard information" which can be

Fig. 3.1.   Cash flow during the process of independent innovation

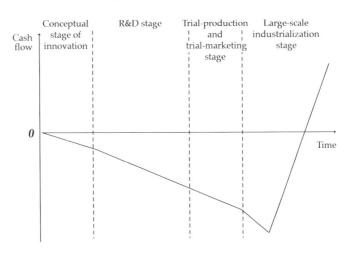

Note: The cash flow of an enterprise includes the cash flow for business operations, investment activities, and financing activities. In order to objectively demonstrate the financing demand for technological innovation, Fig. 3.1 only covers the cash flow for business operations and investment activities.

quantified and transmitted; however, most of them have the highly personalized "soft information." These conditions aggravate the information asymmetry between creditors and innovative enterprises, and have made innovative enterprises face serious financing constraints.

Fourth, innovative enterprises face more risks. Innovative enterprises are uncertain about who will be their competitors and how would they respond due to the unknown and changeable results of innovation. Therefore, innovative enterprises face higher technological risks, market risks and management risks at all stages of the lifecycle. Higher risks have made investors and innovators unable to understand the possibilities of success in innovation, which has fundamentally added financing difficulties to innovative enterprises.

The selection of financing channels at different stages of the lifecycle

From these analyses of the lifecycle and financing demand of innovative enterprises, we can see that its high-risk feature requires that they face more constraints in financing than general enterprises. This has forced innovative enterprises to increase selectable financing channels.

Apart from self-accumulation, the financing channels of innovative enterprises are no more than government aid, risk investment, bank credit and the capital market. In the following studies, we will respectively discuss the influences of these on the financing environment for independently innovative enterprises based on the financing channels at different stages, as are shown in Fig. 3.2. We will focus on elaborating on the suggested targeted policies on the basis of examining the current situation.

Fig. 3.2.   Financing channels of innovative enterprises at different stages

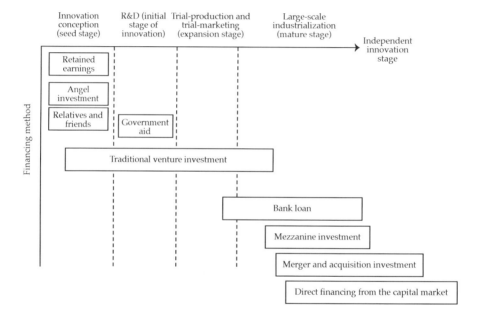

Innovative enterprise's financing and government's institutional support

Economics analysis and realistic considerations of government intervention in business financing

According to economics theory, government does not intervene in markets that can effectively play their own role in the market economy. Government intervention should appear in the markets that lack efficiency or the areas with market failure. The financing difficulties encountered by enterprises due to information asymmetry at the early stage of independent innovation are themselves market failures in financial transactions. At this moment, government needs to become involved in the establishment of a more effective institutional arrangement. In addition, the economics theory believes that technological innovation has spillover effects (technology spillover) and gives rise to positive externality, but these effects may also result in insufficient investment in technological innovation. The investment with technology enterprises as the investees will also be affected by the positive externality of technological innovation, which has made it difficult for innovative enterprises to raise funds. In this case, it is necessary for the government to give priority to the improvement of the financing environment for enterprises.

In recent years, quite a number of ministries, commissions, and financial institutions in China have actively made trials in the investment and financing arrangement in high and new technology industries, including demonstration projects of high and new technology industries, pilot projects of venture capital investment loans, innovation funds for SMEs, 863 Industrialization Base, special funds for the industrialization of agricultural results, the "cooperation platform" set up by the Ministry of Science and Technology, and various financial institutions, which have produced results. However, the financing difficulties faced by innovative enterprises are still serious.

What has happened indicates that the role of government in addressing financing difficulties faced by innovative enterprises is by no means to simply provide funds. More importantly, a rational institutional arrangement should be created in a bid to address the financial distortion brought about by market failure. Some success stories of developed countries, emerging industrialized countries, and even developing countries demonstrate that an effective way is that government needs to establish guiding funds and help build financing channels.

Analysis of the institutional demand of government guide fund in China

We can see from the lifecycle of innovative enterprises that the there is a huge capital demand at the R&D stage after the initial conceptual innovation stage, but enterprises are often in shortage of enough internal accumulation of funds, and thus it is difficult to access financing from banks, or to conduct merger and acquisition (M&A) and get listed. A fund provider is in urgent need to play the role as a bridge. Fig. 3.2 tells us that the only channel for financing at this stage is government aid and venture investment. Government support is not for returns, so risk is not an obstacle. But government's funds are usually less efficient in administrative usage; commercialized venture capital (VC) institutions are highly efficient, but they are not willing to invest in enterprises at this stage considering the risks. The bottleneck faced by enterprises at this stage is the greatest. Statistics show that only less than 20% of Chinese VC has gone to the seed stage and the early stage of innovation in recent years. A large number of commercial VC has been invested in the middle and later stages of innovation. Yet, if we are able to combine government funds without considering risks with the commercial VC institutions with high operational efficiency, then we can develop strengths and avoid weaknesses and explore an effective financing mode. The government guide fund is just the financing arrangement of start-up businesses of this kind.

The government guide fund usually refers to "the fund of funds" (FOF) set up by the government, another government, or designated institutions, and invested in VC firms in the form of stock rights and right of credit. The guide fund is somewhat policy supported and its investment direction is to the enterprises having weakness but with high-growth and high-tech potentials in the financing relationship between normal financial institutions and the financial market. However, the government guide fund is neither invested in VC directly nor involved in the everyday investment decisions of the invested seed funds. Rather, the guide fund leverages on the invested VC firms and relies on market principles to make investment in innovative enterprises. Venture capital investment guide funds for start-up enterprises need to be equipped with professional management that is essentially different from the general investment or loan business, have rich experiences in industrial investment, and have an abundant enterprise network to realize a sustainable development of the guide fund. For instance, the Small Business Investment Act of the United States (SBIC plan) is under the specific management of the Small Business Administration (SBA); Israeli Yozma is under the administration of the specially established Yozma Corporate; Singaporean Temasek is also relatively independent from the government. This kind of government guide fund can effectively integrate government policy objectives with the market operation of VC: first, the demonstration effect of the guide fund can be used to drive private investment of all kinds to set up and invest in VC firms in order to increase its financial strength; second, the influence of the guide fund can be utilized to guide VC firms to invest more in innovative enterprises at the seed stage and the initial stage of innovation.

We will now focus on examining thecorporate vision and operation model of Singaporean Temasek Holdings. Temasek is a whole-state owned holding company of the Ministry of Finance of Singapore and the objectives of its founding are to support government economic policies with investment, invest in the areas that private sectors are not willing to involve, and cultivate world-class companies through effective monitoring and commercial strategic investment so as to make contributions to the economic development of Singapore. As an independent legal entity supported by the government, Temasek has served as an isolation wall to cut off the direct links between the government and enterprises, and to prevent unnecessary government interventions in enterprises. At the same time, the company has become a driver between the government and enterprises and has connected the two while separating their links through capital holdings to guide directions of business. In daily operations, Temasek is able to operate under market principles and under the premise of giving due consideration to the government industrial policy; it regards the market as the orientation, and profit-making as the target and performance indicator. Thanks to Temasek, Singapore has identified a new economic positioning, upgraded domestic industries, and internationalized state-owned enterprises while introducing foreign capital. It is clear that the most successful experience of Temasek is that it has found a perfect balance between government object design and commercial interests, and has followed commercial principles without violating government principles.

Experiences of Zhongguancun Science and Technology Park

Chinese government has also undertaken meaningful experiments in supporting business growth and promoting economic development. In the late 1980s, the Chinese economy witnessed great changes. As learned from international experiences in the high-tech economy, the "electronic street" near Zhongguancun, Beijing, was designated by the state as the first domestic high and new technology industrial development zone and was bestowed with the mission of developing the country through science and education. Zhongguancun was the gathering place for science and technology, intelligence, talents and information. Thirty nine universities and colleges including Tsinghua University, Peking University, and Renmin University of China, along with hundreds of research institutes, including the Chinese Academy of Science, are located in this region. In addition, the development of the electronics industry was shaping the trend with a mix of technology, industry, and trade, which provided a favorable start-up environment similar to that in Silicon Valley in the United States and that in Hsinchu in Taiwan. Under the support of the national high-tech special industrial policies, the regional government kept improving its governance, helping the park continuously develop dynamic vigorous entrepreneurial activities during its development process in following two decades. The total revenue in technology, industry and trade kept a growth rate of more than 20%. By 2006, the total number of high-tech enterprises reached 18,000 with a total revenue of 600 billion yuan. The park also developed into a region covering a total area of 232 sqm, and featured a spatial distribution of one region, multi-parks[2]. It basically shaped the key industry cluster represented by software, integrated circuit, computer, networks, and communication. Of this, desktops have a market share of 40% domestically, laptops reached 25%, software and integrated circuit design industry accounted for one third of the national total, and the export of software accounted for 50% of the national total. Additionally, 95 of the World Top 500 enterprises have set up 148 subsidiaries inside the park. Of these, 65 are R&D companies; in the report of Deloitte Touche Tohmatsu on *Chinese Top 50 High-tech and High-growth Enterprises in 2006,* 22 of these come from Zhongguancun. Among Chinese enterprises listed in NASDAQ, half of them come from Zhongguancun. It is tempting to say that the technological innovation system and its economic size have played a key role in driving the economic restructuring and industrial upgrading in Beijing, as well as promoting the national economy and national scientific and technological progress.

While financially assisting the innovative enterprises in Zhongguancun Science and Technology Park, the government also intervenes in the financing for those enterprises through building direct and indirect financing platforms. The features of the indirect financing platform is to promote the building of a credit guarantee and credit rating system to solve the issue of information asymmetry of traditional financial institutions like commercial banks, and help enterprises access loans; through the direct financing platform, the government help enterprises to risk investment funds through setting up risk investment guide funds. We will elaborate on the experiences in indirect financing platform later on — here we will just make an analysis of the direct financing platform.

In order to help innovative enterprises get the funds from VC institutions, Zhongguancun Science and Technology Park has explored how the government guide integrates government funds with commercial risks. Through the VC seed fund[3] and follow-up investment fund,[4] the government invests funds in start-up enterprises by establishing VC enterprises with equity participation or working jointly with cooperative institutions. But, the government does not engage in specific operations and management of the start-up enterprises, rather, it relies on the market-oriented mode of operation of the cooperative and professional VC institutions, with no governmental administrative invervention. This kind of government guide fund has provided access to financing for early-stage SMEs in Zhongguancun, effectively integrated government policy objectives with the market-oriented mode of operation of VC, driven all kinds of social capital to invest in start-up enterprises, and increased the limited fiscal funds. At the same time, in order to encourage venture capitalists to invest in high-tech SMEs in the park, the park also set up risk subsidies to subsidize recognized VC enterprises according to their real investment in start-up enterprises in the park.

The park has also explored exit options of VC in many aspects. The administrative committee of the park has set up a special fund to assist the restructuring and listing of enterprises in order to motivate mature enterprises inside the park to go public and facilitate enterprises inside the park to leverage on the securities market to regulate their operations and development. The funding costs include: the expenses on accounting and auditing, legal services, sponsorship, asset assessment, financial consultancy, securities broker consulting during the process of restructuring and listing, and the expenses on relevant industrial and commercial registration for alternation. The park has also put in place relevant systems to guide and support enterprises that still require assistance to get listed. The equity exchange is used to carry out the business of property-rights exchange and equity trusteeship of high-tech enterprise and promote business M&A, which has broadened the channel for VC institutions to exit. Second, the transfer system of stock quoted prices, which has operated since 2006 for unlisted limited liability companies inside the park has provided an orderly stock transfer platform to unlisted companies inside the park. This system can help increase the liquidity of stocks, improve the business capital structure, introduce strategic investors and prepare for enterprises to go public as soon as possible. But this system is unfamiliar to a large number of SMEs. Therefore, the administrative committee has adopted policy guidance, online Q&A, and many other channels to promote the system, rolled out regulations on supporting experimental enterprises, and provided them with no less than RMB1 million in special funds for major industrialization and RMB200,000 in funds for restructuring and listing. At the moment, 22 businesses have been chosen as experimental enterprises, and the accumulated trading volume of listed stock has reached nearly RMB15.34 million shares with a turnover of RMB79.74 million. Among the listed companies, one company[5] has started to apply for an IPO, and two[6] have made directional capital increase. This rather innovative and challenging plan has provided more choices for capital exit of VC institutions and for the direct financing of start-up enterprises.

The direct financing system has made innovations in bonds and trust. In early 2003, "Zhongguancun High-tech Enterprises collective Trust Plan" organized a batch of full-fledged SMEs together to raise funds from the public through collectively issuing an trust plan. The Science and Technology Trust Guarantee Company provided a whole-process guarantee to enterprises issuing the collective trust plan. The collective trust has enabled small and medium-sized scientific and technological enterprises to raise capital for development through direct financing. The duration of trust is longer than that of loans, which is more conducive to the business development of SMEs. In 2007, the park also introduced RMB400 million in short-term financing bonds, collectively issued by 11 high-tech enterprises. The innovation of this collective issuance has been recognized by Beijing Municipal Development and Reform Commission and has provided a new mode of thinking for other regions in China to broaden SME financing channels.

The above-mentioned direct financing platform formed inside the park is shown in Fig. 3.3.

Fig 3.3.   The direct financing platform in Zhongguancun Science and Technology Park

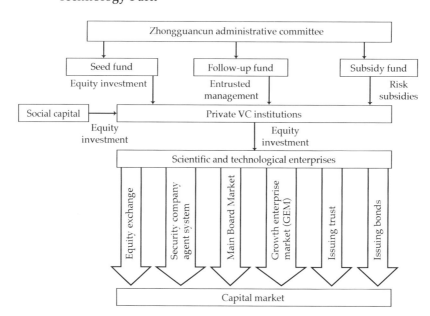

Discussions on the operation mode of the government guide fund in China

As the government guide fund is a new thing in China, the Chinese government needs to learn from international success stories and design a rational management system in line with China's social environment, including a source of capital, type of organization, principle of operation and mode of investment.

(1) Capital source. As the national government VC guide fund, it must be contributed by the central budget or competent departments and ministries and then by policy-supported financial institutions. Policy-based financial

institutions (such as China Development Bank) can provide soft loans for investment when the contributions of the central budget and competent departments and ministries are not sufficient.

(2) Orgainzational form. Optional organization types of guide funds include contractual fund, corporate fund and limited partnership fund. Various issues like legal provisions, fund management and tax avoidance need to be taken into consideration. It is fitting that the guide fund adopts contractual funds in accordance with existing legal provisions in China.

(3) Operation principle. The government VC guide fund provides the leverage to attract social capital to scientific and technological enterprises. The fund should target VC institutions featuring commercial operations, but not directly support start-up enterprises. In order to make sure that the supported VC institutions operate under the market principles, the guide fund shall be managed and operated by professional institutions while government institutions would generally not intervene in the specific operations of the supported seed fund. Rather, government institutions would carry out its supervision by developing regulations, being involved in major decisions, and guiding the seed fund to invest in areas and enterprises encouraged by the central government. The guide fund abides by the principle of giving benefits to the people in the distribution of revenues in order to meet the target of effectively magnifying government funds and guiding private capital. But risk reduction also needs to be taken into consideration for a sustainable development.

(4) Investment mode. In line with the national conditions, the government guide fund may choose the mode of "parent-seed funds." The "parent fund" sets up some wholly-owned special seed funds in light of the needs of national development strategies to support the application, demonstration, and commercialization of major scientific and technological projects in accordance with national medium and long-term plans on science and technology development. It also guide, for example, local budgets, social capital, and offshore funds, to jointly set up regional or specialized seed funds which provide financial services to scientific and technological projects, and to small and medium-sized scientific and technological enterprises.

The guide fund can investe in seed funds in the form of preferred stock. In so doing, concessions of part of the interests can be made to attract social capital. Preferred liquidation order and a series of investment control means can be adopted to control the risks for the guide fund and guarantee its sustainable development. The operation mode of the guide fund is shown in Fig. 3.4.

Fig. 3.4.   Operation mode of guide fund

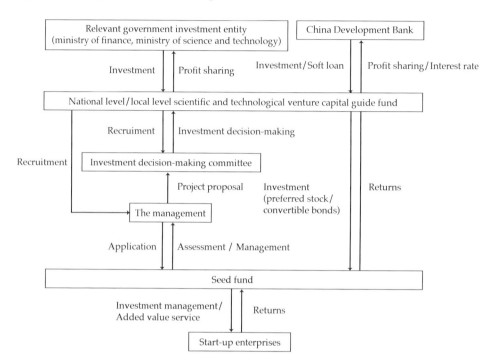

The relationship between innovative enterprise financing and VC investment mechanism

The lifecycle features of independent innovation enterprises have determined that traditional financing channels (including commercial bank loans and financing in the stock market) are not appropriate for them. Fig. 3.2 shows that VC in a broad sense (including Angel Investment, Mezzanine Investment and acquisition investment) actually covers all development stages before the recession stage. At the same time, different types of VC have different emphases at different stages. The following contents will generalize the contributions and shortages of VC to independent innovation enterprises from both theoretical and practical perspectives.

The natural affinity of VC to independent innovation enterprises

Venture capital, together with investment in both real estate and forestry, belong to the assets featuring poor liquidity and high risks among various investment portfolios. Cochrane (2001) made use of the transaction data of VC from 1987 to June 2000, provided by VentureOne, to make empirical studies. It was found that after getting rid of choice preference,[7] the standard deviation of return on VC was 98%, but the standard deviation of Standard and Poor's 500 Index during the same period was only 12%. The risk of VC is obviously higher than general stock investment. These kinds of risks are typically derived from enterprises being invested, including uncertainties of the enterprises being invested, less possibility in asset mortgage, and a higher degree of information asymmetry.

Investors certainly demand higher returns on investment as they take on higher risks. The high expected returns of VC could only come from high-growth enterprises; innovative enterprises might obtain these kinds of returns in 3 to 7 years (Gompers and Lerner, 2002).

Economics theories believe that perfect competition will ultimately make the enterprises of the industry generate average profit of the industry. VC institutions generally prefer that the industry into which they have invested is not a perfectly competitive industry, or their investees are more competitive in the industry they belong to. Therefore, innovative enterprises are favoured objectives for venture capitalists.

The essential differences between venture capital and traditional financing channels: taking bank loans as an example

Commercial bank loans are a popular business financing channel, but the high risk feature of innovative enterprises has made commercial banks shrink back. Venture capital and commercial bank loans have essential differences, displayed in the following three aspects.

First, VC is an equity capital and it generally makes investment in start-up enterprises in the form of stock rights. In so doing, investors can enjoy the high growth and high profits that might be realized by start-up enterprises according to their proportion of shares. But, when it comes to bank loans as a kind of debt capital, commercial banks can only collect the predetermined interest, regardless of the profit of a borrowing enterprise. If the business fails in its operation, the commercial bank might suffer from losses in the loan principal and interests, but the claims order is just before the equity capital.

Second, VC investors are willing to undertake higher risks to gain high expected returns, and thus this is the rich's game. Commercial banks must pursue the security of their loan funds and cannot take on high risks due to the structural features of their assets and debts.

Third, VC focuses on the future development of its invested enterprises, but commercial banks mainly focus on the current financial standing of borrowing enterprises. Venture capital does not require its invested enterprises to provide a guarantee, but it does attach more importance to their commercial plans, their capacity to deliver these plans, and the accomplishment and spirit of entrepreneurs — the "soft information" mentioned before. Apart from examining the report of the project's feasibility of borrowing enterprises, commercial banks put more premiums on their "hard information," such as current financial standing, and can even provide an integrated credit limit to those with strong financial standing. Therefore, enterprises can apply for loans within this limit at any time. For businesses with poor financial standing but a sound development prospect, commercial banks might either turn down their applications or require them to provide guarantees. See Table 3.2.

Therefore, VC is more effective than bank loans for financing technological innovation in line with the great uncertainties and the asset structural features of innovative enterprises at both the early innovation stage and the expansion stage.

Table 3.2.   Major differences between VC and commercial bank loans

VC	Bank Loan
Equity capital	Loan capital
High-risk, high expected revenue	Low-risk, low expected revenue
Mainly invest in intangible assets	Mainly invest in tangible assets
Future-oriented	Present-oriented
Low liquidity	High liquidity
Guarantee is not needed	Guarantee is needed
Revenue—Profit	Revenue—Interest rate
Relatively expensive	Relatively inexpensive

The match between the periodical preference of VC and the lifecycle periodicity of innovative enterprises

Independent innovation enterprises have different characteristics in risk and capital demand at the four stages: the innovation conceptual stage, the R&D stage, the trial-production and trial-marketing stage, and the large-scale industrialization stage. Venture capital institutions also have specific investment preference according to the development stage of the investment objectives (Gompers and Lerner, 2002), and accordingly formed VC of the seed stage, VC of the early stage of innovation, the venture stage of the expansion stage, bridge financing, and acquisition investment. The periodical preference of VC matches the periodicity of financing, which has shaped a sound micro-balance between supply and demand.

Venture capital institutions preferring the seed stage usually first provide a small amount of capital to entrepreneurs in order to decide whether the "idea" merits further consideration and further investment. Generally speaking, these VC institutions do not have a large amount of capital, and many angel investors prefer to make investment in start-up enterprises at the seed stage. VC institutions preferring the initial stage of innovation generally offer their invested enterprises a small amount of VC to help them carry out product development, sample product manufacturing and trial marketing; the invested enterprises at this stage generally cannot make a profit. Start-up enterprises at the expansion stage have sold enough products to consumers and received sufficient market feedbacks; they may have not profited and need more capital support for purchasing equipment, building inventory, and marketing channels. VC institutions focusing on the bridge financing often provide equity capital or subordinated debts to enterprises to help them prepare for listing or M&A. Acquisition investment institutions usually invest in enterprises with stable capital flow and at the large-scale industrialization stage to help early stage venture capitalists to exit.

The control of venture capital institutions over risks

The high risk investment feature of VC does not mean its imprudence in investment. VC institutions control risks through the innovation of method-and-process, and solve difficulties faced by innovative enterprises in financing. These innovations are: examination before making investment, design of investment structure, agreement in investment contract, and post-investment management.

Examination before making investment

Ang, Cole and Lin (2000) pointed out that start-up (innovative) enterprises have a high degree of information asymmetry. Norton (1995) and Nesheim (2000) further noted that the information asymmetry of innovative enterprises is even higher. Therefore, VC institutions might receive more than 100 applications each month, but only a very small proportion can finally obtain investment. As a matter of fact, screening and examination of technological innovation programs before making investment is one of the critical mechanisms for VC institutions to deal with adverse selections. Specifically speaking, before the investment, VC institutions usually accomplish several links, including program source, initial screening of programs, and the due diligence of programs in order to obtain the real information of the alternative investment programs and decrease information asymmetry.

(1) Program source. Generally speaking, the source for VC institutions to access alternative investment programs can be divided into two categories: the active contact of entrepreneurs, and the recommendation of other individuals or organizations. Wang Songqi (2003) found after researching the VC industry in China that active contact of entrepreneurs, recommendation of government authorities and the recommendation of program intermediaries represent the three major channels for VC institutions in China. Studies of Liu Manhong and Hu Bo (2001) on 73 Chinese VC institutions demonstrate that VC institutions usually obtain program information through diversified channels, including recommendation of government competent authorities, recommendation of intermediaries, introduction of friends, introduction of banks, active contact of entrepreneurs, media promotion, recommendation of shareholders, and so on. Of these, the most common and successful channels are active contact of entrepreneurs, recommendation of government authorities and introduction of friends, in that order (see Fig. 3.5).

(2) Preliminary screening of programs. The purpose of preliminary screening is to eliminate programs that are obviously beyond the range of investment. Generally speaking, VC institutions will use the following four standards while carrying out preliminary screenings:

Fig. 3.5.   Source of program for Chinese VC institutions

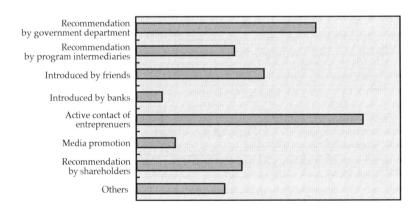

Source: Liu Manhong, Hu Bo, "Investigation Report on China Venture Capital" (research report on national soft subjects, 2001).

The first is the size of financing demand for the program. Most VC institutions will seek to realize a certain degree of risk diversification in their invested program, and set the lowest and highest investment limit for a single program. The average number of investments made by one Chinese VC institution is 10.5 programs. The preferred investment limit is RMB1 to 5 million for one single program, but some VC institutions have made investments in the program valued at more than RMB10 million.

The second is industry and market. The vast majority of VC institutions have a preference for industrial investment.

The third is the location. Venture capital institutions usually need to meet the enterprise's management in order to facilitate their management on invested enterprises, thus in consideration of management cost, some take regional preference as one of the standards for preliminary screening of programs. The survey of Tyebjee and Bruno (1984) indicates that about one fifth of the 46 VC institutions involved in their study had clearly taken regional preference as their standard for preliminary screening of programs, but most were concentrated in the region where the VC institution was located.

The fourth is business development stage. Different VC institutions usually prefer start-up enterprises at different stages.

(3) Due diligence. VC institutions need to undertake in-depth examinations of the alternative programs through due diligence after the preliminary screening to test and verify whether the information provided by start-up enterprises in commercial plans is real or not, i.e. due diligence of technology / product / market, of operation, and of finance and legal. Due diligence is, in effect, an information test and verification as well as information acquisition, and sometimespart of the due diligence goes beyond the range of commercial plans in order to obtain more personalized soft information.

Design of investment structure

Venture capital institutions need to decide which investment instrument will be selected, whether syndicated investment should be chosen, and how to make a staged investment. These have a direct consequence on the effectiveness of VC institutions in managing information asymmetry and risk uncertainties.

(1) Investment instrument selection. Investment instrument selection is one effective means for VC institutions to guard against and manage risks. Making full use of financial instruments, including common stock, preferred stock, convertible bonds, general debt instrument, and financial leasing can help prevent risks brought by information asymmetry to VC.

(2) Syndicated investment. Another common phenomenon in designing VCstructure is the cooperation between VC institutions in making joint investment in VC enterprises (syndication). Gompers and Lerner (2002) made an analysis of the 651 rounds of private capital financing carried out by 271 biotech enterprises between 1978 and 1989 and found that syndicated investment between VC institutions was more frequent than that between non-VC institutions (see Table 3.3).

Table 3.3.   Syndicated investment between VC institutions

Projects	First round of financing	Second round of financing	Third round of financing
Average number of venture investors	2.2	3.3	4.2
Average number of non-venture investors	0.5	0.9	1.1

Source: Gompers Paul A. and Josh Lerner, *The Venture Capital Cycle*, rev. ed. (repr., Cambridge: MIT Press, 2002), 191.

Innovation enterprises have a higher degree of uncertainties and information asymmetry that have forced VC institutions to make investment decisions only based on incomplete information. But introducing the investment project to other investors, the VC institution could obtain more information based on other investors' reactions and ultimately reduce the risks brought by information asymmetry.

(3) Staged financing. Staged financing is one of the most important mechanisms applied in venture capital. VC institutions make a staged assessment to decide whether to make further investment. VC institutions can effectively control uncertainties of venture capital and information asymmetry through staged investment, and reserve the rights to halt further investment due to poor business performance conditions, help investors reduce decision-making errors, and cut costs for wrong judgements.

Agreement in investment contract

Designing and signing the investment contract to help VC institutions and innovation enterprises deal with the distribution of the right to trade mainly includes the distribution of cash flow right and control right. The incentive and binding clauses for entrepreneurs in the contract includes vesting schedule, ratchet clause,[8] liquidation rights, anti-dilution clause and put rights. The purposes of these clauses include: linking the compensations for entrepreneurs to business performance; consistently maintaining the interests of entrepreneurs and investors; and protecting the interests of investors in the event of poor business performance in the future.

Post-investment management

While monitoring the investees, VC institutions remain active in two main aspects: monitoring as well as assistance and tutorship. On top of that, they also assist investees to arrange follow-up financing.

Table 3.4 has summarized the financing difficulties of innovative enterprises and the solutions of VC institutions. It is obvious that venture capital is a more appropriate financing channel compared with the traditional means of financing.

Table 3.4. Financing difficulties faced by innovative enterprises and solutions for VC institutions

Difficulties in technological innovation	Manifestations	Venture capital solutions
High uncertainties	There are high systematic risks due to uncertainties triggered by uncontrollable factors like technology, market or the macroeconomy for innovators.	Maintain the possibility of high proceeds: to make investment in the form of stock rights. Restrict downside risks: to control possible losses through staged investment; to spread risks and improve possibility of success through syndicated investment; to provide help for investees in the market, marketing and follow-up financing, and many other aspects.
	Technological innovators as entrepreneurs usually lack managerial experiences and executive capacity.	Assistance and service: to assist entrepreneurs to enhance the management and executive capacity; to help enterprises recruit and improve the management team. Reserve the control right to change entrepreneurs; to change business managers when necessary; to ensure a benign development of enterprises.
Asset structural issues	Intangible assets stay at the core of the asset structure in the technological innovation enterprises.	No guarantee is needed: venture capital generally makes investment in the form of stock rights. Value assessment: the assessment method of venture capital innovation can effectively manage challenges posed by asset structure

(Cont'd)

Difficulties in technological innovation	Manifestations	Venture capital solutions
Information asymmetry	Innovators know more information about innovation programs and innovative enterprises, but investors can only access incomplete and asymmetric information.	Information acquisition: to conduct preliminary strict screening and prudent investigations of a project; to carry out cross-examinations of technological innovation enterprises by syndicated investment with other investors; to complete a follow-up monitoring and reporting system.
	It is likely to trigger adverse choice and moral risks.	Risk control: to prevent entrepreneurs from using information advantages to harm the interests of investors by using convertible bonds; to restrict technological innovators from using capital through investment contract.
	Technological innovators as business managers have conflicting interests with VC institutions.	Interest consistency: to link the compensation for entrepreneurs to business performance; stock option of the management.
		Restrict the opportunistic behaviours of entrepreneurs: investors reserve the control right to change business managers; gradual delivery of entrepreneurs' share; major decisions need to be approved by the majority.

The support and prospect of public policies for venture capital

Venture capital is the financial field with commercially operated VC institutions as the main body, but foreign success stories tell us that government support is also an important factor. The public policies on promoting venture capital developed by the Chinese government generally include: (1) making direct investment in setting up state-owned VC institutions; (2) participating in VC institutions; (3) formulating relevant laws and regulations to attract foreign capital and private capital to get involved in VC; (4) giving appropriate tax breaks to VC institutions.

The central government promulgated the *Suggestions on Establishing Venture Capital Institutions* in 1999 to encourage and guide VC. The *Provisions on the Administration of Foreign-invested Venture Capital Enterprises* was formulated in 2001 and delivered on March 1, 2003 to attract foreign investment to the field of venture capital. Compared with the central government, the support of local governments are more specific and direct, such as the direct establishment of state-solely-funded VC institutions and the development of local regulations on stimulating VC development. Local regulations are mostly innovative; some even break through the provisions of existing national laws. These local regulations include: (1) allowing VC institutions to exist in the organizational form of limited partnership; (2) conditionally allowing VC institutions to enjoy tax cuts for high-tech enterprises,[9] allowing VC institutions to draw a certain proportion of pre-tax reserve fund for risks, and providing income tax credit to VC institutions for the income for re-investment; (3) setting up venture capital guidance funds and providing capital support to foreign-funded and private VC institutions; (4) allowing VC institutions to make full investment; (5) no upper limit on the proportion of intangible assets of technological innovation enterprises in the shares and the proportion shall be determined by the financers and the investors; (6) encouraging VC institutions and investees to adopt stock options, among other incentive instruments.

Although governments have adopted various measures to promote the development of venture capital and played a key role in this regard, they are not clear about how to define their positions in venture capital, giving rise to the problems of excessive direct involvement and insufficient policy guidance. For example, government direct capital support became equal to the role of direct venture investors, which producing undesired effects on private investment organizations; relevant legal provisions are weak. From the perspective of international experiences, the governments of countries with better VC development, such as the United States and Israel, didn't establish solely-owned VC institutions and were not directly involved in the operation of commercial VC institutions, even at the early stage of the VC development, with the exception of some non-profit community VC projects.[10]

In addition, the institutional environment and market environment for venture capital still need to be improved in many aspects. The *Notice on Preferential Policies on Business Income Tax* promulgated in 1994 states that "newly founded high-tech enterprises enjoy the exemption of income tax for the first two years after going into operation," but the production year and profit-making year are more than two years and have therefore imposed restrictions on the mechanism effect of the tax cut policies. From the perspective of the market system, the shortage of GEM has become a big bottleneck for the development of venture capital in China.

Based on the above studies, the following policy suggestions are proposed:

First, stopping the establishment of new solely state-funded VC institutions with government direct involvement to eliminate the crowding-out effect on private institutions. Existing solely state-funded and even state-holding VC institutions can be gradually eliminated through attracting private capital and foreign equity.

Second, setting up a venture capital guide fund with government capitals, or participating in VC institutions funded by private capital or foreign capital with a certain low proportion (without controlling stake); government capital may enjoy less, late, or even no shares in profit distributions within a certain time limit.

Third, relaxing limits on capital going to the field of VC and increasing the type of investors and the size of capital.

Fourth, working harder to protect intellectual property rights (IPR) and strengthening the social credit system in order to improve the efficiency of venture capital in promoting technological innovation.

Fifth, establishing a multilevel capital market, establishing the GEM in due time, and providing a highly efficient exit channel for venture capital.

Innovative business financing and credit supply of commercial banks

Commercial banks as the traditional financing channel have special decision criteria for the demand of innovative enterprises for financing due to the existence of information asymmetry and uncertainties and the characteristics of their own operation. Their enthusiasm for investing in innovative enterprises is much lower than that of VC institutions. This section will focus on the discussion about the characteristics of commercial banks in innovative business financing and the internal causes as well as feasible methods.

General characteristics of the risk premium of commercial banks and their preference in customer selection

First, commercial banks are risk-averse investors and their attitude to risks lies in the nature of their assets and liabilities, to a large extent. The liabilities of commercial banks are mainly that of depositors and their assets are mainly loans. If commercial banks want to have a prudent operation, they must guarantee the security of their assets and certain liquidity to meet the needs of depositors for withdrawal. Arguably, the assets and liabilities of commercial banks are featured by "hard liabilities, soft assets." In fact, regulators of each country have also required banks to guarantee the security of their assets.

Second, the interest rate demanded by commercial banks has explicit and implicit upper limits. China's commercial banks can now only issue loans with the interest rate regulated by the central bank. Commercial banks are allowed to raise the interest rate on this basis, but there is still an implicit upper limit. Even though the interest rate is liberalized, commercial banks cannot demand the interest rate according to the degree of risk of a bank loan project due to the existence of market competition, and thus the proceeds have implicit upper limits. Petersen and Rajan (1995) proved that banks are unable to demand a higher interest rate to make up for the risks they assumed when it comes to high-risk projects in the competitive market. On top of that, commercial banks are not allowed to hold business shares in countries practicing separate supervision, which has eliminated the possibilityof banks sharing the business growth interests by means of equity investment.

The risk premium return feature of commercial banks has determined that they prefer large-scale loan projects when selecting customers. This can help them obtain the effects of the economy of scale and increase unit loan profits on the one hand, and cut relatively high risks in loans of small enterprises on the other. But, the line of credit of a single loan is generally subject to the restriction of the government in a bid to avoid risks of excessive concentration of investment. For instance, the *Law of the People's Republic of China on Commercial Banks* states, "the balance of loans to a single counterparty shall not exceed 10% of the capital balance of the commercial bank."[11]

The risk premium return feature of commercial banks has also determined that more guarantees are demanded for loans, includingmortgage, pledge and guarantee. If bank clients cannot repay loans on time, banks can dispose the collateral or demand the guarantor to pay back the loans to reduce bank losses. China's regulatory authority regulates that commercial banks are generally not allowed to issue loans on credit.

We can conclude that when choosing loan applicants, commercial banks prefer low-risk clients where the interest rate cannot be increased; when the loan is not great, commercial banks prefer clients applying for a large scale of loans; with other things being equal, commercial banks prefer clients with effective guarantees.

Credit rationing and price discrimination under information asymmetry

The biggest problem encountered by independent innovation enterprises during the process of geting indirect financing is the obstacle for financial institutions to gather, analyze, and process information. With perfect information, the market can allow the role of price in adjustment during the competition process to balance the demand and supply. If perfect information is available during the process of indirect financing, the changes in interest rate will automatically balance the demand and supply of capital. Banks will demand higher interest rates for high-risk projects but lower interest rates for low-risk projects.

But, adverse choice might occur when perfect information is not available, making the price unable to serve as an effective market regulation means. With information asymmetry, innovators (innovative enterprises) know more information about the risk of innovation projects, while banks do not have access to the information. Banks can determine the interest rate level according to the average degree of risks of like projects in the market, but innovators with a lower level of risk do not apply for bank loans, while innovators with a higher level of risks choose to apply. Clients who are willing to accept a high interest rate usually have higher risks. In face of the adverse choice problem, commercial banks will adopt the credit rationing measures noted by Stiglitz and Weiss to address the problem. The so-called credit rationing refers to: (1) two groups of borrowers who have almost all similar conditions while the one group can get loans, but the other group cannot, even if the latter would accept a higher interest rate; (2) the demand of borrowers for loans can only be partially met.

The credit rationing behavior of commercial banks facing information asymmetry can be explained by the following models.

Suppose the loan amount needed by borrowers is B, the interest rate is r, the value of collateral demanded by the bank is C, the risk of loan project is σ, the anticipated proceeds of the project is $E\,(P)$. At the same time, suppose that borrowers totally rely on loans to accomplish their projects. That means borrowers do not use their own funds in project investment, apart from providing collateral.

When it comes to the access of information, suppose the borrower knows the probability distribution of project risk, but banks only know the overall status of risk of like borrowers. If the project succeeds in the end, borrowers can repay the principal of loans; if the project cannot succeed, banks can dispose of the collateral and the project proceeds $E\,(P)$ at the same time belong to banks. There might be a result in between that signifies the project is not successful; the total sum of the project's proceeds and the collateral is bigger than the amount that shall be paid back to banks by borrowers. For the borrowers, the expected project proceeds is the function of project risk σ, that is:

$$E(P) = f(\sigma) \tag{1}$$

In the meantime, suppose $E(P)$ is the increasing function of σ, namely, the expected proceeds will keep rising with the increase of project risks, which is a more realistic hypothesis.

The anticipated profit π of borrowers depends on whether the project will succeed; if it does succeed, the profits for the borrower will be $E(P) - B(1 + r)$; if it fails to succeed, the borrower will not generate $E(P)$, lose the collateral, and its profit will be $-C$, that is:

$$\pi = \max\left[-C, E(P) - B(1 + r)\right] \qquad (2)$$

For any given project risk σ_0, the anticipated profit function of the borrower is shown in Fig. 3.6.

Fig. 3.6.   Anticipated profits of borrowers

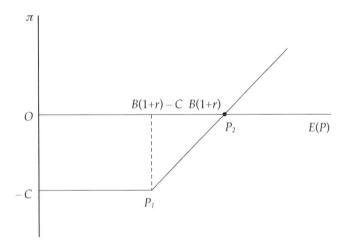

In Fig. 3.6, the vertical coordinate is the anticipated profit of the borrower and the horizontal coordinate is the anticipated project proceeds. In front of point P_1, project proceeds and the collateral together cannot pay back the principal and interest of loans, that is $E(P) + C < B(1 + r)$; the loss of borrowers is $-C$. At the P_1 point, the project proceeds and the collateral together exactly equal the principal and interests of loans; the profits of the borrower is 0, and the point P_2 coordinate is $[B(1 + r), 0]$. Behind point P_2, the borrower can generate positive profits.

For the borrower, the minimum limit it can accept is a positive value of its anticipated profit, which shall fall into the right side of point P_2, that is:

$$E(P) = f(\hat{\sigma}) > B(1 + r) \qquad (3)$$

As such, we can see that each given interest rate $\hat{\gamma}$ has one and only one risk critical value $\hat{\sigma}$, and an enterprise will apply for loans from banks only when the business risk degree is bigger than the critical value $\hat{\sigma}$. In other words, an enterprise with a risk lower than $\hat{\sigma}$ will not choose to apply for loans from banks due to the negative value of the anticipated profit. An enterprise will only choose to apply for bank loans when the risk is bigger than the critical value $\hat{\sigma}$.

This conclusion signifies that a bank, in effect, has turned down some low-risk clients if it raises the interest rate, while clients getting bank loans are high-risk, which has demonstrated the adverse choice problem faced by banks.

As a matter of fact, Formula (3) has also presented the demand curve of enterprises for bank loans and the shape is shown in Fig. 3.7.

Fig. 3.7.   The Demand curve of enterprises for bank loans

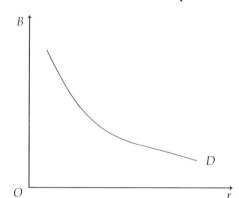

The business loan demand curve is a downward sloping curve. Higher interest rates means less demand for bank loans.

Banks are unable to demand a sufficiently high interest rate for high-risk projects due to the explicit and implicit interest rate upper limit; however, due to the existence of adverse choice, the growth rate of bank proceeds $E(L)$ will be slower than that of the interest rate r, and until it is behind some point r^*, bank proceeds will drop with the rise of interest rate (see Fig. 3.8).

We can see from Fig. 3.8 that the bank supply curve is no longer a typical monotonous rising curve; rather, it is a non-monotonic curve with an optimum interest rate r^* under the condition of information asymmetry.

If the demand for loans surpasses the supply at the interest rate level of r^*,[12] according to the traditional supply and demand theory, market competition will push up the equilibrium interest rate, and banks will increase the credit supply. But with information asymmetry, the supply is less than the demand, but r^* is the market equilibrium interest rate. Therefore, from the perspective of banks, their

Fig. 3.8. Optimum supply interest rate of banks

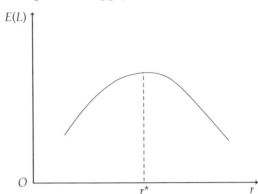

expected return will shrink as the risk might be higher when they issue loans with an interest rate higher than r^*. As such, banks would rather choose credit rationing but not expand the size of credit with a higher interest rate.

In line with the demand of enterprises and the supply of banks, the credit rationing behavior of commercial banks under the condition of information asymmetry can be further analyzed (see Fig. 3.9).

Fig. 3.9 can be used to explain the discrimination treatment of commercial banks to different kinds of borrowers. As shown in Fig. 3.9, suppose D_1 represents the loan demand curve of a group of enterprises with lower risk degree and without information asymmetry, and D_2 represents the loan demand curve of enterprises with higher risk degree and information asymmetry. Under this condition, banks will adopt the competitive equilibrium interest rate r_1 for the first group of clients and the rationing equilibrium interest rate r^* for the second group of clients. As a result, the size of loans provided to the first group

Fig. 3.9. Analysis of the credit market under information asymmetry

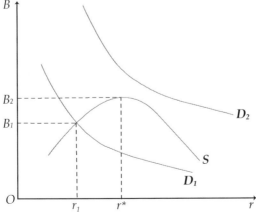

of clients is B_1 and the loan amount issued to the second group of clients is $(B_2–B_1)$. Thus it can be seen that commercial banks have practiced the price discrimination policy under different information conditions. This has reflected a reality in the current market: banks usually provide loans with a lower interest rate to large enterprises with more transparent information and stable cash flow, but practice credit rationing and demand higher interest rate for small and medium-sized technology enterprises with serious information asymmetry and higher risks (Li Yang, and Yang Siqun, 2001).

Solutions to the difficulties in getting loans and financing for enterprise's independent innovation

We have analyzed that the traditional businesses of commercial banks avoid issuing loans to innovative enterprises from the theoretical and practical perspectives, but commercial banks are also trying to develop new business varieties to expand the business scope under an increasingly competitive banking industry. Diversified operations of commercial banks have become a global trend. During this process, commercial banks are now making their way into the financial sector in a prudent manner through business and product innovation. Both Chinese and foreign financial institutions have made some successful explorations. We will make an analysis of the effective innovations of the two different channels based on the two cases, namely the Silicon Valley Bank and the "Gazelle Plan" of Beijing Zhongguancun Science and Technology Park.

The investment and credit integration mode of the Silicon Valley Bank

Silicon Valley Bank was founded jointly by several banks in 1983 with a registered capital of USD5 million. Silicon Valley Bank has remained at the center of entrepreneurial activities in the valley and has been regarded as a strong force for supporting the dynamic entrepreneurial activities in the valley. Silicon Valley Bank was founded at a time when the U.S. commercial banks' business re-integrated with the investment banks' business. After going through the process of integration–separation–integration, American commercial banks started to develop the investment banking industry again. Especially in 1970s, the securitization of the capital market accelerated this trend. In the 1980s, the innovation of commercial banks and the diversification, specialization, centralization, and internationalization development of investment banks exerted an impact on the business strategy of Silicon Valley Bank and made it, as a commercial bank, have many investment bank characteristics.

The operation philosophy of Silicon Valley Bank in supporting business innovative financing is by integrating investment and credit. In order to cut down risks, the bank only provides services to the company supported by VC enterprises and tries to find more VC enterprises for cooperation. Silicon Valley Bank will ink an agreement with its clients, requiring them to use their technical patents as the security guarantees. In accordance with the agreement, the technical patent of a company will belong to the bank if the company cannot repay the money; if a company cannot continue its operation, the proceeds on the selling of the patent will also be used to repay Silicon Valley Bank first and then to reimburse VC companies. This has forced VC companies to cooperate with the Silicon Valley Bank. This requirement has undoubtedly promoted related VC companies to have a close link with the bank, which has made the bank learn more about the operation status of the enterprise and thus reduce risks. Silicon Valley Bank, at the same time, provides VC institutions and its investees with direct banking services and it usually establishes banking outlets adjacent to VC institutions. Silicon Valley Bank is also a shareholder and partner of more than 200 venture capital funds. All efforts abovementioned have made Silicon Valley Bank and venture capital build a network of relationships together so that they can share information and engage in mutual cooperation.

The operation mode of integrating the investment and credit of Silicon Valley Bank has two types of investment: "leveraged" investment and "equity" investment. The "leveraged" investment is mainly to draw part of the funds of its clients. A large majority of the venture capital funds come from the sales of bonds and stocks, but Silicon Valley Bank will draw part of the funds of its clients as the capital for venture investment as a means to cut the amount of the raised capital and expenses from its fund raising. Then the bank will invest the capital in the form of lending in start-ups, obtain an interest rate higher than that of common investment, take back the principal, and pay corresponding interest rate. For example, the Silicon Valley Bank adopted the "leveraged" investment when selecting Quintessent Communication Company in the information and electronic technology industry as a client and an objective for venture capital, offering loans with a higher interest rate to the company as a price for risks; it helped the company develop through consultation and assistance. When "equity" investment is adopted, the Silicon Valley Bank will sign an agreement with VC enterprises on share stock rights or subscription rights in order to reap benefits from the exit. For example, Silicon Valley Bank adopted the "equity" investment when making venture investment in Neurogenetics Corporation, a biotechnological company, and closely cooperated with Advent International, a venture capital firm, to strengthen corporation supervision, provide consulting, banking services, and find more investors for the company.

Thanks to the development over the past more than two decades, Silicon Valley Bank has made remarkable achievements. The bank has 9,500 clients, nearly 1,000 employees and an asset of USD4.4 billion. Its branches can be found in all regions with concentrated high-tech companies across the U.S. By 2005, Silicon Valley Bank had provided financial services to 40,000 startups. Nearly one third of the technology and life science companies that launched IPOs in 2000 and 2001 are the clients of the Silicon Valley Bank; more than half of the VC companies of the United States are served by Silicon Valley Bank.

"Gazelle Plan"[13] of Zhongguancun Science and Technology Park

Zhongguancun Science and Technology Park, as a product of China's reform and opening-up, was founded in the late 1980s. As the first high-tech industrial development zone, the park possessed the science, technology, intelligence, talent, and information force of the capital city. Under the support of the national special industrial policies of a high-tech industrial development zone, the regional government kept improving government conducts and maintained the park to be a dynamic region for starting up businesses, and witnessed a gross income growth rate of over 20% technology, industry, and trade in the following two decades.

Zhongguancun Science and Technology Park has a large batch of "Gazelle Enterprises" which undergo rapid development, and which are involved in numerous national strategic sectors such as software, integrated circuit, biomedicine, new material, environment, and new energy. In 2003, the park recorded a total of RMB50 billion in revenue for technology, industry, and trade— an increased average of nearly 60%. However, these Gazelle Enterprises lack the capital for industrial development and have quite a narrow financing channel; this had become a bottleneck for Zhongguancun to overcome. Given this situation, Zhongguancun Science and Technology Park rolled out the "Gazelle Plan" aiming at providing financing solutions to Gazelle Enterprises in the park, helping them jump higher and leap faster.

The design principle of the "Gazelle Plan" is to integrate credit assessment, credit incentives and a restraint mechanism with guarantee lending businesses, then concentrate financial resources and build a high-efficient, low-cost guarantee lending channel through government guidance and motivation to overcome the bottleneck for bank loans and to introduce VC institutions.

The Administrative Committee of Zhongguancun Science and Technology Park built the guarantee and credit rating systems in line with the information asymmetry between financial institutions and enterprises seeking finance in order to strengthen the information transfer. In 1999, the park founded "Beijing Zhongguancun Science and Technology Guarantee Co., Ltd," a policy-based credit guarantee institution with a registered capital of RMB423 million contributed by the local government. This company immediately became the guarantee institution with the closest enterprise connection in Zhongguancun. As the operation unfolded and the requirements for credit ratings became higher in 2003, under the advocacy of Beijing Zhongguancun Science and Technology Guarantee Co., Ltd, the Administrative Committee of Zhongguancun Science and Technology Park, high-tech enterprises inside the park, and relevant intermediary organizations jointly established "Beijing Zhongguancun Enterprises Credit Promotion Association"— a non-profit social organization legal entity. The association introduced, for example, the credit rating report, credit investigation report, and debt receivable management consulting report; it also absorbed enterprises with certain credit ratings inside the association as members, guided the credit consciousness and behaviour of enterprises, built credit information platforms and made public the list of credible and non-credible enterprises. The building of this kind of socialized credit system has made the "soft information" transfer of SMEs more rapid and effective; commercial banks could simultaneously access the soft and hard information of innovative SMEs, which is an important premise to changing credit rationing and price discrimination.

According to the "Gazelle Plan," the Administrative Committee of Zhongguancun Science and Technology Park will present a certificate for "Gazelle Enterprise" when enterprises need loans; for these credible enterprises, the Beijing Zhongguancun Science and Technology Guarantee Co., Ltd. will shorten the examination and approval time and simplify the guarantee measures. The cooperative banks inside the park practice the principle of "providing loans at the sight of guarantee" and without raising the interest rate; the Administrative Committee will also offer an interest subsidy of 20%. Small and medium-sized science and technology enterprises can also get loans from banks even if they do not get involved in the "Gazelle Plan," but the counter guarantee conditions are rigorous , the process and formalities are complex, and the loans need to raise by 30% from the 5.31% benchmark interest rate, if through their own credit. However, after joining in the "Gazelle Plan," the enterprises will have a fixed interest rate for loans and the enterprises meeting the conditions of "Gazelle Plan" for many consecutive years will enjoy an even higher interest subsidy. At the same time, their accumulated credit can build a sound image for enterprises, which is certainly an advantage in project bidding.

Beijing Zhongguancun Science and Technology Guarantee Co., Ltd and Beijing Zhongguancun Enterprises Credit Promotion Association have enlivened all parties in the indirect financing channel of scientific and technological SMEs. The first is the interactions between enterprise intermediaries, the credit promotion association, and the science and technology guarantee companies. The credit promotion association designated 11 strong credit intermediary companies to rate the enterprises inside the park; all assessment reports must be submitted to the credit promotion association for filing. Due to the lack of influence from the credit system, the consciousness of enterprises inside the park in purchasing these credit products was initially limited. In order to encourage and promote the credit system of the park, the Administrative Committee developed a mandatory regulation that these credit reports were required in order for the enterprises to access the assistance of various special funds, and apply for guarantees and bank loans. At the same time, using credit reports in the guarantee business, such as loan guarantee, performance guarantee, and the guarantee for sales on account, avoids the preliminary examination of the guarantee company. Additionally, the Administrative Committee also adopted a more direct financial support method to press ahead with the use of credit products by the enterprises inside the park. It set up the special support fund for enterprises to buy intermediary services, and contributed 50% of the actual amount gratis when enterprises purchased credit intermediary service, finance intermediary service, and an IPR agent, among other intermediary services. These measures have helped many enterprises understand and correctly use the credit report and credit investigation system. The second is the interaction between the guarantee institution and commercial banks inside the park. The "Gazelle Plan" requires the involved enterprises to accept the rating of the credit intermediary institution developed by the credit promotion association and also to become a member of the association. They can then enter into the fast track of guarantee examination and approval, and into that of cooperative banks. Cooperative banks will deliver loans at the sight of guarantee and carry out the benchmark loan interest rate. These enterprises can also enjoy 50% of loan interest subsidies from the Administrative Committee. This measure has not only become a green channel for SMEs inside the park to seek indirect financing from commercial banks, but also alleviated the embarrassing situation that the commercial banks inside the park are in no position to issue loans to many SMEs.

See Fig. 3.10 for the indirect financing platform formed inside the Zhongguancun Science and Technology Park.

At present, the financing chain of high-tech SME, the credit intermediary institution, credit promotion association, science and technology guarantee company, and commercial banks in the park has worked well. The science and technology guarantee company has provided the financing guarantee support valued at a total of RMB11.6 billion to 2,800 enterprises. 15% of the total enterprises in the park are beneficiaries, and the guaranteed loans amount

Fig. 3.10.   The indirect financing platform inside the Zhongguancun Science and Technology Park

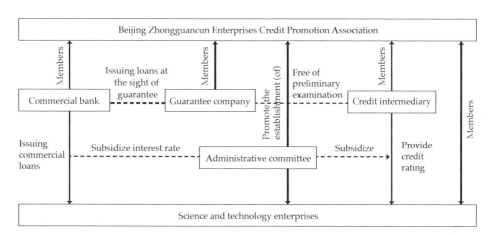

account for 10% of the loans for enterprises. The "Gazelle Plan" and the three green channels have also made remarkable achievements. In 2006, 254 enterprises were provided with special guarantee loans valued at RMB1.703 billion. The newly-founded enterprises with a production value of more than RMB100 million in the park in 2006 are this kind of Gazelle Enterprise. The virtuous cycle of this kind of indirect financing system has resulted from the combined effect of credit rating, credit incentives and restraint mechanism. The high efficient and low-cost innovative SME financing platform has been built through the efforts of the local government in capital assistance and cultivation of social morality. Fig. 3.11, Fig. 3.12 and Fig. 3.13 respectively demonstrate the effect of the "Gazelle Plan" on alleviating the financing difficulties of innovative enterprises in the park.

Expected institutional innovation of financial institutions

We have seen from the above cases the two different paths in improving the financing environment for innovative enterprises. The locations of the two cases have distinctive features. The Silicon Valley is the cradle of technological innovation enterprises of the United States and Zhongguancun is the area with the high concentration of high-tech enterprises in Northern China. Innovative enterprises in these two areas are all facing difficulties in getting debt financing from traditional commercial banks. The problems of the Silicon Valley are solved by the institutional innovation of financial institutions; the Silicon Valley Bank integrates the leveraged investment with the equity investment and

Fig. 3.11. The number of enterprises obtaining VC in Zhongguancun Science and Technology Park between 2001 and 2005

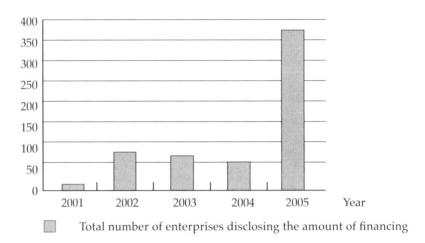

Total number of enterprises disclosing the amount of financing

Source: Zero2IPO Research Center.

Fig. 3.12. The total amount of the guaranteed loans obtained through the "Gazelle Plan" between 2005 and 2006

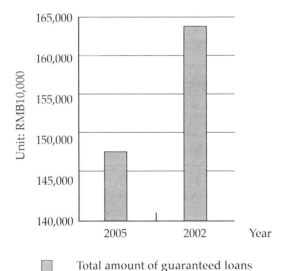

Total amount of guaranteed loans

Source: "Statistical Results of the Questionnaire on the Data and Analysis of the Business Financing Market in Zhongguancun Science and Technology Park in 2006" (Financial Branch of Beijing Software Industry Association).

Fig. 3.13.   Surveys on the knowledge of the green channel for loans in the park

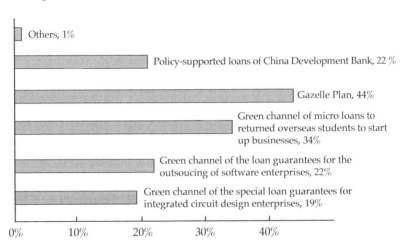

Source: The same as Fig. 3.12

closely connects with VC companies. At the same time, it also provides direct banking services to both the enterprises invested by VC institutions and the VC institutions themselves, not only serving small VC innovative businesses but controlling risks under an acceptable level. No single financial institution in Zhongguancun has made explorations in institutional innovation for innovative enterprises there. Rather, it is up to a government body (the Administrative Committee of Zhongguancun Science and Technology Park) to coordinate various institutions in the park, including the credit promotion association, the science and technology guarantee company, and banks. In addressing the issue of information asymmetry, innovative financial services are realized by relying on information transfer (credit rating system) and information improvement (credit guarantee system). Apart from these two development paths, many scholars have also proposed developing small and medium-sized financial institutions and private financial institutions to address the financing difficulties in small-scale enterprises' independent innovation. They believe that small banks and private finance have more strong points in dealing with information asymmetry and the researchers of this book consider this to be the third path for financial innovation for improving business independent innovation. There are both possibilities and a necessity to promote these three methods in the Chinese financial market at the moment.

Institutional innovation of financial institution: investment and credit alliance

The model of the Silicon Valley Bank can be summarized into "investment and credit integration" and "investment and credit alliance" which is the close cooperation between bank credit business and the investment business of VC companies. The principle of "security, liquidity and profitability" of bank capital has made many independent innovation enterprises unable to meet the lending criteria of banks; thus, there is a contradiction between the "high risk" of enterprises and the "stable return" of banks. But this contradiction can be alleviated to a large extent through integrating the two platforms of investment of VC companies and bank loans, which will reduce the difficulties in science and technology commercialization, and in development financing of small and medium-sized science and technology enterprises.

The specific approach is that a VC company signs the "Investment and Credit Alliance Agreement" with a bank, and the bank provides a certain line of credit. Within the line of credit, innovative enterprises, under the support of VC companies, can entrust a bank to apply for loans within the line of credit if they are in shortage of capital—they do not need to go through red-tape examinations, and the entrusted loans can be directly used as the short-term liquidity loans for innovative enterprises. Through drawing on the experiences of the Silicon Valley Bank, banks can become actively involved in the investment projects of VC companies, and not necessarily wait for the enterprises in need of financing to come to them.[14] This alliance is not the entrusted loan in a general sense; rather, it is a meaningful experiment of commercial banks involved in venture capital. Under this innovative mode, banks have cut the costs for accessing acquisition and confirmation of information through the professional investment management of VC companies. They have also built a retiform contact mechanism through serving borrowing enterprises and VC companies to track the development of enterprises and greatly reduce the information asymmetry of borrowing enterprises. Meanwhile, banks can also accumulate and cultivate a batch of potential excellent customer bases and provide them with more extensive financial cooperation space. For the part of VC companies, long-term investment and management have made them fully aware of the operating management and capital demand of the enterprises into which they invested. In so doing, VC companies can effectively control capital risks and provide enterprises with substantial value-added services. For businesses, this platform can help them avoid complicated procedures when dealing with banks, and also get short-term liquidity for their development.

This kind of innovation mode has already been put into practice in China.

In early 2004, Shanghai Venture Capital Co., Ltd. and China Merchants Bank established the first "investment and credit alliance" specialized in supporting Shanghai Venture Capital Co., Ltd. and its invested enterprises with real market potentials and repayment capacities. According to the "investment and credit alliance" plan, China Merchants Bank provided RMB150 million in a line of credit to Shanghai Venture Capital Co., Ltd. China Merchants Bank can be entrusted to issue loans to the high-tech enterprises invested by Shanghai Venture Capital Co., Ltd. so long as Shanghai Venture Capital Co., Ltd. believes that they have capital demand. Shanghai Venture Capital Co., Ltd. can take professional management as the guarantee and generate more profits by means of debt-to-equity swap, which has been widely welcomed. Until now, Shanghai Venture Capital Co., Ltd. has provided support for nearly ten enterprises with a financing amount of RMB50 million, and not a single case has become overdue or generated bad debt. Therefore, financial resources have been effectively integrated.

Institutional innovation of financial institutions: credit rating and guarantee system

We have found from the case of Zhongguancun the important role of transferring information through a credit rating system and of strengthening information through a guarantee system. These two channels are invaluable for financial institutions and innovative enterprises with information asymmetry.

The position of the credit rating system in the market economy has been gradually recognized. In particular, for independent innovation enterprises with high risk, great uncertainties, a high degree of information asymmetry, and a large amount of intangible assets, the evaluation of a social independent rating agency can decrease their market transaction costs, enabling easy financing. The evaluation of enterprises by credit rating institutions covers not only the balance sheet, income statement, cash flow statement, and other financial standings, but also the analysis of business operation environment, enterprise operation strategy, management quality, internal control, crisis management, intangible assets, and litigation report, among others. The result of this rating includes both "hard information" and "soft information" which will better help them obtain financing. In this sense, using this scientific and objective social rating agency as the channel for information transfer is currently an essential measure to improve financing for independent innovation enterprises.

At the early stage of development of the credit rating system, the government's encouragement is essential. Government can, through directly building an enterprise credit information database, designate companies to carry out commercialized operations, and coordinate the industry and commerce, tax, and other government departments to cooperate and deliver data to the credit information organization, and move forward with the establishment of a credit rating system. The credit rating system in Zhongguancun is a success story of government guidance.

Setting up a credit guarantee system is an international practice to support SMEs, a financial intermediary behaviour integrating credit certification and asset liability. This is the result of the government's comprehensive application of market economic means and macro-economic regulation and control, and an important financial service means to strengthen information transfer between enterprises, control risks and assist SMEs. In the credit guarantee system, professional credit guarantee institutions can effectively eliminate the obstacles of insufficient collateral for the financing of innovative enterprise, complement deficient business credit, and cut risks undertaken by financial institutions.

China attaches great importance to the practices in the credit guarantee and has achieved more results in this regard. In 1992, SME mutual guarantee funds emerged spontaneously in various areas. In 1999, the central government formally started to implement the pilot program of the SME credit guarantee system and generalized it as "three principles," featuring "a mix of supporting development and preventing risks," "a mix of government support and market operation," "a mix of providing guarantee and improving credit," and a "one main body, two wings, three levels" framework.[15] "One body" refers to the mainstay of the guarantee system while the "two wings" refer to mutual guarantee institution and the commercial guarantee institution with SME as their service objectives; these two institutions are important supplements for the public guarantee institution founded by the government. The "three levels" refer to the central, provincial, and local municipal government levels. The local municipal government level is responsible for the direct guaranteeing business, the provincial government level mainly re-guarantees the institutions guaranteed by local municipal government level, and the central government level is the state SME re-guarantee institution.

For all this, there all still many problems pending solutions. For instance, the "one main body, two wings, three levels" framework lacks clear-cut legal protection for two distinct problems: the mixed operation of a government-supported guarantee institution, commercial guarantee institution and mutual guarantee institution, and the problem of the absence of classified management and operation, and lack of criteria and grounds in risk control, responsibility sharing, and the internal and external supervision. Therefore, the top priority should strengthening the monitoring and risk control on the basis of levelling legislations, and making clear the relationship in rights and interests between the guarantee corporation, government, and enterprise.

Institutional innovation of financial institution: small and medium-sized financial institution and private financial institution

Developing small and medium-sized financial institutions and correctly guiding informal finance are one of the means to solve the financing difficulties of Chinese SMEs. Although the central government has repeatedly emphasized that large banks need to support the development of SMEs,[16] the operation features of large banks have made them impossible to actively heed the call of the government. In contrast, small and medium-sized financial institutions have a stronger relationship with innovative enterprises due to the easy access of both hard and soft information" of innovative enterprises. However, the Chinese government has imposed strict restrictions on small and medium-sized financial institutions, including classification in the scope of their operation, which has resulted in limited access to financial resources under the monopoly of large banks. At the same time, the improper intervention of government at all levels in small and medium-sized financial institutions, the absence of a credit rating system and the unsound credit guarantee system have restrained the growth of small and medium-sized financial institutions. Under this connection, relevant policy support needs to be provided to small and medium-sized financial institution.

When it comes to the development mode of small and medium-sized financial institutions, there is a need to relax restrictions on private economic and institutional innovation behaviour, regardless of whether scholars advocate the joint involvement of government capital and private capital, and the cooperation between commercial financial institutions and cooperative financial institutions, whether it relies on the growth of private endogenous financing system. Considering that there are few financing channels and low financing efficiency of SMEs and innovative enterprises in China, many enterprises finished their early stage internal accumulation by depending on the original capital provided by informal finance, thereby qualifying them to apply for loans from formal financial institutions. How to accurately examine the effect of informal finance is an important aspect of addressing the financing difficulties of innovative enterprises.

According to Marshall's dilemma,[17] the original advantages in information acquisition and costs of informal finance (private finance) will gradually decrease with the growth of the number and size of participants. Therefore, even if we recognize the effect of informal finance institutions, there is still a need to help them evolve into small and medium-sized financial institutions. The existence of a large amount of small and medium-sized financial institutions is conducive to modernizing the financial system and enhancing the efficiency of social funds.

Innovative enterprise's financing and capital market operation

According to the analyses of the lifecycle of innovative enterprises, independent innovative enterprises, after entering into the mature stage, may go for direct financing from the open market apart from applying for loans from commercial banks. Of course, the traditional experiences of enterprises in financing cannot be used to explain how much financing space the capital market could give these innovative enterprises. This section will analyze relevant issues about the financing of independent innovative enterprises through issuing stocks in the open market from both theoretical and practical perspectives.[18]

Independent innovation and financing through getting listed: theoretical explanation for information asymmetry and difficulties in going public for financing

Information asymmetry and high costs for financing

Enterprises seeking financing through openly issuing stocks do not want to provide all their information to investors out of considerations for protecting their own interests. The interests of stock investors may be damaged due to information asymmetry. In order to maintain order in the capital market, the government formulates strict regulations demanding listed companies to disclose all kinds of hard and soft information. China Securities Regulatory Commission regulates that listed companies must publish their relevant financial statements regularly and major events irregularly to the market. Beyond that, the senior management, directors, and major shareholders of listed companies must report the stock trading of their companies to the stock exchange. Even so, information asymmetry may still exist. Investors' assessments of the invested company are also made based on information such as the size and popularity of the company, except the published financial information. Therefore, small and less popular innovative enterprises are often not recognized by investors, which have resulted in higher costs and even failure in. In this case, information asymmetry of independent innovation enterprises has basically eliminated the possibility for them to issue stocks in the general market for financing.

Information asymmetry, capital structure and signaling theory

An enterprise may choose debt financing (such as bank loans) and equity financing (such as going public to issue stocks) when it needs to seek financing for some investment opportunities. With the principle of maximizing shareholders' interests, business managers may select the financing with the lowest costs. According to the signaling theory of Ross (1977), investors possess less information than the managers about risks of the invested projects and the enterprises, due to information asymmetry. An enterprise will only choose to issue stocks for financing when the stock market price is higher than its real value. Otherwise, it will go for debt financing or wait for an opportunity of stock price overestimation. This is a rational choice of business managers under the conditions of information asymmetry. Rational investors will regard the stock issuing behaviours of an enterprise as the signal of an overestimated share price of the business, and therefore regard stock issuance as negative news, as they know the existence of information asymmetry and understand the choices of the business managers.

The signalling theory further proves that technological innovation enterprises with a higher degree of information asymmetry need to pay more financing costs for stock issuance.

Listing requirements and features of independent innovative enterprises

Innovative enterprises do not have much demand for capital at the early stage of development. For instance, the capital demand of an innovative enterprise is generally a few million yuan at the early stage of innovation. But the capital demand is generally around RMB10 million at the expansion stage (Liang Laixin, 2003). Because the process of innovation has a strict order and the accomplishment of the former stage is the essential condition for the start of the next stage, it has determined the characteristics of staged growth of capital demand for innovation.

Statistics of the Organization for Economic Co-operation and Development (OECD) in1996 pointed out that the capital size of a technological innovative enterprise at the R&D stage in OECD countries is usually between USD500,000 and USD1,000,000; these figures will become USD2 million–6 million at the growth stage. Generally speaking, technological innovative enterprises can only generate profits on operation at the stage of large-scale industrialization.

In contrast, stock exchanges have put in place higher requirements for enterprises to go public in capital size and profitability. Generally speaking, stock exchanges usually have the minimum requirements in pretax profit, market value for listing, and business record. Table 3.5 summarizes the listing requirements of Chinese and foreign stock exchanges. We can see from the table that stock exchanges have a higher threshold for small and medium-sized technological innovation enterprise, making it more difficult for them to get financing from the stock market.

Apart from capital size, the cost for listing is also a major obstacle for technological innovation enterprises to issue stocks for financing. Chinese businesses need to pay the initial listing fee, listing fee and annual fee for listing to the exchange; the amount varies from hundreds of thousands of yuan to several million yuan. In addition, a company needs to pay very expensive fees for listing to the sponsor, underwriter, market marker, accounting firm, law firm, and other intermediary organizations. Using Hong Kong Growth Enterprise Market (GEM) as an example, science and technology enterprises need a total of millions to tens of millions of Hong Kong dollars for various listing fees. Most small and medium-sized science and technology enterprises are incapable of taking on such expensive costs.

Table 3.5.   Listing requirements of some stock exchanges

Stock exchange	Pretax profit	Market value for listing	Business record
New York Stock Exchange	The total sum of pretax profits is no less than 10 million U.S. dollars in the first three years after getting listed.	The total market value globally is no less than 500 million U.S. dollars.	Three years
Stock Exchange of Singapore	The accumulated pretax profits in the recent three years surpass 7.5 million Singapore dollars, and the pretax profits of each of the three years exceed one million Singapore dollars.	According to the issue price, the total market capitalisation is 80 million Singapore dollars at least.	Three years
Hong Kong Exchanges	The profits in the last year shall not be less than 20 million Hong Kong dollars; the profits of the previous two years shall not be less than 30 million Hong Kong dollars.	The expected market value of new applicants must not be less than 100 million Hong Kong dollars, and the expected market value of securities held by the public shall not be less than 50 million Hong Kong dollars.	Three years
Shanghai Stock Exchange	Making profits for the previous three consecutive years.	The total capital stocks shall not be less than 50 million yuan.	Three years

Source: Sorting based on the listing rules of different stock exchanges in China.

Management system and information disclosure are also obstacles for independent innovation enterprises to go public for financing. Listed companies must maintain a certain transparency due to the involvement of public interests. To do this, two conditions must be met: the first is a scientific corporate governance structure; the second is a sound accounting and auditing system. For small and medium-sized innovation enterprises, the core management of most science and technology enterprises usually only possess a technology background and lack managerial knowledge, and are in need of a sound accounting system. These factors have made the information disclosure quality of technological innovation enterprises unable to meet the requirements for going public, resulting in the relatively lower efficiency in information transfer. This has further influenced the possibility for technological innovation enterprises to go public for financing.

The above analyses indicate that all technological innovation enterprises, after entering into the stable development stage, can choose stock market, especially various GEMs (second board markets), to raise funds. But, going public for financing is quite difficult for technological innovation enterprises at the seed stage, initial stage and expansion stage. Some countries have launched GEM in order to solve financing difficulties faced by technology enterprises.

Financing space provided by GEM to innovative enterprises

GEM, also known as the second board market, refers to the market with lower listing standards than the main board market and the stock market, which specializes in providing growing SMEs with places for listing and financing services. Compared with the main board market, the conditions for listing in the GEM have been reduced greatly (see Table 3.6). Generally speaking, the GEM is appropriate for the technological innovation enterprises already entered into the large-scale industrialization stage. But, the technological innovation enterprises which are still at the innovation conceptual stage, R&D stage, and trial-production and trial-marketing stage are still unable to use the GEM for equity financing.

GEM as a crucial channel for venture capital exit has significant meanings. Venture capital needs to exit from the investees for two reasons. First, venture capital is a value-added—based investment. In addition to offer capital to the investees, venture capital institutions generally provide value-added management and services to the investees, including getting involved in the board of directors of the invested enterprise, helping enterprises recruit core administrative staff, helping enterprises improve commercial plans and

Table 3.6.   Listing requirements of some GEM stock market

Stock exhcange	Requirements for operation	Market value for listing	Business record
NASDAQ	Net tangible assets are no less than18 million U.S. dollars.	No less than 75 million U.S. dollars	Two years
Hong Kong GEM	The turnover in the last year is no less than 500 million Hong Kong dollars, or the total assets are no less than 500 million Hong Kong dollars in the previous fiscal year, or the new issuance size is no less than 500 million Hong Kong dollars.	No less than 150 million Hong Kong dollars in market value before listing	One year
EASDAQ	The total assets are no less than 3.5 million Euros.	No less than 50 million Euros of market value for issuance	No specification
Kuala Lumpur Second Board Market	The paid-up capital is no less than 10 million Ringgit, the average pretax profits of the first three years after listing are no less than 2 million Ringgit.	No specific rules	Three years

Source: Sorting based on the listing rules of different stock exchanges.

development strategies, assisting enterprises with follow-up financing, and so on. As the investees become mature, the value of the management services provided by VC institutions decreases and the required rate of return of VC institutions is bound to keep decreasing. As such, venture capital needs to exit from the investees and go to other innovative enterprises for a new round of capital cycle. The second reason is the main organizational form of venture capital. In western countries, the independent and limited partnership venture capital institutions are managing more than half of the nation's total venture capital.[19] But, these limited partners funds have set an approximate deadline of ten years, in general. The limitations on the terms of partnership have forced venture capitalists to exit from the investment and end the partnership.

The exit channels for venture capital include Initial Public Offerings (IPO), M&A and liquidation. Empirical data demonstrate that among the three channels, IPO can bring about the highest investment return to VC institutions (Gompers and Lerner, 2001). Sheng Lijun (1999) found after comparing different venture capital exit channels that the average returns realized through IPO exit is three folds of the returns realized through other channels (see Table 3.7). Therefore, the first exit channel for VC institutions is to help VC enterprises go public.

Due to the higher listing conditions in the main board market, the enterprises invested by venture capital usually cannott meet the conditions, so GEM has become the first choice of VC institutions and the investees.[20] The existence of GEM has a remarkable influence on the venture capital industry, seen in Table 3.7. Fig. 3.14 has demonstrated the relationship between the IPO number of enterprises that have gained venture capital in the last five years and the total amount of venture capital investment in the U.S. We can see from the Fig. 3.14 the changing trend of the IPO number is same as that of the venture capital amount across the United States; the changes in the latter are comparatively later than the changes in the former, which has reflected the time lag for IPO to influence venture capital.

Table 3.7.   Comparisons of the exit channels of venture capital

Exit channels	Average shareholding period (year)	Average return multiplier
IPO	4.2	7.1
Acquisition	3.7	1.7
Buy-back	4.7	2.1
Second sale	3.6	2.0
Clearing	4.1	0.2

Source: Sheng Lijun, *Venture Capital: Operation, Mechanism and Strategy* (Shanghai: Shanghai Far East Publishers, 1999), 98.

Further statistical analyses indicate that the two have an obvious positive correlation. Taking the IPO number of the year *T* as the explanatory variable, the venture capital amount of the year (*T* + 1) as the explained variable, and making a Unary Linear Regression analysis can help find the evident explanatory capability of the explanatory variable to the explained variable.

Gilson and Black (1999) found after comparing the venture capital industry between the United States for one part, and Japan and Germany for another, that dynamic GEM has an inevitable connection with the advanced venture

Fig. 3.14.   The Relationship between IPO market and venture capital

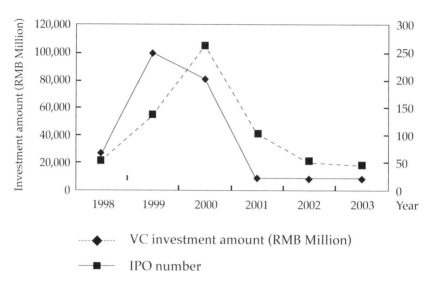

----◆---- VC investment amount (RMB Million)

————■———— IPO number

Note: IPO Number refers to the number of venture-backed IPO.
Sources: *Venture Economics and Venture Capital Journal* (1998–2003).

capital industry. The success stories of the U.S. and the relatively unsuccessful experiences of Japan and Germany demonstrate that a dynamic GEM is essential to the development of the venture capital industry.

The vibrancy of GME has become a wind vane for the financing market. Although not all venture capital transactions can exit through IPO, the vibrancy of the IPO market influences the expectations of investors for future earnings, which still affect the decisions of investors in investment. Given the major influence of venture capital on the financing of innovative enterprises, the vibrancy of the GEM market will directly make a difference in the vitality of technological innovation.

Conclusions

This chapter starts from information asymmetry to analyze the theory and practice of the financing of independent innovation enterprises, and discusses the features of the financing demand of enterprises at different development stages and their matched financing path by analyzing the lifecycle of independent innovation enterprises. At the same time, this research study also demonstrates the applicability of small and medium-sized financial institutions for innovative enterprises based on the theoretical analysis of both formal and informal finance, and puts forward relevant policy suggestions in line with the historical evolution of China's supporting system for the financing of independent innovation enterprises.

The study concludes that financing difficulties are protruding at both the seed stage and the initial innovation stage of independent innovation. At these stages, they have a long way to go before getting bank loans, carrying out M&A and getting listed. Only government and venture capital could serve as a link between different stages to accomplish innovation. However, institutional efficiency is lacking when it comes to the administrative operation of government funds alone, while pure commercial venture capital is not willing to get involved in that stage. Therefore, the government venture capital fund combined withgovernment funds and venture capital from commercial venture capital institutions becomes a possible choice. This method has effectively integrated government policy objectives with the market-oriented operation of venture capital, amplified fiscal funds, and guided other social capital, which is an important innovation to make up for the insufficient government investment and financing system. At the moment, an appropriate operation may be that the Ministry of Finance and the Ministry of Science and Technology take the lead to build a syndicated investment with China Development Bank (a nationwide government venture capital guide fund) and designate a professional institution to manage and operate the fund.

When an innovative enterprise develops from the early to the mature stage, commercial venture capital institutions may contribute financing. As VC is a kind of equity capital, it can take on higher risks and focuses more on independent enterprises. Venture capital has effectively solved the financing difficulties faced by independent innovation enterprises when seeking financing via traditional channels. The means it used include the selection and examination before investment, the design of the investment deal structure, the encouragement of investment contracts to businessmen, the distribution of control power, and the management and service after the investment. Chinese local venture capital institutions are now in a state of moving forward, slowly, after the rapid development in the early 1990s. The government needs to gradually put an end to state-holding venture capital firms, allow pension funds and insurance funds to enter into the venture capital market through professional funds, and keep making continuous amendments in taxation and legal aspects so as to realize further development of venture capital.

Independent innovation enterprise enters the mature stage of large-scale industrialization after finishing R&D, trial-production, and trial-marketing. Commercial banks, as the traditional financing channel, gradually get involved, but still practice credit rationing to small and medium-sized technology enterprises with serious information asymmetry and higher risks and demand a higher interest rate. Financial institutions may be expected to realize the three-dimensional institutional innovation in breaking difficulties faced by innovative enterprises in getting loans. The first is the "investment-credit alliance" that is a close cooperation between the credit operations of banks and the investment business of venture capital firms. Banks provide certain lines of credit to venture capital firms; within the line of credit, the banks can offer entrusted loans and there is no need to go through red-tape examinations when the VC-supported innovative enterprises encounter financial difficulties. The second is to fully develop a credit rating and credit guarantee system, reduce information asymmetry from the perspectives of information transmission, confirmation and assurance, reduce the blindness and homogeneity of financial institutions in investment, improve market order and environment, and enhance financing efficiency. The third is to develop small and medium-sized financial institutions and private informal finance, give full play to the advantages of small and medium-sized financial institutions in information acquisition and processing, relax barriers on market entry, liberalize the interest rate, guide the formal development of informal finance, and build diversified small and medium-sized financial service systems.

After entering into the mature stage, independent innovation enterprises could also apply for direct financing in the open market apart from applying loans from commercial banks in the traditional financial channels. But effective capital market theory, signaling theory, and pecking order theory all point out the influences of information asymmetry in the open market, and stock exchanges have set relatively high requirements for enterprises to get listed in capital size and profits. At the same time, the expenses on listing and the management system of innovative enterprises also represent the obstacles for the enterprises to go public for financing. Therefore, it is also more difficult for independent innovation enterprises at the mature stage to successfully get financing in the open market.

4
Chapter

Technological Innovation Through Resource Integration

Song Hua, Yi Zhihong, and Yu Kangkang

Introduction

The complex and ever-changing environment, the still-fiercer competition and diversified market demand require enterprises to engage in an all-directional competition. Great production efficiency, extremely high quality, and even flexibility are inadequate for enterprises to maintain their market competitiveness. Continuous innovation is increasingly becoming an inexhaustible source and driving force for enterprises to survive and keep developing. This not only calls for enterprises to carry out technological innovation but also forces them to engage in systematic, continuous and comprehensive innovation centering on technological innovation. Peter F. Drucker noted in his book *Innovation and Entrepreneurship* that innovation is a specific instrument for entrepreneurs to demonstrate their entrepreneurship and a new capability given to resource to create wealth. Therefore, innovation in itself creates resources. In other words, through innovation, enterprises can obtain new resources, and consequently obtain core competitiveness and form a sustainable competitive edge. Enterprises must fully integrate existing internal resources and capacity, and make full use of the external valuable and scarce resources and capability, if they want to enhance their independent innovative capability and strengthen their competitive edge. One of the reasons behind the weak capability of Chinese enterprises in independent technological innovation is that they have failed to optimize the allocation of and utilize their internal and external resources and capability. This chapter adopts the single case research method to explore a new trend of technological innovation. Technological innovation based on resource integration, through the literature review of relevant theories and practices of technological innovation, provides an inspiration for enterprises to improve their technological innovative capability and maintain continuous competitiveness through analyzing the core contents and key channels of this mode.

Literature Review and Theoretical Summary

Austrian American economist Joseph Schumpeter pointed out in his work the *Theory of Economic Development* in 1912 that "innovation" is the critical dimension of economic change and the "creative destruction" process. According to the definition of Schumpeter, technological innovation is to establish a new production function and introduce the "new combination" of production factors and production conditions, which have never existed before, into the production system. This "new combination" includes the following contents: (1) introducing new products; (2) introducing new technologies; (3) opening new markets; (4) controlling the source of raw material supply; (5) realizing new organization of industries. After Schumpeter, innovation theory began to develop in two directions. One is the technological innovation school represented by Mansfield and Schwartz, who have made in-depth studies on technological innovation from the perspectives of the relations between both innovation, and the imitation, promotion and transfer, and shaped some representative theories. The other is the institutional innovation school represented by North, who combined innovation and institution to study the relations between institutional factors, business technological innovation, and economic benefits, and emphasized the importance of both the institutional arrangement and the environment for economic development (Ding Juan, 2002).

It could date back to Schumpeter's classical definition of technological innovation that the focus of technological innovation was limited to technology, and regarded "market" and "organization" innovation as an important contents of technological innovation. As the theory and practice continues to develop, both Chinese and foreign scholars have successively put forward some new thoughts and theories focusing on technological innovation, innovation process, innovation system, and so on so forth (Dodgson, Gann, and Salter, 2002). The fifth generation technological innovation process model emerged between the 1950s and 1990s (see Table 4.1) and explained the process of technological innovation from different perspectives. We can see from the trend that most scholars have given up their simplified understanding that technological innovation is the combination of "technology" and "innovation" and have denied the simplified and linearized cognition model as the understanding of technological innovation expands. Instead, they have regarded technological innovation as a complex, multi-dimensional, nonlinear, networking and interactive system and process.

Table 4.1.   Development of technological innovation process model

Model	Time	Feature	Content
Driven by technology	1950s–1960s	Simplified, linear, technology-driven (Jeseph Schumpeter)	Technological innovation is technology-oriented linear and spontaneous conversion process, market passively accepts technological results, and is reflected as the technology-driven process.

(Cont'd)

Model	Time	Feature	Content
Driven by demand	1960s–1970s	Simplified, linear, demand-driven (Kamien and Schwartz, 1975)	The model emphasizes that market is the source of innovation ideas, market demand provides opportunities for product and process innovation, and plays a key role in innovation.
Interaction between technology and market	1970s–1980s	Based on interaction (Rothwell and Zegveld, 1985; Steinmueller, 2000)	The model emphasizes that the whole process is triggered by the interaction between technology and market demand, and the technology-driven model and demand-driven model have different effects at different stages of product cycle and innovation process.
Technological innovation integration	1980s–1990s	Based on Chain-linked Model (Kline and Rosenberg, 1986)	The model emphasizes the integration of the interface between R&D and manufacturing, and the close coordination between enterprises, suppliers and lead users.
Integrated network of technological innovation system	1990s	System integration and network (Dodgson, Gann, and Salter, 2002)	The model regards the whole process as a complex network of an enterprise's internal and external communication path, which connects various internal functions and highlights a closer strategic links between enterprises.

Source: Mao Wuxing, "Studies on the Capability of Enterprises in Total Innovation Management," doctoral dissertation of Zhejiang University (2006).

There are two standards to evaluate how technological innovative capability can bring competitive edge to enterprises: applicability of "technology" and applicability of "development" (Helfat et al., 2007). The applicability of technology is defined as: how a capability plays its role effectively, taking no consideration of how a capability makes an enterprise survive; the applicability of development or the applicability of externality involves the selected environment. Dynamic capability can help obtain the applicability of development and build the environment, to a certain extent. The Dynamic Capability Theory represented by Teece (2007) holds that the modern economic environment is featured by the diversification of geographical locations and organizational resources of innovation and manufacturers, and the sustained competitive edge requires not only irreproducible resources but also unique and hard-to-replicate dynamic capability. These capabilities can be continuously created, extended, upgraded and maintained and can keep the unique resource foundation of related enterprises. They also require enterprises to adapt to the ever-changing customers and technology opportunities, and to improve their ability to build their business survival environment.

Enterprise is an open and dynamic target system (process), so it needs to take into consideration not only the static structure and connection but also the dynamic factors and external communication. Therefore, enterprises must continuously go beyond technologies and markets to seek and explore opportunities both locally and distantly (March and Simon, 1958; Nelson, and Winter, 1982). Innovation activity can be deemed as a form of exploring new products and process, but local exploration is only one part of relevant studies. A large proportion of new products are introduced from external resources under the rapid development environment. Therefore, exploring and developing activities should not be limited to local areas. Enterprises must find key resources from the center to the periphery of their business ecological environment, and must include potential partners, i.e. the clients, suppliers and complementors, who actively engage in innovation activities (Teece, 2007). This kind of model, which centers on technological innovation and unites other related parties in the periphery, including enterprises, universities, research institutions, governments and financial institutions, is defined in this chapter as "wheel radiation." In these types of innovation groups, each individual must have its own "ecological position," and innovative individuals with discontinuous gradient must have more communication with the world of information and energy if they wish to jump to a higher gradient level.

In 1997, Iansiti put forward the concept of technology integration, believing that technological integration management is more capable of coping with discontinuous technological innovation. The idea and testing of a new product and the belief in the new product are jointly established by all internal divisions of an enterprise, and by the partners and clients of the enterprise. This kind of brand-new integrated mechanism has ensured the realization of discontinuous and crossing innovation, and it includes not only the necessity of investment in research activities and the recognition of the customer's demand and technology, but also the understanding of potential demand, industrial and market structural changes, and the response of suppliers and competitors. As such, enterprises could engage the technology opportunity and recognize the demand of clients at the same time, which, in turn, could bring about opportunities for further commercialization (March and Simon, 1958; Nelson and Winter, 1982). Empirical research indicates that the probability for one innovation to be commercialized successfully is closely related to the understanding of the developer about the demand of its clients (Freeman, 1974). Customers are sometimes the first to realize the potential application of new technology. If the new technology of a supplier cannot understand the demand of its customers, the developed products will not succeed. However, the supplier is also the driving force for end product innovation. With the rapid innovation of the components-supplier, the success in the downstream competition is also well-positioned to influence the upstream enterprises in obtaining leading innovative capability (Teece, 2007). This chapter defines this kind of integrated innovation mode, focusing on the supplier to the client and the upstream and downstream enterprises, as "chain integration."

Janszen pointed out that innovation is a complex, self-adaptive system and various factors in innovation need to be taken into consideration from the systematic perspective. Relevant research include: national innovation system (Nelson, 1993; Edquist, 1997), regional innovation system (De la Mothe and Paquet, 1998) and innovation network (Freeman, 1991; Eisenhardt and Martin, 2000; Fleming and Sorenson, 2003). Prahalad and Hamel also noted in the book *Corporate Core Competence* (Prahalad and Hamel, 1990) that business innovation is a system which must be established from the industrial level so that enterprises with a strong innovative capability will have a continuous, competitive edge. The industrial innovation system is rooted in a series of closely linked major innovation sources. These innovation sources drive other relevant innovations through dynamic technology transfer and feedback mechanism, and have both upgraded the industrial level and promoted industrial development as a whole. The concept of the industrial innovation system effect indicates that a host of industries are linked to one network structure that is built on the basis of the inter-dependence and mutual complementation of dynamic and strong technical economic connections. Teece (2007) believed that these kind of mutual complementary innovations are very important, especially in the industries with innovations with accumulation as the feature or "platform." But these kind of mutual complementarities have not only brought about the expansion of size and scope; more importantly, they have brought about the coordinated specialization between various industrial plates. This chapter defines the network which regards industry as the basic level of classification and coordinates with various business innovations, as "group aggregation."

Combining wheel radiation, chain integration and group aggregation will form the business innovation network which integrates internal and external resources, and this network will bring about direct or indirect innovative performance to enterprises. Many scholars also believe that the application of external technologies is expected to bring about strategic profits, such as avoiding internal development costs (Noori, 1990), achieving rapid growth (Capon and Glazer, 1987; Granstrand et al., 1992), adding to technical knowledge (Cohen and Levinthal, 1989; Huber, 1991) and enhancing business technical capability through the process of seeking and applying external technologies (Chatterji, 1996; Jonash, 1996). As such, product and process innovation has led to higher performance. However, Cai Kunhong and Wang Jianquan (2005) have proven that obtaining external technologies could only significantly add to business performance under the regulation of internal R&D. That is to say, the technological innovative capability of enterprises is based on the integration process of internal and external resources, and enterprises obtain technologies to match internal development activities through technology alliance (Industrial Technology Bureau, 2003). In so doing, enterprises could continuously launch and implement new innovations for quite a long time and keep realizing the economic benefits of innovation. Based on the above theory, this text builds a new technological innovation model on the basis of resource integration (see Fig. 4.1), and its major characteristics include: (1) in

Fig. 4.1.   Technological innovation model based on resource integration

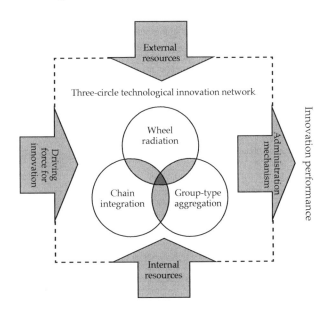

the aspect of innovation source, to extend the definition of the driving force for innovation from supply and demand directions; (2) in the aspect of innovation content, to focus on the integration, synergy and dynamic adaptation of internal and external resources; (3) in the aspect of innovation spatial scope, to include various key strategic resources in regional economy, industrial cluster and supply chain into the innovation network for integration; (4) in the aspect of the evaluation of innovation benefit, to consider the benefits brought about by the cooperation network of external alliances. We will confirm the applicability of this model through case studies, to follow in this text.

Research Method and the Selection of Sample Enterprises

This research adopts the single case research method which is to select a typical single sample enterprise in the industry by examining multiple secondary units of analysis in order to establish and verify theory building. In the case study, many scholars pointed out that multiple case studies are more reliable than a single case study, but a single case study which can challenge or extend the existed theories is still significant. In particular, a single case study is effective for a unique or even extreme case. Beyond that, using a single case study is easier to analyze various intricate situations and the status of business operations through careful multi-unit analysis of a single enterprise to propose a relevant theory.

The industrial background of this study is the Chinese coal industry and Shenhua Group is chosen as the research objective. Shenhua Group is a coal-based integrated energy company, the largest domestic coal producer and marketer, the world's second largest publicly-listed coal company, and boasts the largest-scale high-quality coal reserves in China. Its principal businesses include: coal production, marketing, electricity generation, heat production and supply, and relevant railway and port transportation services. In recent years, Shenhua Group has taken development as the objective and relied on scientific and technical progress to speed up scientific and technical innovation, and thus has explored a road of rapid development in line with its own features. In the first half of 2007, the operating income of Shenhua Group grew by 29.8% to RMB38.331 billion and its operating benefits grew by 25.0% to RMB16.452 billion. In the main business income (see Fig. 4.2), the coal business keeps growth. In the first half of 2007, the coal production and sales volume recorded 76.6 million tons and 97.8 million tons respectively. The raw coal and salable

Fig. 4.2.   The main business revenue structure of Shenhua Group in 2006

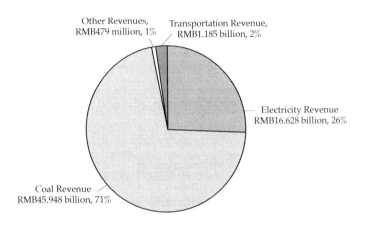

Source: 2006 Annual Financial Report of Shenhua Group.

coal of Shenhua Shendong Coal Group Corporation Ltd. surpassed 100 million tons. The overall efficiency, the unit yield and digging, and other indicators, hit a new high. The electricity business of Shenhua Group realized rational layout and stable development. By 2006, the total installed capacity and the sales amount of electricity of the entire group reached 11,960 megawatt and 51.71 billion KWH respectively. Between 2004 and 2006, the annual compound growth rate reached 41.7% and 20.7% respectively. The transportation business also developed rapidly. Shenhua Group has a railway- and port-integrated transportation network, including five railway lines (Baoshen, Shenshuo, Shuohuang, Dahuai, Huangwan) with an operating mileage of 1,367 km. The special-purpose ports, Huanghua Port and Shenhua Tianjin Coal Dock, have provided full guarantee for the production and sales of coal. In the first half of 2007, 39.8 million tons of coal was transported through Huanghua Port and 9.2 million tons were transported through Shenhuan Tianjin Coal Dock.

With the integration of the production structure adjustment, the modern corporate system, and operation mechanism reform, Shenhuan Group's development also reflects the results of scientific and technological innovation. Shenhua Group takes the rational layout of the coal, electricity, road, port, and oil industries as the new starting point, formed a set of scientific and technological innovations and R&D strategies, and founded three technical research centers. One of these centers is the Coal Liquefaction Technical Research Center, which serves respectively the coal, electricity, and oil (coal chemicals) sectors of the group. In the development and production, Shenhua Group adopts advanced technical equipment according to the actual demand, carries out systematic synthesis and domestic transformation through introduction, digestion and assimilation, and employs the development method with the integration of enterprises, universities and research institutes, which have formed the "Technology of High-yield and High-efficiency Mine in Shendong Mining Area." This represents both the advanced technology in the Chinese coal industry and the advanced technologies and scientific and technological achievements in the electricity, railway, and port industries. The maturity and development of the scientific and technological innovation mechanism of Shenhua Group have provided a basic technical support for the formation of its core assets and capacity, and also offered the ultimate driving force for continuous upgrading of its production structure. Given this situation, Shenhua Group is a representative for carrying out the single case study in exploring the technological innovation mode of the large state-owned enterprises in the coal industry. The field investigations and in-depth interviews in Shenhua Group were conducted in July 2007; the objectives for interview include the senior management of the group and the manager of its science and technology division.

Shenhua Group's Technological Innovation Model based on Resource Integration

The original innovation achievements of Shenhua Group obviously cannot be totally realized through its existing internal resources, and the group should creatively and comprehensively utilize external resources needed for innovation. The realization of independent technological innovation requires this enterprise to fully integrate existing internal resources, and to fully absorb and exploit external valuable and scarce resources. An enterprise must effectively integrate internal and external innovation resources if it intends to improve its independent technological innovative capability and strengthen its competitiveness. Shenhua Group is an enterprise with diversified energies and its main businesses are the key basic industries for national energy guarantees, which involve multiple industries, and possess broad range, high threshold and a huge technology demand. However, Shenhua Group is different from other state-owned key enterprises as it does not have many technology research and development entities. This has made it create a new independent technological innovation path with the integration of enterprises, universities and research institutions, taking market as the platform, technology demand as the orientation, cooperation for win-win results as the bond, and strategic alliance as the guarantee, all while trying to adjust and integrate external resources.

During its development process, Shenhua Group played an active, initiating, leading, and creative role as the independent innovation subject, created a new technological innovation model with wheel radiation, chain integration and group aggregation as the core, and obtained innovation performance better than its peers through its established technological innovation network.

Driving force for technological innovation

The driving force for the technological innovation on the basis of resource integration in Shenhua Group comes from two aspects: first, the government's promotion of a highly efficient and energy-conserving production of the coal industry; second, it is the driving force from the enterprises, which are the independent innovation subject.

A well-known American strategic management scholar, Michael Porter, believes that the development of a nation's competitiveness must undergo three stages: factor-driven stage, investment-driven stage and innovation-driven stage. The development of Chinese competitiveness is still in the factor-driven stage and investment-driven stage, which faces dire and pressing constraints of resource bottlenecks and the demands on the ecological environment. China's development is bound to advance from the investment-driven stage to the innovation-driven stage by promoting the development of energy conservation and high efficiency production mode, sticking to the civilized development road by encompassing the growth of production with an affluent life and sound ecosystem, and building an energy-efficient and environmentally-friendly society. With increasing industrialization, urbanization and modernization, China's demand for coal products continues to rise (see Fig. 4.3). The market competition among coal enterprises has been eased and the focus of competition has moved from price to resources. According to *Some Opinions of the State Council on Promoting Healthy Development of the Coal Industry* and the *11th Five-Year Plan* for the coal industry, China will basically build a new type of coal industrial system featuring high resource utilization, guaranteed security, strong economic benefits, sustainable development, and less environment pollution. China's central government has implemented a series of measures: selecting Shanxi province as the experimental regi1on for implementing the policy measures of the coal industry's sustainable development; extending the objectives for shutting down, from illegal coal mines and the coal mines that cannot guarantee safety in production, to those not supported by coal industrial policies and are considered as having an irrational layout, resource damage and excess pollution; working harder to integrate coal resources; planning to construct 13 state-level coal bases involving 14 coal producing provinces with a reserve of 690.8 billion tons, accounting for 70% of the national total, and planning to realize an annual output of 1.7 billion tons in 2010; reforming the tax and fee system for coal; and cutting down taxes for coal enterprises to a rational level, among others. This policy environment is conducive to the fast and stable development of the coal industry, and can drive Shenhua Group to work harder in technological innovation and improve the productivity and recovery rate.

Fig. 4.3.   China's demand for coal between 2003 and 2006

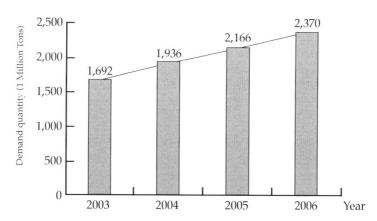

Source: Promotional materials about the interim performance of Shenhua Group in 2007.

The economic function of a government is to carry out macro management, macro regulation, and control of economic activities of the entire society. In order to realize technical progress of the entire society, the government needs to be the initiator for technological innovation, and more importantly, to provide policy support and build the environment for technological innovation. Enterprises are profit-oriented, have clear-cut demand, technical competence and economic strength, and represent the main force of technological innovation. According to the survey of 78 of Top 100 Chinese coal enterprises, RMB7.59 billion was invested in R&D in 2006, up by 50.9% from the year before. The general manager of the science and technology development department of Shenhua Group highlighted in the interview the necessity and urgency of regarding enterprises as the main force of independent technological innovation. Enterprises need to be the main force of independent innovation, which is the core of building an innovation-oriented country. The demand, input and commercialization are all indispensable and only enterprises possess all the three at the same time. Specifically: (1) production and operation cannot continue if problems that occurred in the process have not been solved; (2) an investment in human, financial, and material resources need to be made; (3) all projects within the enterprise need a seamless connection without problems in transformation.

In the light of enterprise internal demand, Shenhua Group has kept investing more in R&D and has made organizational improvements in recent years (see Fig. 4.4). For example, regarding the problem of channel accretion in Huanghua Port, Shenhua Group conducted sufficient studies and verification on the mechanism and influencing factors, and innovated the fast embanking of open, unshielded sea areas, large-scale neritic backfilling, extensive waterway excavation, and other advanced technologies. As another example, during the R&D process of coal liquefaction, the group found there was a risk of equipment unable to operate for a long period when the group introduced technology from abroad at the initial stage, and thus immediately decided to set up a special liquefaction research center to conduct independent R&D of coal liquefaction processes and equipment to guarantee the development of technologies.

Fig. 4.4.   **R&D costs of Shenhua Group between 2004 and the first half of 2007**

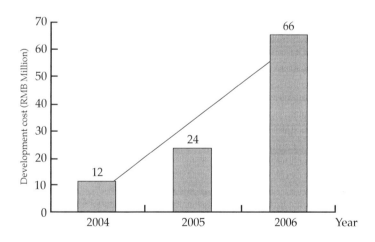

Sources: 2005, 2006 annual reports and 2007 interim report of Shenhua Group.

As a comprehensive energy company, Shenhua Group's independent innovation features a new philosophy, great demand and high starting point; more organizations but few personnel; a new model and broad cooperation; and high economic returns. However, the group's technologies are not current and it only has three technical centers for coal, electricity and coal chemical respectively, with a mere 200 researchers. It is very difficult for the group to carry out independent innovation on its own. This difficult situation has further propelled Shenhua Group to keep exploring an effecitive independent technological innovation path in line with its own development. Therefore, it is necessary for the group to give top consideration to the role of resource integration in improving business capacity through independent technological innovation. It is equally important to coordinate in many aspects, leverage on colleges and universities and research institutes, mutually complement with free R&D forces and form a joint technology development and innovation network to realize the two main strategic targets of independent innovation mentioned by the general manager of the science and technology development department of Shenhua Group in his interview: first, to provide effective technical support to Shenhua Group; second, to lead the technical progress of the industry. As a state-owned enterprise and as a leading Chinese player in the industry, this responsibility is necessary. Technology will become more and more important with the group's development, and independent technological innovation is expected to continue to provide a core competence to Shenhua Group.

Three-circle technological innovation network

Wheel radiation at the center of the innovation group

"Innovation group" is the relative concentration of a technological innovation group in the regional technological innovation ecosystem. During the technological innovation process, Shenhua Group takes the science and technology department as the hub, the technology center of the industrial plate as the support, the high-quality and professional technicians as the core, the post-doctoral scientific research workstation as the platform, and makes use of external technology resources to establish a high level of the technological innovation system. Under the premise of having command over the core stages (systematic integration and management of innovative technologies) and core technologies for technological innovation, the group has leveraged on social innovation resources to work harder to integrate and develop external scientific and technological enterprises or organizations and establish strategic alliance integrating the enterprise, universities and research institutions, and the innovation network with the combination of internal and external resources of the group. Shenhua Group has organized and coordinated the relations of enterprises in various phases and has driven the development of the enterprises, organizations and scientific research academies in relevant industries through the coordination of internal resources and the integration of external resources, by setting up industrial standards, sharing technologies and resources, and controlling core phases, which have leveled the overall performance of the innovation group (see Fig. 4.5).

Fig. 4.5.   Wheel radiation of Shenhua Group

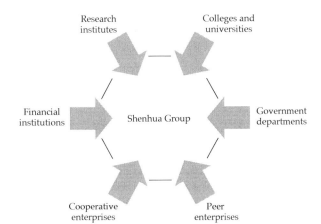

The technological innovation-related group is the base of the innovation group while the technological innovation principal group is the core. As the heart of the wheel radiation, Shenhua Group has integrated various external resources and capability, but its core technology is still limited to internal operations rather than outsourcing to other enterprises, and its key phases are also controlled through monitoring the implementation process. The general manager of the science and technology department of Shenhua Group, highlighted that they must control social resources under the preconditions of the "two firm controls": the first is to firmly control the core phases of technological innovation; the second is to firmly control the core technologies; and Shenhua Group has to make full use of social resources.

For example, during the coal liquefaction development process, Shenhua Group cooperated with relevant companies and institutions to develop the coal liquefaction catalyst and to direct the coal liquefaction processes with the proprietary intellectual property rights; it established the pilot plant with a daily handling capacity of 6 tons of coal by relying on the domestic technology strength. But, coal liquefaction technology is an emerging knowledge-intensive technology in China and involves numerous disciplines, such as coal, petroleum, chemical engineering, machinery, control and materials; Shenhua Group does not have sufficient proficient researchers in all these fields. Therefore, the group brought in domestic experts in coal liquefaction from relevant companies and institutions, and technicians from foreign petrochemical engineering enterprises and research institutes. However, the core technologies are still in the possession of the Shenhua Group technicians. As another example, during the R&D process, Shenhua Group developed the plan on implementing the hydraulic support localization, made clear the implementation principles, overall plan and process arrangement of localization, set up a technical appraisal team consisting of academics and industry experts, and studied the technical feasibility, economic rationality and technical plan for localizing R&D, but it was still the Shenhua Group who determined the major technical parameters and standards for localization equipment.

For the peripheral partners in the wheel radiation, Shenhua Group conducted diversified research on technological breakthroughs and technical exchanges by means of entrusted development, joint development and entrusted exam, for example. Also cited in his interview were several key cooperation programs for the Shenhua Group: built a strategic partnership with China Aerospace Science Technology Corporation and Shanghai Municipal Government to engage in mutual exchanges of needed products; cooperated externally through innovative enterprises, including 15 state-owned enterprises; established technological innovation strategic alliance under the lead of the Central Iron and Steel Research Institute, the Central Agricultural and Machinery Institute, China National Chemical Engineering Co., Ltd., and Shenhua Group. Therefore, the win-win target was realized through these channels and by fully leveraging external scientific research resources.

Specific analyses of the peripheral structure of the wheel radiation of Shenhua Group mainly include the following aspects:

(1) Radiate to research institutes and higher learning institutions. The mode featuring the cooperation among industries, universities and research institutes is a strong support for improving enterprise independent innovative capability, the source of knowledge and technology. Colleges, universities, and research institutes as the source of knowledge and technology can create new knowledge and technologies, effectively promote the spread of knowledge, information and technology, and cultivate talents for enterprise independent innovation. For instance, Shenhua Group and Southwest Jiaotong University jointly developed the "Shuohuang Railway Transport Complex (Simulation) Training System" and realized the standardization and automation of the training for technicians.

(2) Radiate to government departments. Governments provide public services to enterprises for technological innovation. The communication and cooperation with related departments and ministries as well as industry

associations have created a favorable external phenomenon for the future development and independent innovation of Shenhua Group. Shenhua Group filed an application to the working coordination guidance group for cooperation among industries, universities and research institutes, including the Ministry of Science and Technology, the Ministry of Finance, the Ministry of Education, the State-owned Assets Supervision and Administration Commission, All-China Federation of Trade Unions, and China Development Bank, to establish the "strategic alliance of clean and high efficient development and utilization of coal and the innovation of liquefaction and poly-generation technologies," and provide technical support and services to several major industries including coal supply, oil supply, power supply, and transportation capacity.

(3) Radiate to peer industries and promote the flow and optimum distribution of information, technology and knowledge resources. In the project of direct coal liquefaction to make oil, Shenhua Group cooperates with its peer industries, integrates part of the research results of Japan, the United States, and Germany and makes innovation on this basis.

(4) Radiate to enterprises with cooperation. During the R&D process of high strength hydraulic support localization, Shenhua Group signed a research contract of the 2.4m and the 6.3m hydraulic supports with domestic equipment manufacturing leaders, and most preliminary research results have been applied to the manufacturing of these two kinds of support.

(5) Radiate to financial institutions, provide financial support for technological innovation, share innovation risks, and so on. Shenhua Group has established strategic partnerships with a number of banks, including the Industrial and Commercial Bank of China, Pudong Development Bank, and China Merchants Bank, and has leveraged on the resources of financial institutions to deliver financial services and conduct credit risk preventions.

Chain innovation guided by supply chain

The supply chain management is to integrate various internal and external enterprise commercial activities or processes into a consistent and high-efficient business model, and it includes logistics management activities, production operating activities, marketing activities, financial activities and relevant information. Shenhua Group has maximally integrated the resources of the upstream and downstream of the entire chain through the operation mode of supply chain and has continuously analyzed, found, and solved the plentiful technical difficulties in production, construction, operation and management. This mode is typically demonstrated by using the technological innovation demand of Shenhua Group as the starting point to drive the midstream and upstream enterprises in the supply chain (and the upstream of the upstream enterprises), comprehensively realize key technological innovations, and ultimately realize the all-round innovation of raw materials, components and subsystems. It means Shenhua group has realized the dynamic interaction mechanism of technological innovation and management innovation, and has also formed a strong dominant power in the industrial chain while obtaining scale expansion.

What might trigger innovation ideas include the changes of the specification and property of the raw materials provided by suppliers, the changes of customer demand, the opportunities brought about by the introduction of new technologies from colleges, universities, scientific research institutes, and new actions of competitors. But, these information resources and knowledge resources are mostly fragmentary and are not systematized before integration. If we want to leverage the maximum value of these resources, we must carry out integration, abandon useless information and knowledge, and organically integrate valuable information and knowledge to make these resources more flexible, rational and systematic. In this chain, it could be said that the innovative capability of Shenhua Group comes from the establishment of the chain for innovation, and has driven the innovation and development of upstream and downstream enterprises through breakthrough innovations of production processes.

The practice of Shenhua Group in conducting R&D of key equipment localization is a case in point. The coal industry has a higher dependency on the equipment manufacturing industry and its technical level and market competitiveness depend primarily on the technical equipment level of the manufacturing industry. Shenhua Group is the largest user of domestic high-end excavating equipment, but most high-end excavating equipment was imported years ago. In early 2004, Shenhua Group decided to make a breakthrough in hydraulic support to carry out localization research of excavating equipment and listed it a the key scientific research project of the group to start to organize technological breakthroughs and technical exchanges. First, it is to carry out in-depth research on the design method and concept of introduced equipment by means of attraction, and engage in re-innovation of design technology through the problems of the introduced equipment in production and application; second, it is to conduct special research on the high-strength steel plate welding by means of independent R&D; third, Shenhua Group has entrusted China Aerospace Science and Technology to research and develop the electronic hydraulic control system and has leveraged China Aerospace

Science and Technology to solve this core technology; fourth, Shenhua Group requires that manufacturers of hydraulic support must strengthen the quality guarantee system and introduce a monitoring system in the whole designing and manufacturing process of hydraulic support.

During the process of setting up programs and cooperation, the effectiveness of work is guaranteed through specific measures. The first is the internal management. To establish a technical center, technical commission and expert consultant association, research institute and so on, relies on the post-doctoral working station to train people, and build a complete management system from setting up a project to appraisal, examination and approval, acceptance check, and post-evaluation, which strictly follow standard procedures. The second is external management, including the selection of project-assuming companies, strictly following the project bidding requirements, and building strategic partnership through competitive negotiations.

Shenhua Group has assisted and boosted the technological innovation of upstream enterprises through management, demonstrated in the following aspects:

(1) Innovative proposal of "performance of the first independently-innovated product." Shenhua Group adopts the method of replacing purchases with R&D; in other words, the equipment satisfies the objective of localized R&D, major technical parameters, and technical standards proposed by the enterprise. Then the manufacturer will allocate capital to organize technological breakthroughs, design and production. After the tests and industrial trials of the first sample machine in accordance with the requirements proposed by the user, the user will pay the market price to buy the equipment and declare that the R&D company owns the business achievements of this series of products and is qualified to bid in the purchase in the market later on; if the trial failed, the costs for the R&D will be assumed by the manufacturer. As such, Shenhua Group has facilitated suppliers to realize improvement and innovation of components or equipment through providing relevant technical parameters and technical requirements.

(2) Shenhua Group provides relevant technical support and creates conditions for field trials after the type tests of the sample hydraulic support. Shenhua Group is involved in the whole R&D and production process of the hydraulic supports of each manufacturer, and works together with their R&D staff and technicians to make breakthroughs and solve key technical problems, which have greatly accelerated the R&D process. After 35,200 and 38,500 times of test of the lifetime of the 4.5m and the 5.5m hydraulic support developed by Zhengzhou Coal Mining Machinery Group Co., Ltd. and Beijing Coal Mining Machinery Co., Ltd., Halagou Coal Mine and Daliuta Coal Mine of Shendong

Mining Area offered two fully-mechanized mining working surfaces for the comparative application of localized equipment and imported equipment under the same condition in a bid to test the efficiency of localized equipment.

(3) Shenhua Group provides financial, technical and market support during the R&D process of equipment localization. In particular, during the cooperation process with China Aerospace Science and Technology Corporation in the R&D of hydraulic support, Shenhua Group broke the routine to provide funds for the R&D of 30 industrial trial supports. This kind of risk-sharing mechanism in finance, technology and market has greatly accelerated localized R&D process.

Group aggregation based on multiple industrial chains

The industrial chain takes competitive enterprises and competitive products featuring better prospects, higher technological content and stronger relevance in products as the core, product technology as the connection, and capital as the bond, links both the upstream and downstream, stretches downward, and connects the front and the back. Shenhua Group has creatively integrated a number of industrial chains related to coal and has formed a manufacturing service network with coal as the basis and other business plates as assistance and extension through the integration of coal supply, electricity supply, road and port capacity and oil supply (see Fig. 4.6). In addition, the innovation of Shenhua Group is not apparent only within each phase, rather, it focuses more on the coordinated specialization in and between links of phases, which has broken

Fig. 4.6.   **Group aggregation of Shenhua Group**

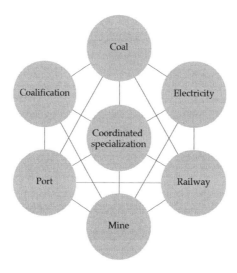

the division of sectors and territory blockades, and eliminated inter-industrial conflicts and internal frictions. In this way, it has brought about the assembly effect between different industrial chains, driven the innovation performance of the entire network, and maintained steady and sustainable growth.

The interactive combination of the various business plates of Shenhua Group has formed the synergistic effect on the basis of monomeric efficiency; this value creation mode ensures long-term excess profits for enterprises. The success of the group in internal development of new technologies and new products signifies the emergence of a new production field, which can often lead to a series of new spillover effects, such as the production of new equipment, the demand for new materials, and the building of new after-sale services, among others. The group has set up a number of multi-sector and interdisciplinary production or business units, while the diversified business has enabled the group to share the expenses on R&D in a broader production field forming a virtuous circle as shown:

(1) The synergistic effects of coal, electricity and oil. Shenhua Group has a number of integrated industrial chains, from coal producing, coal washing and coal transporting to the downstream coal power generation, coal chemical industry and coal liquefaction. Shenhua Group practices the vertical integration operation and produces various relevant products to make coal enterprises enhance resource utilization efficiency, cut down costs and increase income. The coal and electricity integration layout can both help the company find stable customers of coal and provide a timely supply of the fuel for the power plant. The joint R&D and system innovation of coal, electricity and the development of various related businesses have brought about to Shenhua Group synergistic effects. Of these, the electricity business and coal business complement each other's advantages, the electricity business has provided stable and large-scale market to the coal business, while the coal business has provided reliable fuel supply to the electricity business. Shenhua Group can also obtain more high-quality investment opportunities in the electricity business.

(2) Synergistic effects of highway, port and mining. The self-owned railway of Shenhua Group has made the concentrated allocation of goods and materials possible, and the point-to-point transportation has helped the group accomplish mining, transportation and washing efficiently, which ensures effective transporting time and helps raise coal value. Taking the prompt goods sales as an example, Shenhua Group's coal pit price is RMB110 per ton, but the average price at the Huanghua Port is RMB380.8 per ton. The closed-loop system formed by the mine, highway and port has greatly reduced external interventions; in particular, it has avoided the influences exerted by the adjustments in the

maximum proportion of transportation of various goods and materials by the state-run railway, thereby reducing uncertainties of the product accessing the market.

(3) Synergistic effects of coal, electricity, highway, port and mine. The coal business provides materials to the electricity business, and the electricity business in turn supports coal production and operation; the coal business improves the power generation business through port transportation. By relying on the strong resource supply and transportation capacity, Shenhua Group has gained high-quality investment opportunities in the electricity business in coastal areas which experience power shortages, and have formed double-advantages in costs and electricity prices.

Performance evaluation of innovation network

Shenhua Group has pressed ahead with technology R&D through the building of the three-circle technological innovation network. When the interview with the general manager of the science and technology development department of Shenhua Group came to the end, he listed the direct and indirect benefits of the application of a number of technological achievements:

Shendong mining area project has realized an economic benefit of over RMB10 billion; one of the Shenhua railway projects–Shuohuang railway project, has changed the railway operation mode; Huanghua Port sedimentation project could gain an annual economic benefit of RMB5 billion as the sedimentation mechanism is clear; the Shenhua excavation project, which completely uses localized equipment, has rejuvenated the equipment manufacturing industry; the Shenhua coal quality improvement project has expanded the Shenhua coal scope of application; the Brown coal purification technology has been developed, and now converts low-content coal into steam coal; the Shenhua water saving power generation project has solved the problem of influence from water on power generation; the Shendong working area enhancement project has increased the supply side and made the resource recovery reach 14%.

These innovation achievements have been applied to practices and increased the production and sales volume of the leading businesses (see Fig. 4.7 and Fig. 4.8) such as the coal business and the electricity business of Shenhua Group, which have been reflected in the financial performance (see Fig. 4.9), including the steady rise of the operating revenue, earnings before interest, taxes, depreciation and amortization (EBITDA), and the operating income of the coal business and the electricity business.

However, the innovation performance of Shenhua Group is not just demonstrated by a single enterprise but involves upstream and downstream

Fig. 4.7.   Coal business status of Shenhua Group between 2004 and the first half of 2007

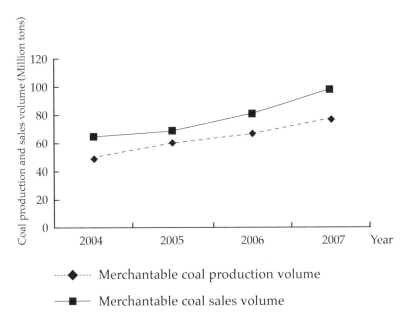

Sources: 2005, 2006 annual reports and 2007 interim report of Shenhua Group.

Fig. 4.8.   Electricity business status of Shenhua Group between 2004 and the first half of 2007

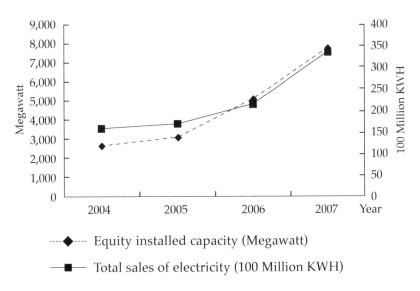

Sources: 2005, 2006 annual reports and 2007 interim report of Shenhua Group.

Fig. 4.9.   Financial standing of Shenhua Group between 2004 and the first half of 2007

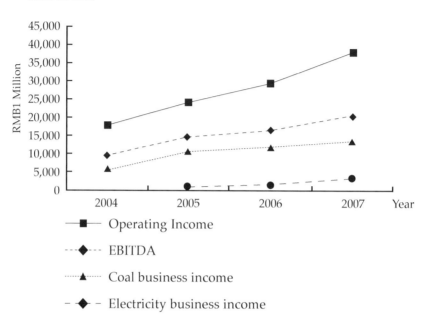

Sources: 2005, 2006 annual reports and 2007 interim report of Shenhua Group.

enterprises, the entire innovation network, and even the overall performance improvement of the industry. In his interview, the general manager of the science and technology development department of Shenhua Group generalized the technological achievements and also mentioned the social benefits generated by the independent technological innovation of Shenhua Group; the technological innovation of Shenhua Group has driven up the industry's technical level. For instance, the high production and high efficency mode of Shendong are replicated by other enterprises, the safety control of coal mines are promoted nationwide, and the performance issues of the first independently innovated equipment have been solved. The strategic alliance model will continue to be promoted because the model of inviting cooperators according to business demand is more competitive. The technological innovation of Shenhua Group has addressed the energy security and coal industrial structual issues, which has pressed the entire industry to advance in this regard, and has guided the technological innovation of the whole industry with a particularly remarkable benefits in coal conversion and seawater desalination.

In the innovation network of Shenhua Group, cooperation with other research institutes can create new knowledge and new technology, effectively promote the spread of knowledge, information, and technology in the industrial group, and realize the flow and optimal allocation of information, technology and knowledge through the crosswise (peer competitors) and lengthwise (upstream and downstream enterprises) links. This interactive whole includes three levels: innovation enterprises, basic industry and support network. The three levels are not mutually independent, rather, they are closely connected based on economic transaction, industrial relations and knowledge sharing. Any level can make a difference on other levels. The high degree of coordination of various factors in the network structure of three levels may also realize the improvement and enhancement of functions of each sub-network system and even the entire innovation network system while adding to its own profits.

Conclusions and Future Research Directions

The case study of the Shenhua Group has testified to the applicability of the technological innovation mode based on resource integration in practice. This mode requires a business to cross the border of enterprise while leveraging the business advantages in independent technological innovation; form a long-term, stable and mutually beneficial innovation network through integrating internal and external resources by means of wheel radiation, chain integrated and group aggregation; make internal capacity and external resources play synergistic effects, and finish the dynamic process of knowledge flow and resource activation to realize the ultimate goal of improving a business' independent technological innovative capability and maintain a competitive business edge. The specific conclusions from the analysis are as follows:

First, the driving force for technological innovation is the combination of internal and external forces.

The Shenhua case tells us that independent technological innovation is not driven by a single force; rather, it is driven by the joint forces of national policies and the enterprise as the main force of independent innovation. The central government correctly positions and understands the overall industrial structure and development environment, and promotes business technological innovation and industrial progress in science and technology from many aspects, including the promulgation and delivery of industrial policies and specific measures, the setup of major scientific research projects and scientific research institutions, and the support in operation capital and technical talents. However, enterprise is the mainstay of the market economy and the engine and main force for technological innovation. R&D institutions are well-positioned to promote innovation, but the combination of the new idea and the realization of market value of innovation does not happen in research institutes outside the enterprise. The business creates the demand for new ideas through entering into the fierce competitive market to contact its customers. For government scientific research institutes and research-oriented universities, the driving force for continuous innovation only exists when R&D activities are closely related to the business production activities. Therefore, the mainstay of independent technological innovation must be enterprises, and only by the joint force of national policy and business innovation could an enterprise realize an interactive integration of science, technology and economy, and effectively boost the development of technological innovation. Given the growing international pressure and impact, enterprises need to independently move forward with the improvement of technological innovative capability and have a good command of core technologies and proprietary intellectual property rights, in line with the support of national industrial policies, in a bid to meet the ultimate goal of enhancing their comprehensive strength and core competitiveness.

Second, the core of technological innovation is the integration of internal and external resources.

The traditional resource classification method believes that business resources principally refer to internal resources. But, under the concept of "competition and cooperation," external resources have become part and parcel of business resources. Enterprises regard the obtaining of external technologies as an important channel for innovation (Duysters and Hagedoorn, 2000). Clearly recognizing the core business resources is the foundation and precondition for effectively allocating and utilizing resources to improve business independent innovative capability. Business resources are all the things that can potentially or practically influence the creation of business values, including not only the resources possessed or controlled by the enterprise but also those that cannot be easily controlled by the enterprise. Resource integration is to integrate, group and recombine various relatively independent resources into a new whole and a new resource system featuring high effectiveness and efficiency in line with certain needs. The result of resource integration is to realize a larger-scale collection of things, of which the benefits and efficiency are larger than that of an individual thing or the repeated addition of individual things. This kind of resource integration does not target only industries; rather, it targets the whole enterprise ecosystem (including community and institutions participating in the form of organizations, and business clients and suppliers in the form of individuals). In this connection, related groups include upstream and downstream enterprises, rule makers, a main body of standard setting and educational institutions, among others. Shenhua Group, as a government-owned enterprise, has realized cross-industry and cross-department resource integration and greatly integrated resources at different levels. At the macro-level, the national, local and industrial resources have been integrated; at the micro-level, the internal resources of different departments within the enterprise and the resources of relevant upstream and downstream enterprises with cooperation have been integrated. In so doing, Shenhua Group's technological innovation is not in a general sense, rather, it is the system with the coordination of internal and external resources and the entire management system. This exactly echoes the empirical research conclusions of Cai Kunhong and Wang Jianquan, in that obtaining external technologies will not significantly improve business performance, but its contributions to business performance will be added as a result of the integration with internal R&D capacity.

Third, the form of technological innovation is the innovation of the entire network.

Modern business management stresses that the realization of comprehensive performance not only lies in the breaking down of the barriers between different functions inside an enterprise but also needs to eliminate various barriers among an enterprise for one part, and its peers, the industry, and the government for another, to realize a comprehensive integration of commodity flow, logistics, information flow, and capital flow. With the upgrade of product structure and the application of high and new technologies, technology R&D has become more and more difficult; it is hard for an enterprise to rely on its own knowledge and capital strength to support large-scale technology development. Therefore, modern technological innovation is no longer a partial and individual innovation but the innovation of the entire network, which requires the concentrated production organization form of an individual enterprise to give way to the decentralized organization form of the business network. This kind of organization form needs to fully leverage on external resources and strength to improve the whole competitiveness through wheel radiation, chain integration and group aggregation, displayed in three areas. First, it must jointly carry out technology development with colleges and universities, scientific research institutes, government departments, and cooperative enterprises, fully leveraging on the mutual complementarities between social forces and free R&D strength by means of entrusted development, joint development and entrusted examination and verification, and demonstrating the wheel radiation under the preconditions of possessing core technologies and key phases. Second, it must establish a longitudinal cooperative alliance with upstream and downstream enterprises, and promote upstream and downstream technological innovation through providing technical support, establishing technical index, playing the monitoring role and adopting other channels, and establish the chain integration featuring shared risks, benefits and innovation. Third, it must possess coordinated specialization of relevant business plates within the industry, forming the external environment to meet diversified demand and dynamic changes, and at the same time improving the group aggregation of innovation network performance formed by a number of industrial chains. In this connection, entrepreneurs and managers could create special values through integrating the coordinated specialization assets inside the business (Teece, 2007).

In conclusion, this chapter has revealed a new innovation trend through the Shenhua case study–the technological innovation mode based on resource integration, and has reflected the innovation model and path for large, state-owned enterprises to explore breakthrough development during the reform process. However, studies in this chapter are carried out based on a single case study, and thus lack the testing of extensive sample data. Additionally, the conclusions and judgments made from the case study also require extensive data verification.

5

Chapter

The Operation of the Innovation-Oriented Service Supply Chain

Song Hua

Introduction

Customer service has increasingly become an important factor for the management of the business supply chain. An enterprise could not only effectively enhance its competitiveness but also better satisfy its customers and realize customer loyalty and good business performance through various service activities which meet the demand of its customers. In this sense, how to provide better services has become an important subject for researches in the business and theory circles of the modern supply chain. However, the rise of the Service Supply Chain (SSC) in recent years has posed a great challenge and many scholars suggest that the SSC has essential differences from the service in the manufacturing supply chain. First, many previous studies explored the performance or influence of supply chain services from the manufacturing perspective and gave little consideration to the unique role of service factors in the management structure and operation of the supply chain. Both the structure and the operation of the SSC have essential differences with the traditional supply chain service.; Second, the concept of the past supply chain service regarded service as the management factor derived from the product supply chain and failed to recognize the huge value-added effect of human resource, information sharing, and capital management during the service process. This value-added ratio cannot be demonstrated or measured by the service performance of the product supply chain. Third, the management elements of supply chain service have some similarities to that of the SSC, but they have many different characteristics which determined that enterprises need to form an operation model that transcends the traditional supply chain. Therefore, this chapter adopts case study method to systematically analyze and explore the innovative operation regularities and model of the SSC in line with the Chinese equipment manufacturing industry. It also rationally breaks down the management elements of the SSC to probe into the SSC theory, move forward with the development of the Chinese business supply chain management, and formulate the innovation management model for key Chinese equipment manufacturing enterprises, transcending traditional products and the manufacturing competition pattern.

Literature Review: Transformation from Product Supply Chain to SSC

With the increasingly fierce competition among enterprises and the development trend towards process and network management, enterprises have emphasized delivering products and services in a highly efficient manner. To make this happen, a critical means is to design and coordinate the supply and distribution networks, i.e. the supply chain management system. According to the definition of the United States Council of Supply Chain Management Professionals (USCSCMP), supply chain management refers to the behaviors planning and managing various activities, including procurement, supply, transfer, and all logistics activities, and in particular, refers to the behaviors and management activities organizing and coordinating the relations among supplier, circulator, Third Party Logistics (TPL), supplier, and customer. In essence, supply chain management is to integrate the supply and demand management inside an enterprise and between enterprises. However, past supply chain management research were primarily conducted from the production and manufacturing perspectives, and lacked sufficient recognition and exploration of the SSC. In fact, in the practices of modern business supply chain management, the performance resulting from service-oriented activities has taken up 24% of the proceeds of the entire supply chain management and 45% of the profits. The values created by service activities have surpassed the product supply chain to become the direction for further development and reform of supply chain management.

Characteristics of SSC

What are the characteristics of a SSC? Chinese and foreign researchers have different understandings toward it and their studies can be generally divided into the following categories.

The first category understands the SSC as the links and activities related to services in the supply chain. On this basis, their studies try to find a way that can balance the optimal services and the lowest costs to operate the SSC. In recent decades, enterprises have been committed to finding a balance between the best customer services and the lowest operating costs. At present, many enterprises have recognized that service operation can be taken as a center of profit that can satisfy their customers and shareholders at the same time. For that matter, enterprises need to plan a blueprint for the activities and links of service delivery, reverse logistics, assessment of suppliers, and the forecast of demand and stock, among others. (Kevin Poole, 2003) According to the definition of Dirk de Waart and Steve Kemper (2004), the SSC consists of the entire procedures and activities of planning, removing and repairing goods and materials to support a product's after-sale services. Based on this, researchers have offered a five-step method to guide the success of the practices of the SSC and balance the costs faced by enterprises to advance stock and service while realizing the optimal order delivery speed and quality. Some other scholars also carried out special research on the services and after-sale services in the global supply chain (William E. Youngdahl and Arvinder P.S. Loomba, 2000; N. Saccani, et al, 2007). Domestic scholars have focused more on studies on the logistics SSC. For instance, Tian Yu observed that the basic structure of the logistics SSC is "suppliers coordinating logistics service suppliers–suppliers integrated logistics service–manufacturers and retailers." Shen Chenglin argued that the logistics SSC is a new-type supply chain with supplier-integrated logistics service as the core, whose role is to provide all-around logistics services to the logistics demander.

The second category regards the SSC as the supply chain of the service industry or service sector corresponding to the supply chain of the manufacturing industry or manufacturing sector. It compares the similarity and differences of the two in a bid to find the supply chain management method that is suitable for the service industry. Initially, Henk Akkermans and Bart Vos (2003) made a comparison between the reinforcement effects in the supply chain of the telecommunication industry and the bullwhip effects in that of the manufacturing industry, and found the source of the reinforcement effects in the supply chain of the service industry. Later on, American scholar Lisa M. Ellram (2004), compared the supply chain management in the manufacturing sector in the book *Understanding and Managing the Service Supply Chain* and stressed the importance of the supply chain management of the service sector, systematically elaborated the differences between the two sectors in supply chain management, and established, on this basis, the SSC model, applicable to the service sector. Recently, Kaushik Sengupta and other scholars (2006) put forward and tested this finding with the empirical research method, employed the factor analysis and regression equation method to analyze the differences between the influence of factors (including information sharing, product or service customization, long-term relationship, advanced planning system, supply and distribution network structure on the operation, and financial performance) on the two sectors. Yang Mingming noted in the book *Hong Kong's Port Service Supply Chain* that port supply chain means taking ports as the core enterprises, effectively integrating various service suppliers and customers, and delivering correct commodities, in correct quantities, to correct locations, to realize the lowest system cost.

From the perspective of the development trend, the first category of studies focuses on the generality of the supply chain of different industries and builds a universal model on this basis; the second category carries out systematic studies on the SSC of different service industries, such as the logistics service industry, in line with industrial characteristics, so as to better guide the practices of the industry. This chapter develops a third category of understanding based on these two categories, which is to regard the SSC as the service-led integrated supply chain. When a customer makes a service request to the service integrator, the service integrator will immediately respond to the request, provide the customer with systematic integrated services, disintegrate the customer's service request when necessary, and outsource part of the service-oriented activities to other service providers. As such, starting from meeting the service request of customers, different service providers at different service positions disintegrate customers' service requests step by step, and different service providers cooperate with each other to form a supply relationship while the service integrators are responsible for integrating various service elements and the management of whole process. We call this the Service Supply Chain. From this point of view, SSC not only includes service and is limited to the service industry, but also depends on whether the supply chain is set up by focusing on the demand of customers and with resource integration and service integration as the guide. To a certain sense, SSC has carried on the customer-centered philosophy in modern supply chain management and driven the management and operation of the entire supply chain in a bid to meet the diversified demand of customers through service integration.

The model and structure of SSC

The SSC is different from the manufacturing supply chain, but they still have similarities in the constructing of the two. In this sense, researching the structural similarities and differences between these two supply chains, in a comprehensive and detailed manner, have quite a significant influence on the design and structure of the SSC.

Comparing manufacturing supply chain and SSC

From the perspective of the structure of the supply chain, a representative structural model is H-P model which was designed by Lee and Billington (1995) and then was cited and described by Hewlett Packard (H-P) (see Fig. 5.1). The H-P model demonstrates that supplier, manufacturer and customer are connected by the commodity flow in an uncertain environment, and many inventory locations have provided a cushion for these uncertainties. As pointed out by Davis (1993), it is these kinds of uncertainties that have added to the difficulties in managing a supply chain. The uncertain environment in this model still exists in the SSC, but it is reflected in workload instead of inventory because uneven access to information might also have some reinforcement effects. Therefore, the service supply still needs to balance the improvement of service level and the costs brought about by an increased workload.

Fig. 5.1.　Hewlett-Parkard model

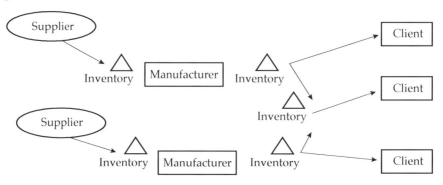

Another model is the widely accepted Supply Chain Operations Reference Model (SCOR) promoted by the Supply Chain Council. The SCOR model has described the process and activities of supply. SCOR has reproduced the operation process and combined the description and definition of the process with the best practices and technologies (see Fig. 5.2). The SCOR model organizes by centering on the five major management procedures, namely plan, source, make, deliver, and return. In the SSC, activities including plan, source, make, and deliver still exist in some node enterprises but they are probably no longer the necessary phases and must processes for the leading enterprises of the integrated service.

Fig. 5.2.　SCOR model

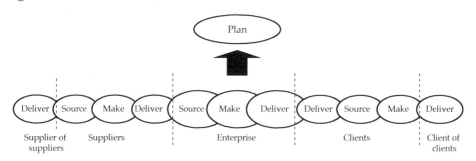

Source: www.supply-chain.org.

Another typical model is designed by Croxton et al. (2001) and put forward based on the Michael Porter's Value Chain Model (1985). Relevant studies about this model are named Global Supply Chain Forum Framework (GSCF); that is, a supply chain consisting of operation process, management, and chain framework (see Fig. 5.3). Customer Relationship Management (CRM) and Supplier Relationship Management (SRM) are also included in this chain. In this model, the supply chain includes all players, from raw material providers to the ultimate consumers. The product flow process goes through this chain framework and considers the return process at the same time. The efficiency and performance of the entire supply chain have been improved through information cooperation and integration in the supply chain. The information flow in this model is the key part of the SSC; the CRM, Demand Management, and SRM continue to be used in the management of the SSC, but Skill Management and Service Delivery Management (SDM) are what characterize the SSC.

Fig. 5.3.   GSCF

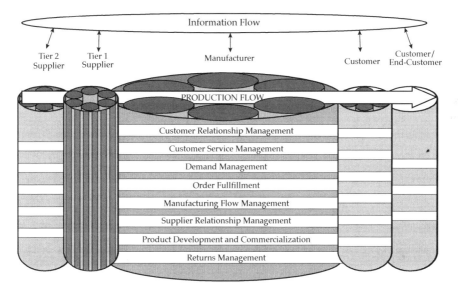

Source: K. L. Croxton , Carcia Dastugue, D. M. Lambert, and D.L. Rogers, "The supply chain management process," *The International Journal of Logistics Management* (2001).

Model building of SSC

Based on the analysis of these three models and the characteristics of the SSC, the Service Supply Chain Framework (SSCF) built in this text takes the service as the node, takes the workload as the cushion, and takes the indirect service supplier, direct service supplier, service integrator, and end customer as the mainstay, including Demand Management, CRM, SRM, Skill Management, SDM, among other leading activities, and integrates the SSC of physical flow and information flow (see Fig. 5.4).

As shown in the Fig. 5.4, the services delivered by indirect service providers converge to direct service suppliers and then from direct service suppliers to service integrators who will deliver the whole set of services to end customers;

it might be also possible that service integrators have been equipped with the service capacity of direct service suppliers to directly accept services delivered by indirect service suppliers. No matter which path it is, the delivery and convergence of the services from all nodes depends on information flow and sharing., and of course, some parts of the services still depend on the physical flow of products. In this model, service integrators dominate the building and management of the entire SSC. Service integrators are capable of understanding the changes and updates of demand and continue to develop and explore new possibilities for value growth through predicting customers' demand and CRM. After that, through assessing the suppliers' performance and SRM, they will integrate the resources and capability of both direct and indirect service suppliers, create value-added services based on basic products and services, provide improved integrated services to customers, and build the entity and information-flow network with service as the node.

Fig. 5.4.   SSCF model

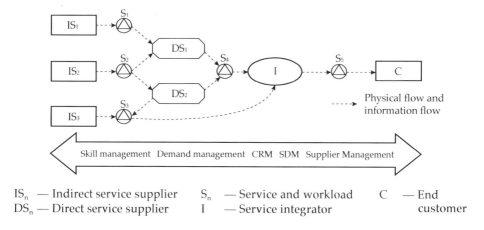

IS$_n$ — Indirect service supplier   S$_n$ — Service and workload   C — End
DS$_n$ — Direct service supplier   I — Service integrator   customer

In specific analyses, information flow plays a key role in recognizing the demand and sharing information, and it is also reflected in the service level, working conditions, working border, service skill and performance feedback. Information flow is the basis for any effective supply chain and can reduce risks and uncertainties faced by all types of supply chains. The SSC puts more emphasis on the sharing of information. Apart from the technical support of the information system and network platform, the high-efficient and long-lasting operation of the entire SSC also depend on capacity management, demand management, relationship management, SDM, capital management, and the integration and coordination of numerous functions. Skill management that is based on the employee's technical skills and service quality is one of the elements for service integrators to realize competitive differentiation. But, quick response to the demand of customers also relies on the prediction and planning for the ever-changing demand of customers. Relationship management includes two aspects, namely CRM and SRM. CRM needs to have an all-round understanding of customer demand, and must concentrate resources and capability to meet these demands, including customer segmentation and customer relationship. As a driving force for the operation of the SSC, CRM should also include rapid response and adaptability to changes to ensure that the customers' demands are met in a timely and comprehensive way. The realization of customer value also needs the cooperation and coordination of all participants in the entire chain and the assistance of SRM, including various management activities from suppliers' selection, evaluation, and coordination to management. In the SSC, the selection and evaluation of suppliers based on the definition of the contents and scope of work, and also involves clear service level agreements (SLAs) to cut uncertainties in the service delivery process. All in all, in order to realize the common development progress of the entire SSC, service integrators must effectively coordinate the competition and cooperation relationship between different nodes to make all parties realize their own development on the basis of benefiting the whole supply chain. In so doing, the delivery of integrated service can be timely and effective.

However, the SSC is still exposed to many problems and challenges. The increase of customer satisfaction will also add to costs, and the balance between the two still needs the optimal marginal critical state. The Bullwhip Effect, which emerged in traditional studies on the supply chain, is replaced by reinforcement effect in the SSC. The resolution to the effect rests on workload rather than inventory and thus the reinforcement effect can be weakened to a certain extent through focusing on skill management, quality management, and information sharing. Additionally, there is a need to further study whether service can be managed level-to-level, like category management, and whether cyclic service is needed, as in reverse logistics.

Research Method and the Selection of Sample Enterprises

This research adopts the single case research method, which is to select a typical single sample enterprise in the industry by examining multiple secondary units of analysis in order to establish and verify theory-building. In the case study, many scholars pointed out that multiple case studies are more reliable than single case studies, but a single case study which can challenge or extend the existed theories is still significant. In particular, a single case study is effective for a unique, special or even extreme case. It is easier to analyze various intricate situations and deeply reflect the status of business operation through careful multi-unit analysis of a single enterprise in order to propose a relevant theory.

The industrial background of this study is the fan industry. The fan industry plays a key role in the national economy and belongs to the major equipment manufacturing industry. The principal products include air blower, compressor, ventilator, and energy recovery turbine installation, among others, which are widely used in major industrial sectors, such as the petroleum, chemical industry, metallurgy, electricity, coal, building, textile, and transportation industries. These products are general machineries that are largely demanded and broadly used. In the compressor and fan field, there are only a small number of production and operation enterprises including Shenyang Blower Works (Group) Co., Shaanxi Blower (Group) Co., Ltd., Shanghai Blower Works Co., Ltd., Chongqing General Industry (Group) Co., Ltd., Zhejiang Shangfeng Company, and Wuhan Blower Company. Among which, Shaanxi Blower Co., Ltd. (hereinafter referred to as Shaanxi Blower) is a rapidly changing enterprise. The foundation stone of Shaanxi Blower Co., Ltd. was laid in 1968. In 1975, the company was founded and put into operation. The company used to be managed by the Ministry of Machine-Building Industry and Shaanxi Machine-Building Department, and was then transferred to Xi'an Municipal Government in 1985. It was restructured into a limited company in 1996. In 1999, the company launched the establishment of Xi'an Shaangu Power Limited Liability Company with the support of the group company's production, operation, and quality assets. Before 2000, Shaanxi Blower was a Chinese traditional blower equipment producer and operator, and was weak both in competition and the capability in shielding against market changes. But as of 2000, its business performance and competitiveness have changed dramatically. Toward the end of 2005, its assets totaled RMB3.45 billion, RMB2.53 billion more than at the end of 2000 (RMB918 million); the net assets stood at RMB932 million, 4.16 times that of 2000's net assets of RMB224 million, and surpassed the total assets of late 2000. In 2000, Shaanxi Blower was still a traditional manufacturer with an employment of 3,326; it recorded an output value of RMB340 million and a profit of RMB14 million. In 2005, Shaanxi Blower had a total of 2,969 employees, 11% fewer than that of 2000. It realized an output value of RMB2.5 billion and a profit of RMB344 million, 7.35 and 24.27 times of those in 2000 respectively, and the profit posted in the year of 2005 alone was more than the whole-year gross output value of 2000. The company recorded RMB113,900 in per capita sales revenue in 2001, but the figure rose to RMB2,803,900 in 2005, up 25 times. In addition, from the perspective of the competitive position in the industry, Shaanxi Blower registered RMB2.5 billion in the gross industrial output value, RMB2.18 billion in sales and RMB880 million in industrial added value in 2005. During the same period, the second largest industrial player, Shenyang Blower Works (Group) Co., posted a total of RMB1.73 billion in the industrial output value, RMB1.78 billion in sales and RMB490 million in industrial added value. The root cause for this change of Shaanxi Blower is the "two changes" operation idea raised in 2001. The first change is to "put customers at the core, carry out value-driven operations, and strengthen customer-orientation," occupy the high end of the value chain through full studies of the end market, and enhance the competitiveness of solutions in line with the demand of users. The company seeks to gain profits through providing solutions with high-technology contents and high added value; the company also creates profit space via systematic purchases from suppliers. The second change is "brand operation and the highlight of the systematicity and comprehensiveness of the operation." It is just because of the proposition and implementation of this business mode that has brought about the revolutionary development of Shaanxi Blower. This case is unique in the Chinese fan industry. Therefore, it is practical to adopt the single-case, multi-angle and multi-aspect research method under this background.

Our field surveys and in-depth interviews were conducted between January 15 and 17, 2007. Our interviewees include the company's top leaders, middle and lower management staff. The departments of the company we surveyed include the procurement and supply department, the products service center, complete equipment center, product sales department, contract management center, and the technology department, among others. All materials for these analyses come from field interviews.

The Operation Mode of the SSC of Shaanxi Blower

As the social economy keeps developing and the competitive environment keeps changing, manufacturing enterprises have focused on building a supply chain based on the core capacity. They have adopted the flat organizational structure consisting of production units which can be rapidly reconstructed, replacing the pyramid-like multi-layer management structure with fully self-governing and distributed cooperative works. They have given emphasis to human creativity and service, and transformed the competitive relations between enterprises into the "win-win" relations with competition and cooperation.

Flexible operation stresses the integration of internal and external functions based on information openness, sharing and integration. This integration has not only broken internally established business and information barriers, but also removed various barriers among enterprises, and between enterprise and customer. The "two changes" idea of Shaanxi Blower has rebuilt the operation and management model of the business supply chain based on this thinking.

Starting point of the supply chain operation—sales and marketing

For manufacturers, one of the most important functions in supply chain management is the sales and marketing activities. According to Shaanxi Blower, traditional manufacturers focused on the manufacturing, development and marketing of physical products. After passively accepting an order from a customer, an enterprise would carry out logistics functions, such as manufacturing, stocking, installation and delivery, and pay little attention to the supply chain service in operation. However, Shaanxi Blower has, since 2001, adjusted the traditional operation mode, actively expanded the customer base and customer demand, and organized business operation activities by focusing on the overall demand of customers. Specifically, apart from providing self-produced main engines to customers, Shaanxi Blower is also responsible for complete equipment (including system design, system equipment delivery, system installation, and debugging) and project contracting (including infrastructure, plant and peripheral facility construction), to provide a broader range and systematic solution. For instance, while providing the Top Pressure Recovery Turbine (TRT) to Baosteel, Shaanxi Blower was also responsible for the construction of complete equipment, plant, infrastructure and periphery facility; apart from providing traditional TRT main engines, they implemented the "turn-key" project and provided the blast furnace gas waste heat and pressure recovery for power generation function. One of the benefits for customers is that there is no need to carry out special project management and coordinate among the main engine manufacturer, supporting equipment manufacturer, and project executor. Beyond that, the coordination has been greatly reduced, and any problems arising would be handled by Shaanxi Blower. The second benefit for customers is to control the project investment and investment cycle, which has effectively addressed the issues of rising costs and an uncontrollable cycle resulting from an unmatched system and insufficient experiences in implementation. The departments of Shaanxi Blower for the implementation of this activity include the marketing department, the complete equipment department, and the project contracting center. After learning the overall demand of customers, these departments delivered the contract of intention to the management center for examination, coordination and management (see Fig. 5.5), which leads the operation of the SSC of Shaanxi Blower.

Fig. 5.5.   Operation model of the service supply chain of Shaanxi Blower

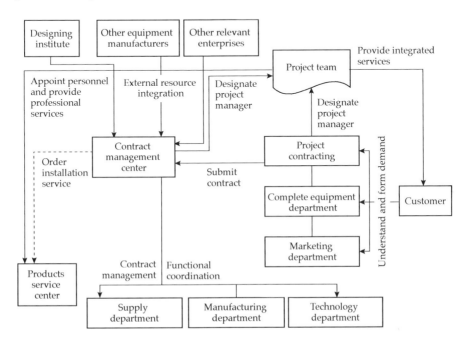

Internal coordination and integration of the supply chain

After getting to know the customer demand and forming the integrated service demand order, how to effectively examine contract orders and control operating risks, especially how to coordinate and integrate various cross-functional departments and activities, have become the keys for the internal supply chain operation. The function of contract management center is to integrate supply chain management departments, which is displayed in three ways: The first aspect is to design and forecast, including forecasting and drafting the annual plan, seasonal plan and monthly plan. The annual plan is generally to arrange the contracts for the next year (including placing orders, producing, and yielding) and to make a forecast in line with the social and economic situation, customer factors and other various parameters; after the formation of the annual plan, the contract management center will start to draw up and balance the seasonal plan. The seasonal plan is generally drafted and balanced by taking into consideration the product condition, loan, customer importance, supporting condition, and other factors. Once the development of the seasonal plan is finished, the monthly plan will be formed accordingly. The monthly plan is worked out by the contract management center and works with other functional departments to realize a production-supply-marketing plan. Then the departments of procurement and supply, production, and planning will form their respective binding operation plan. At Shaanxi Blower, the monthly coordination meeting is hosted by the head of the contract management center to facilitate the complementation of the monthly plan.

The second aspect is to implement the contract. The matters of various equipment contents and small order size of different single pieces have made it more difficult in contract organization and management. Therefore, there is a need to interpret the contract, and the marketing contract in particular, before fulfilling the contract signed with customers. In general, relevant tasks need to be made known to each department after interpreting the contract. The contract management center will take the lead to conduct training programs before fulfilling the contract if necessary, to assist each department with fully understanding the customer demand and any issues in contract implementation.

The third aspect is to guide, manage and coordinate the marketing departments, including the project contracting department, the complete equipment department, and marketing department. Only by realizing full internal communication and that between different marketing departments could an enterprise better meet the customer demand.

The contract management center generally has the right to evaluate and review any contract signed, form instructional opinions and standards for the marketing, and set a relevant date of delivery. A marketing department can ink a contract by itself under this regulation, but the time exceeding this prescribed period shall be approved by the contract management center. The contract management center will also offer feedbacks regularly to the marketing department for communication. For example, the peak time for order is generally between March and April. During that period, the contract management center will rearrange activities and bring forward the order for June to May, sometimes even 2–3 months earlier, in order to balance market irregularity. At the same time, the contract management center will coordinate with marketing departments and require them to feed back the market information once every ten days, enabling coordination.

Procurement and supply in the supply chain

After the contract order is examined and verified by the contract management center, the first activity is to purchase and supply—often a key link in supply chain management. Sound procurement and supply activities support the production and follow-up activities; the quality and value position of procurement and supply will directly determine the competitiveness of enterprises. The materials procured and supplied in Shaanxi Blower is complex and falls under four categories. The first category is raw materials, mainly steels and non-ferrous metal; the second is auxiliary materials, such as machining, chemical materials, accessories, and props; the third is auxiliary equipment for entering into the plant and, during the general assembly process, it needs additional equipment, such as motors and seals, which take up half of the parts and components; the fourth is consumable items and other tools.

In line with these, Shaanxi Blower has adopted different procurement and supply systems. Generally, the design determines the technical parameters, what to buy, and when to buy; the procurement and supply times are generally decided according to the time of manufacture, supply, transportation, and customs clearance, and it is generally required one month in advance. The procurement quantity of steel is small and this purchasing is mainly carried out by the purchase department of Shaanxi Blower. The raw and auxiliary materials adopt the zero-storage procurement model: the order is outsourced to another enterprise for procurement and Shaanxi Blower takes charge of price examination and review. The reason for outsourcing the procurement of auxiliary materials is mainly that there are many qualities and prices of these materials. The company would otherwise be loaded down with those trivial details, and the value-added ratio of those materials is low. In contrast, outsourcing procurement has not only simplified the management activities of the company but could also greatly enhance its competitiveness if managed and controlled well.

However, the effective set-up and implementation of outsourced procurement management is important for supply chain procurement management. Shaanxi Blower has properly carried out quality control over outsourcing companies in due time, the core materials are mainly examined and reviewed by it, and the technology department has prepared a technical agreement requiring that outsourcing companies have their own quality tests and the Shaanxi Blower conducts spot checks. Date of delivery, quality, after-sales, market exploration, and other standards are determined by Shaanxi Blower, and some small products are also outsourced to external enterprises due to the overcapacity. Training programs are also organized for supporting plants and the supplier manages the storage in supply logistics. That means outsourcing enterprises have set up storage in Xi'an, providing timely delivery and distribution to Shaanxi Blower.

Accessories and props mainly seek procurement tenders and price-compare procurement. Sometimes when the customers designate products, it mainly uses price-compare procurement. For instance, the company goes for unified bidding for pipelines and valves and the number is fixed each year; transmission is outsourced, the demand is made known to the supplier through bidding, and a unified contract will be formed. The management is controlled in four ways: advance payment, progress, delivery, and quality assurance. In other words, the supplier should operate for 72 hours as a trial; the company will adopt payment control, which means before the delivery the company will deduct retention money and the final payment will be made after the product service center recognizes the one-year operation. For frequently used parts and components, Shaanxi Blower does not set up storage; rather, parts are directly delivered to the installation spot. But, the supply department will check, test, examine, and review the materials based on quality control, supply and other logistics performance indicators to decide future procurement and supply contracts.

Project team building and project implementation management

As Shaanxi Blower has implemented the "turn-key" project and provided integrated services, the management is more complex and the activities are extensive and complicated. The operation strategy is bound to be impaired if there is not strong project team building and supervision. After examining and developing the project implementation plan, the contract management center takes the lead to form a cross-functional and cross-activity project team that will be dispatched to the customer site for whole-process tracking and project implementation. Shaanxi Blower has also put in place the two-level project manager system in order to guarantee a smooth delivery of the project and effective coordination of various resources. First, after the project contracting center signs a project service contract, each will be equipped with a project manager who has the right to make decisions in field activities and management. The project contracting center is responsible for formulating the manager manual to guide and regulate the behaviors and activities of project managers; secondly, the contract management center examines and reviews customer orders and assigns another project manager to follow up the entire process of the project implementation. Project managers are generally senior technicians and administrative staff who have retired from Shaanxi Blower and are familiar with the situation and technologies of the enterprise. The scope and target of management are different between the project managers designated by the contract management center and those of project contracting. The latter is mainly to carry out field monitoring and management of the project, while the former conduct dynamic tracking of the project implementation and also assist the on-site project managers in coordinating and managing various functions internally. In this sense, the monitoring and management effects of this kind of two-level project manager system are significant. Therefore, the service quality has been greatly improved and a whole-process, round-the-clock dynamic supply chain management has been made a reality.

Customer service

Another important functional department in the operation process in the supply chain of Shaanxi Blower is the products service center. It originally engaged in the maintenance of the mechanical equipment of the company, but after the implementation of the "two changes," the maintenance work has been outsourced to other professionals. As a result, the center integrated its personnel and formed a specific whole-process service department. The major functions of this department in the supply chain process include: (1) working with other departments to form a project team to help install and debug equipment when fulfilling the customer contracts to ensure the task performance in project implementation; (2) providing follow-up equipment maintenance, and that all follow-up services will be provided by the product service center if any equipment abnormality occurs within the contract's guaranteed period; (3) providing whole-process equipment maintenance and even whole-process monitoring. The product service center of Shaanxi Blower not only provides services within the guaranteed period but also extends the services to the entire operation process. Furthermore, the department also provides comprehensive services for any equipment and product exceeding the guaranteed period and takes this as new added-value services and the growth point for profit; (4) providing tailor-made services to customers. The product service center of Shaanxi Blower not only provides whole-process services of its own equipment but also extends the service chain to any equipment and product of its customers. In other words, if the customer demand arises, relevant services will be provided whether the product equipment is manufactured by it or not.

The Structure and Key Elements of the SSC of Shaanxi Blower

A unique feature in the development of SSC is to stress on the customer demand as the driving force behind process, instead of simply relying on products to push the running of the operating system. The value of customers is determined not only by the value of a single product and service but also by the costs for integration, connection, and search of multiple products and services demanded by customers. In this sense, the key for an enterprise to succeed lies in the shift of the focus from its own product and service to its customers and the solutions they seek. The surveys on Shaanxi Blower conclude that its operation and management mode have demonstrated the SSC management. The company's SSC has showcased the following characteristics:

1. Management system of comprehensive demand and customer relationship

The comprehensive demand and CRM of Shaanxi Blower are carried out in three aspects.

The first is the systematic sales service practiced by Shaanxi Blower after 2001.

The second is the continuous maintenance service. Shaanxi Blower has outsourced the maintenance of its own equipment to professional companies in order to better optimize resources, which has guaranteed the quality of its own equipment maintenance service and cut down maintenance costs. In order to take full advantage of its own resources and professionals, the company has organized its human resources to invest in more value-added services, and has leveraged its own professional advantages to provide maintenance services in a swift, timely and highly efficient way. In particular, Shaanxi Blower has leveraged its equipment collaboration network to integrate the supporting resources of other plants and deliver systematic services to extend its customer service to the entire product lifecycle and cut down the total costs for customers.

The third is the professional remote equipment management. The safe and stable running of turbomachinery is essential for the normal operation of the customer's whole system. Shaanxi Blower implemented "afterwards remedy," which means the maintenance for equipment is provided after a mechanical failure and is bound to impact smooth production and operation. And the company has changed the afterwards remedy to "beforehand monitoring" after the implementation of the "two changes" program, to ensure a smooth, year-round operation of the customers' equipment. Shaanxi Blower has developed the rotating machinery process monitoring and fault diagnosis system which enables the company to carry out real-time remote monitoring of the field unit operation via the Internet and to learn the running state and changing trend of the equipment in time. For the part of customers, this system can make them have an accurate picture of the unit's running state, rationally arrange the time for maintenance, upgrading, and the quantity of standby redundancy as preventive measures, and ensure safe operation of the unit. Shaanxi Blower provides 24-hour on-duty service, conducts real-time monitoring of the running unit, and dispatches experts to make the judgment of the severity of the unit in trouble and regularly deliver the monitoring report on the operation, allowing its clients to focus on their principal works. Meanwhile, the analysis, online observation, and forecast of experts can provide the marketing team of Shaanxi Blower with beforehand and accurate information about the demand of customers for maintenance, upgrading, and standby redundancy.

2. Logistics service deliver management

Inventory works as a cushion for enterprises to deal with market fluctuations, but the shortcomings of the traditional inventory management mode are mainly reflected in the failure of irrationally using and managing suppliers' resources, the huge amount of tied up capital, the increase of costs for inventory and business operational risks. In this connection, Shaanxi Blower has established strategic partnership with its major raw material suppliers who are entrusted to manage the inventory. Suppliers deliver raw materials based on the production planning of Shaanxi Blower, who uses the raw materials according to the plan and makes a unified settlement. Any tied-up capital has been reduced through zero raw material inventory management. According to statistics from the procurement and supply department, Shaanxi Blower realized a direct benefit of RMB1.33 million from the zero raw material inventory management in 2004.

In addition to strengthening the raw material inventory management, another feature of Shaanxi Blower's logistics delivery service is to optimize customers' spare inventory. The complete fan equipment provided by Shaanxi Blower has a high degree of specialization and represents the core device of the equipment during the manufacturing process, and thus requires high reliability. Once the equipment goes wrong, the entire system will totally stop running, which may even cause destructive events. In order to prevent that from happening, clients usually store some quick-wear parts as spare parts, which will result in a large amount of tied up capital, a large number of reserves, and difficulties in safekeeping. To solve this issue, Shaanxi Blower provides its clients with zero storage of spare parts services. For example, Shaanxi Blower provides Laigang Group with spare parts service, so Laigang Group no longer need to purchase any spare equipment in advance. These spare parts will be supplied in a timely manner if a unit breaks down. Laigang Group's capital will not be tied up and it is unnecessary to worry about the halt of production triggered by an urgent demand for spare parts which are not stored. In this way, their storage costs are reduced. Shaanxi Blower can provide these kinds of services to many clients due to the setup of the joint storages of spare parts, and only a small amount of stock is needed as a result of the high degree of serialization of product, little difference in parts, and virtually no difference in semi-finished products in particular (even if individual reserves are not available, it can be prepared in advance by leveraging on remote monitoring and diagnose information). Once a client requests it, Shaanxi Blower could transfer spare parts from storage or manufacture a small amount of the spare parts before delivering to the client by road or by air. The benefits generated by the client from this service will be distributed through mutual negotiation, which means Shaanxi Blower and its clients share the benefits generated from the joint storage management. According to statistics from Shaanxi Blower, the order quantity of spare parts services generated RMB124 million between 2002 and 2004 with an average annual increase of 45%. The joint inventory management mode of Shaanxi Blower, which is built on the basis of supplier trusteeship, can not only improve its clients' operation performance and enhance the operation capacity of the supply chain, but also gain new service profits.

3. Supplier relationship management (SRM)

Whether an enterprise can effectively implement the supply chain integration service is closely related to the whole-process management of external resources. Shaanxi Blower provides its clients with the whole set of supply chain services, which is bound to involve the coordination and management of many suppliers. For example, in a project valued at RMB12 million, Shaanxi Blower only provided RMB500,000 while the rest was provided by external coordinating enterprises. Therefore, it would be hard to guarantee a smooth implementation of the integrated services if an effective network relationship management system cannot be established. Shaanxi Blower builds teams of subcontractors nationwide and, before fulfilling the signed service contract with any client, it usually goes to the subcontractors for investigation and survey; in addition, its clients can recommend subcontractors and Shaanxi Blower can assess and screen alternative subcontractors based the standards it developed. Shaanxi Blower practices the whole-process management of outsourcing enterprises, which means the management will start at bidding and will include not only relevant business departments but also the financial department, discipline inspection department, and the bidding department of Shaanxi Blower to get involved in the assessment process of each tender offer, in order to reduce risks in the project implementation process and fully evaluate each outsourcing enterprise. When evaluating an outsourcing enterprise, Shaanxi Blower not only examines the technology, integrity and culture of the team of the outsourcing enterprise, but also makes and controls the budget in advance, and then analyzes the rationality of the costs. According to relevant sources with Shaanxi Blower, the quality of the whole process is very likely to be damaged if it were simply to compare the quality and the price during the tendering period. Therefore, the demand of customers can be met and the high quality of project services can be provided only with specific activity analysis, the control of each link, and price of the process.

In the SRM, organized supplier coordinative institutions are pivotal for stabilizing and developing supplier relationship. Shaanxi Blower has built its external resources collaboration network–supplier strategic collaboration network. It integrates competitive suppliers through this permanent organization and convenes an annual meeting for multi-directional communications which include promoting its culture and strategy, conducting some system technology R&D and even some specialized seminars including technology and market, and integrating supply chain coordination and communication. On top of that, regular meetings for coordination are held on a monthly basis, mainly on-site meetings, and the participants include not only the quality, supply, technology and finance departments of Shaanxi Blower, but also some enterprises of the collaboration network. They join together to report the current situation and make known their respective difficulties in finance, delivery method, and others. In these two collaboration networks, Shaanxi Blower also invites suppliers to critique its functional departments, and evaluate its working process and the work performance to help it discover its own problems and inadequacy, thus improving its work efficiency.

4. Compound capacity management

The formation and operation of the SSC also depends on the integrated management of all kinds of capacities (compound capacity). In particular, the delivery of the fan industry's system service partially depends on the capacity of supporting manufacturers and external organizations. Shaanxi Blower proposed to take the supporting manufacturers to enhance its capacity meeting the demand of the market. In September 2003, it organized 56 supporting enterprises to establish the "Shaanxi Blower Complete Set Technologies and Equipment Collaboration Network" with a membership including Siemens, Emerson, GE, and many other world-renowned companies. Collaborating partners can realize resource sharing and jointly improve the quality of technologies through the operation of this cooperative network; at the same time, this cooperative network can help Shaanxi Blower advance human resources. As cases in point, Emerson offers human resources (HR) training for 50 supporting enterprises of Shaanxi Blower each year in Singapore;

Formosa Plastics offers HR training for 20 partners of Shaanxi Blower each year in Taiwan. Shaanxi Blower also cooperates with colleges, universities, and research institutes by means of commissioned research, joint development, and entrusted review to strengthen its own R&D power; it also engages in close technical cooperation with its peers around the world during the process of joint development of the domestic market, and collaborates with supporting manufacturers in research on system technology and related technology to integrate supporting products and self-made products. For instance, Shaanxi Blower and Zhejiang University worked together to conduct the "top pressure stability analysis and control test research in the blast furnace top gas pressure power generation device" and was granted with a national patent; Shaanxi Blower also cooperated with MAN Turbo to develop a large supporting compressor for an air separation facility, and this cooperative behavior has met the demand of the market for various services.

5. Capital and financing management

Capital and financing management is one of the most critical elements in the operation of the SSC and represents the developing trend of modern supply chain management. Shaanxi Blower has begun to develop toward supply chain financing management while providing integrated supply chain services to its clients. Specifically, Shaanxi Blower has, since 2004, established a strategic partnership with many banks, including the Industrial and Commercial Bank of China (ICBC), Pudong Development Bank, and China Merchants Bank (CMBC), and has capitalized on the resources of financial institutions to provide its clients with financial services and guard against credit risk. First, Shaanxi Blower has established an accounting center to carry out unified management of independent accounting units, practice centralized capital storage and repay the large amount of loans; banks set up special counters in Shaanxi Blower to provide specialized services to help it effectively manage capital. This is because Shaanxi Blower has a large amount of capital flow and a change of 1% in the interest rate means tens of millions of yuan in benefits or losses. Therefore, there is a need to strengthen the rational integration of capital available in order to generate the maximum benefits from financial management.

Second, Shaanxi Blower has established a contract management center and a financial center to manage supply chain financing and operate a financing warehouse business. Financing warehouse is the development of SSC in the financial field and is a comprehensive way of managing commercial distribution, logistics, information flow, and capital flow, using capital to revitalize commercial distribution and logistics while using these to drive capital flow. The core idea of a financing warehouse is to seek opportunities in integration, mutual complementarity, and interaction of various flows to improve business efficiency and cut down operation capital and risks. According to the director of the contract management center, Shaanxi Blower has launched the financing warehouse based on movable property management by combining resources of banks. This enables it to leverage on its own capital and bank credit lines to carry out financing based on the chattel (such as equipment) mortgage and warehouse warrant (of products such as steel), mortgage or pledge. After the accomplishment of the project, Shaanxi Blower will obtain a relevant contract amount and financing charges.

For example, Rizhao Steel Co., Ltd. used to entrust Shaanxi Blower with the construction project of 6 blast furnaces, but it could only prepay 15% of the contract amount due to its insufficient capital, and the rest had to be paid by gains. This kind of contract cannot be performed based on the common management method. But Shaanxi Blower, on the basis of the recognition of the client and the need for strategic cooperation, adopted the method of financing for the construction. In the early stage, various departments carried out field and legal evaluations and research on guarantee, and started to manage after the delivery of the loans and with the approval of Xi'an municipal government. In terms of financial risks management, Shaanxi Blower generally worked together with commercial banks to formulate relevant financing plans and control the returned money, stage by stage, on the basis of assessing project risks. In other words, after the project was put into operation, it was re-evaluated to see whether the first batch of loans could be repaid on time before dealing with the second batch of loans several months later. After the implementation of this project, banks issued loans to Shaanxi Blower with the equipment and warehouse warrant as the collaterals, and then Shaanxi Blower provided financing to Rizhao Steel. In addition, during the implementation of the project, Shaanxi Blower analyzed risks according to the situation of suppliers, and all subcontractors shared relevant risks while obtaining deserved financial benefits. As such, both Shaanxi Blower and the subcontractors have formed the benefit sharing and risk sharing operation mode.

Conclusions and Future Research Directions

It is easy to find from the above case study of Shaanxi Blower that the formation and operation of the SSC have provided Chinese various industries, especially the equipment manufacturing industry, with a brand new competition mode. In particular, amid the fierce international competition, the SSC will become a channel for enterprises to make breakthroughs. This is because the SSC is not a product manufacturing supply chain. Therefore, the competitive performance does not just depend on a single product, production technology, or manufacturing process, rather, it is the management and operation using services to drive the whole supply chain in an effort to meet the diversified demand of customers through the integration of services. Because of this, Youngdahl and Loomba named this kind of operation mode as a "service factory" which plays a role as a consultant, showroom, and dispatcher between marketing, external clients, and R&D. But this study believes that compared to the "service factory," the SSC can better generalize this model of Shaanxi Blower, which integrates through service integrators the suppliers of all kinds (other equipment and system supplier), other direct or indirect service providers (such as a building design institute, logistics enterprise, and so on), distributors of the products and services, financial institutions, and end customers, and provides a comprehensive service system based on capacity management, demand management, SRM, CRM, SDM, and capital and financial management. We can find through this case study that the realization of this target or mode must depend on the following elements:

First, service integrators must be equipped with a high level of credit asset and capability. As the SSC emphasizes a high degree of resource and capacity integration management, an organizer of the SSC must be equipped with strong credit assets and capacity. This must not only meet the demand of clients for the most direct and the simplest product and equipment, but must also leverage on the capacity of core enterprises to offer high-level credit assets. This includes not only whether the products and services could deliver core value to its clients and could meet any demand raised by its clients (including tangible and intangible), but also whether the knowledge and knacks of various parties are utilized to drive the whole network into a brand new and unexplored "blue ocean market." Only when an enterprise has these credit assets and capacities, can the SSC be established. This is demonstrated not only in the Shaanxi Blower case study but also in the past theoretical studies. Sako, in view of the sustainable relationship between Japanese companies, also noted that credit is in itself a kind of asset and has expanded the relationship through a three-level credit mechanism (i.e. contract credit, capacity credit, and business reputation).

Second, service integrators can seek integration opportunities and comprehensive benefits among different individuals, organizations and elements. The main feature of SSC is that it has broken through traditional industrial and territorial constraints and realized huge value-added profits through the coordination and integration of all elements. This added value not only comes from earnings in the product operation (i.e. the earnings generated from the product supply chain), but more importantly reflects in the reducing of low efficiency resulting from social transaction costs through the service integration. Shaanxi Blower's story is a case in point. Traditionally, its clients did not enjoy the rights of being party A as it is in the contract, due to system and institutional factors of some economic subjects (such as the building design institute) during the real operation process. Later on, since Shaanxi Blower provided solution-based project contracting, it helps its clients save a great deal of energy, cost, time, and various consumption on management, and realize added value indirectly. In this connection, the operation of the SSC can realize direct economic returns brought by integration, and enable participants of the network to reduce social transaction costs as a result of service integration. All these are the management targets of service integrators.

Third, service integrators can establish sound social capital between individuals and organizations through its own behaviors and operations. Social capital is an intangible resource, compared with physical capital and human capital, and takes trust, regulation and network as the carriers, including not only system, regulation, networking and other organization structural features in social relations, but also citizens' trust, prestige, social reputation and other personality networking features. It can be seen from the Shaanxi Blower case study that a key element formed by service integrators is the strong social capital established through various rules, systems and long-term accumulated credit. This social capital is not just the process-based trust and network built between individuals or between different sectors inside the enterprise, it also built the effective group social capital between the enterprise and related stakeholders in the network (horizontal) and even between the enterprise and management departments (vertical). To achieve group social capital requires a business to make changes in communication and engagement.; The effective horizontal bridging (the process-based linking between different departments of Shaanxi Blower) and vertical bridging (the linking between different management layers of Shaanxi Blower, such as the coordination mechanism between the contract management center and other functional departments) could be formed inside the enterprise on the one hand; on the other hand, horizontal bridging (the linking between Shaanxi Blower and cooperative enterprises such as other equipment factories, organization and institutions) and vertical bridging (the trust and collaboration between Shaanxi Blower and administrative department of state-owned assets) also need to be established between the enterprise, organization and institutions. Only when these bridged networks are built, the group social capital can be realized and the SSC can achieve sustainable development.

Fourth, the effective operation of the SSC is built on the control and integration of product supply chains by service integrators. The organization and operation of the SSC includes various direct or indirect economic subjects and institutions, and its management contents include not only SRM, customer relationship management, and customer demand management, but also capacity management, service delivery management, and cash flow financing management, so the management difficulties and risks of service integrators are rather high. Against this background, it would be difficult for the SSC to continue operating effectively if service integrators cannot be directly involved in or even manage product supply chains. The success of Shaanxi Blower's SSC is built on the good design and organization of product supply chains, and if Shaanxi Blower lost the technical management of the fan equipment and the control over the physical product supply chain, it would be hard to form the SSC. Shaanxi Blower has established the SSC on the basis of the organization of product supply chains, which has made the two kinds of supply chains well-coordinated and mutually facilitated. This can be reflected in the compensating of the SSC to the technological innovation and product supply chain. This means Shaanxi Blower, through providing maintenance and whole-process service to customers' equipment, has more information about the technical condition of equipment and customers' demands, which in turn improves the internal equipment technologies of Shaanxi Blower.

Fifth, financing and capital management is an important management target of service integrators. Traditional product supply chains focus more on the management of commercial distribution, logistics and information flow, while the SSC stresses the role of capital flow in coordinating and driving logistics, commercial distribution and information flow. At the same time, capital flow is also the leading source for the SSC to realize added value. During the organizing process of the SSC, effectively addressing the difficulties in capital and financing in the network, especially when the involved parties face financial stress or when the capital obtained based on real estate mortgage cannot meet the demand of production and operation, and at the same time banks and financial institutions cannot provide capital out of the consideration of risk control, will become the key for the performance of the SSC. An important factor for the success of Shaanxi Blower's SSC is that it has not only formed a service network and provided integrated service products, but also provided comprehensive services in the management of logistics, bill flow and capital flow as a financing warehouse.

In conclusion, this chapter has revealed the organization and operation mode of the SSC through the Shaanxi Blower case study, which has reflected the innovation model and path of the breakthrough development of the equipment manufacturing industry amid Chinese economic transformation and business changes. However, the study in this text was carried out based on a single case study and lacks large-scale sample data for testing. In addition, the conclusions and judgment of the case study also needs to be tested and verified in other equipment manufacturers and each conclusion and judgment should be subjected to an in-depth analysis. These are the shortcomings of this research, which need follow-up.

6
Chapter

Performance of Enterprise Informatization under Institutional Innovation

Song Hua, Wang Lan, and Wang Xiaoliu

Introduction

The arrival of the knowledge-driven economy has created enormous challenges for management, especially management of Chinese enterprises. At present, Chinese enterprises are changing from a planned economy to a market-driven one. Innovation of businesses in this new knowledge-driven economy is an evolving process. China needs economic development and so must adopt modern management systems that are equipped to handle large volumes of high quality goods on the most competitive terms in the world. This requires efficiency, standards and intellectual capital, and will require a lot of innovation in the lower levels of management in order to keep up with these developments. Informatization is the major area in which Chinese enterprises can meet these challenges. How these kinds of changes will influence the performance of innovation management is the subject of research.

Demand for medical products is increasing at the same time as the Chinese pharmaceutical industry is coping with rapid development and innovative changes. At the same time, there is also increased competition at home and abroad, given economic globalization and the opening of the medicine circulation field. As a result, Chinese medicine circulation enterprises are experiencing many challenges. For example, the average percentage of circulation expense to sales volume of China's medicine business is 12.5% while the profit ratio of sales is less than 1%. The average profit ratio of sales of the pharmaceutical wholesalers in the United States is 1.5% while the percentage of circulation expense to sales volume is only 2.9%. It's therefore clear that competitiveness enhancement is a burning issue for Chinese pharmaceutical enterprises. These enterprises must cut circulation costs, find efficiencies in their distribution networks and level up innovation services. Information technology and networks can greatly improve information sharing, reduce circulation costs, optimize inventory management, and comprehensively enhance management efficiency. Consequently, improving management through informatization and networking has become the common choice of Chinese medicine circulation enterprises. Enterprise informatization has not only brought about opportunities for Chinese medicine distribution enterprises to improve circulation efficiency but has also provided them with opportunities to gain competitive advantages as late comers in the international market.

But Chinese medicine distribution enterprises are not doing well in implementing informatization. Although they have invested a great deal of human and material resources in Good Supply Practice (GSP) certification and in reforms towards informatization, they have not achieved the expected results. As a result, some enterprises have given up on informatization and some have even idled their equipment. We therefore have conducted in-depth and systematic research into informatization as it relates to medicine distribution enterprises in order to assist medicine distribution enterprises move forward with business informatization in a practical and effective manner and thus improve their business performance and competitiveness.

Literature Review

Characteristics of enterprise informatization

The term "informatization" was first used by a Japanese scholar in the 1960s (Yoichi Ito, 1994). "Informatization" is a technical term with distinctive oriental and Chinese features. Most European and American academicians use "information," "information technology (IT)," "information system (IS)" or "e-business" as the key words for issues connected with the relevant areas. The Chinese concept of "informatization" is, however, somewhat different and so this paper will investigate the characteristics of informatization from the view of Chinese authorities and academia.

At the National Informatization Working Conference in 1997, the Leading Group for Information Technology Advancement under the State Council stated: informatization refers to the application of modern information technology, the in-depth development and wide-ranging utilization of information resources and the acceleration of the national modernization process in agriculture, industry, science and technology as well in all aspects of social life under the unified state plan and organization. We can see from this definition that informatization is a process including at least two parts: the process of applying modern information technology and the process of developing and utilizing information resources; thus, interpreting informatization only as the application of information technology in related sectors is incomplete (Wu Jiapei, 1999).

When it comes to our research objective focusing on "enterprise informatization," we discovered that there is not yet a unified and authoritative definition. Scholars still remain divided over the understanding of its characteristics and have different focuses in their definitions. Some representative definitions are as follows:

Enterprise informatization refers to the process by which enterprises use modern information technology to effectively develop, utilize and manage information resources. Informatization involves all links of an enterprise and represents a long-term and complicated work. Enterprises need to gradually put in place an automatic system in which all the following are considered essential: production control, a management system with financial cost management and an investment and financing decision-making and marketing system with e-commerce. The top priority at the moment is to establish and improve the management systems with financial cost management at their cores and then progressively advance the development of e-commerce (Liu Li, 2000).

Yang Xueshan (2001) pointed out that enterprise informatization refers to the process wherein an enterprise comprehensively applies information technology in all aspects of its business including its production and operations, its management and decision making, its research and development and its sales and marketing and it which it also establishes an application system and network, adjusts or restructures its organization and business mode of the enterprise through effective development and utilization of information and knowledge resources all in order to reach business development targets and enhance business competitiveness.

Li Xueying (2003) believes that enterprise informatization means leveraging information technology and equipment to effectively develop information resources, comprehensively integrate internal and external information, coordinate logistics of human flow and capital flow with information flow, and finally level up the management of business production and operation.

Jin Jiangjun (2004) noted that enterprise informatization is adapting management philosophy in order to integrate an enterprise's existing production, design, manufacturing and marketing links, and to provide, timely, accurate and effective information to its management and decision makers, so that they can quickly respond to market demand; the nature of informatization is to strengthen an enterprise's core competitiveness and to provide capacity to coordinate with other enterprises."

Based on the foregoing, the authors observe that enterprise informatization includes two mutually influential concepts: the first is the application of information technology and the development of information resources; the second is the interaction between information technology and other components of the enterprise, namely, the changes and adjustments of business mode, management mode, process and organizational structures occurring during the adoption and integration of information technology either voluntarily or involuntarily (Feng Rende and Chen Yu, 2003). The role of information technology in business performance is just transmitted through this kind of interaction (Christina Soh et al., 1995). Therefore, the second level can also be regarded as the influence that information technology has on the enterprise's performance.

The target of enterprise informatization is not just to realize the "automation" and "electronization" of a business' operation and management by using information technology The more profound objective is to promote the integration of information flow, commercial distribution, logistics and capital flow within an enterprise itself and between that enterprise and its supply chain partners This is all to enable the business to better respond to, and meet market demands.

Enterprise informatization and performance

After the rise of the paradox of information technology and productivity, many economists, management scientists and experts in information system have done a great deal of work analyzing the theoretical and empirical relations between information technology (IT) and productivity. The main content and conclusions of their research are discussed below.

Between the 1980s and early 1990s, there was a weak or no relationship between IT investment and productivity. Since the 1990s, however, IT investment and productivity have shown an obvious positive correlation.

In 1987, the Nobel Prize laureate, Robert Merton Solow, published a brief article in the special column for book reviews in the *New York Times*. In that article, he firstly addressed the productivity paradox in the application of information technology and concluded that that paradox is why the information technology revolution has resulted in a fall in labor productivity and Total Factor Productivity (TFP). In the same year, a chief economist of Morgan Stanley, Steven Roach, expressed a similar point of view. As a result of inadequate statistical data and samples of IT investment in enterprises, early studies were unable to prove that IT investment can generate returns (Brynjolfsson and Hitt, 1996, 2000); Franke et al. studied the business data of the service industry involving insurance and banks and concluded that there is a weak or no relationship between IT investment and productivity (Franke, 1987).

Since the early1990s, however, IT investment and productivity have showcased an obvious positive correlation. In the 1990s, especially after 1993, researchers beganusing data based on much larger samples (Detailed IT investment data of large American companies was provided by specialized market research firms and questionnaires were employed to obtain basic data on relevant companies from information system directors and executive staff; in addition, reliable financial data was gleaned from various open databases). Using this data and a production function-based econometric model, researchers were able to calculate elasticity IT investment and IT capital output. In an article "Paradox Lost: Firm Level Evidence on the Returns to Information Systems Spending," Brynjolfsson and Hitt (1966) used data from 367 large enterprises based on the years from 1987 and 1991. They adopted the basic model $Q = F (C, F, S, L; I, t)$ to conduct an empirical study which concluded that: information systems contribute greatly and significantly to the output of enterprises; between 1987 and 1991, the average return on investment (ROI) of computer capital of sample enterprises was 81%, which exceeded that of other capitals. In addition the business output brought about by the investment in labor of information systems was several times more than that brought about by the investment in labor of non-information systems. As a result, it is reasonable to believe that the productivity paradox had disappeared in 1991, at least in the sample enterprises. The studies of Brynjolfsson and Hitt (2000) found that data indicates that investing in information technology can significantly increase productivity. This means that the returns on IT investment are reflected not only in the improvement of business labor productivity but also in the growth of multifactor productivity (MFP), and that this influence can last for 4–7 years. (Brynjolfsson Erik et al., 2000). Gilchrist et al. (2001) studied the sample data of American manufacturers and concluded that IT investment plays a significant role in improving labor productivity.

Additionally, the role of information technology in improving business performance keeps increasing with the accumulation of time and experiences. As noted above, Brynjolfsson and Hitt (2003) based their research on data of 527 American enterprises covering the years from 1987 to 1994, they used the MFP calculation method and concluded that, in the short term (data of one year), informatization (computerization) can contribute to a measurable growth of productivity and output that is basically consistent with investment cost. In the long term (data analysis of 5–7 years), however, the contribution of computerization to productivity rises to over 5 times, and the observed data demonstrate that the contribution of computerization to productivity often comes over a longer term and following a larger investment in organization capital.

Conditions for enterprises to realize informatization performance

To understand why some enterprises invested heavily in IT but did not reap good results, Weill, in 1992, studied the case of IT investment failure from the firm level and raised the concept of "IT transformation effect." On this basis, many scholars started to advance relevant theories (Process Theory and Discrepancy Theory) to account for how IT investment translates into the increase of production efficiency, the realization of enterprise value and the improvement of organizational performance, among others (Weill P., 1992).

In their 1995 dissertation entitled "Information Technology as a Factor of

Production; the Role of Differences among Firms," Brynjolfsson and Hitt (1995) reached the same conclusion as did David (1989). They concluded that an enterprise must spend a certain amount of time making some basic changes, including in its organizational strategy and structure, if it wishes to realize increases in productivity through computer investment.

Christina Soh and M. Lynne Markus (1995) systematically made a comparative analysis of the model of Lucas, Grabowski and Lee, of Markus and Soh, of Beath, Goodhue and Ross, and of Sambamurthy and Zmud. In the article "How IT Creates Business Value: A Process Theory Synthesis," they proposed a process model about how, why and when IT investment yields considerable increases in business organizational performance. To be specific they found that different organizations invest in IT and generate different degrees of IT assets due to the different effects in the management process. If a high quality of IT assets can be rightly used, then that business could generate substantial IT effects; if the substantial IT effects are not affected by the competition process, then the organizational performance would be improved. The process theory put forward by Christina Soh and M. Lynne Markus includes three sub-processes. The first process model is competition process; the second is IT utilization process; the third is IT transformation process (see Fig. 6.1).

During the first process (competition process), whether "IT effect" can result in the improvement of organizational performance will also be affected by controllable external factors which might result in the improvement or failure of organizational

Fig. 6.1.   Process model of IT investment transformation into enterprise performance

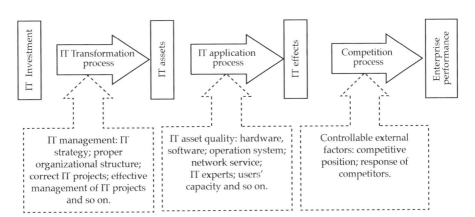

Source: Prepared based on the article of Christina Soh and M. Lynne Markus.

performance; these external factors include such things the business' competitive position, whether its competitors could respond quickly and, of course, luck.

During the second process (IT utilization process), "IT effect" refers to the intermediate result generated during the IT utilization process accompanying business internal necessary conditions (though not sufficient conditions). In other words, for a specific organization, the ultimate performance results—improvement in decision-making the improvement of coordination, flexibility and others, are different from business finance and customer satisfaction. The essential condition for IT effect is the high quality of IT assets, specifically the application platform, the IT infrastructure and the users' capacity (Markus and Soh, 1993). IT application platform refers to the utilization of IT inside an organization, i.e. the applied software; IT infrastructure involves the basic hardware system, operation system, network service and IT experts (Weill, 1992); users' capability is a very important IT asset meaning that users must know how to use their system and facilities effectively or else the IT application platform and infrastructure cannot play their potential role.

During the third process (IT transformation process), IT asset is the result of IT investment (expenses). While IT investment is an essential condition, it is not the only condition required. Not all organizations can transform IT investment into IT asset with the same efficiency (Weill, 1992). In a business, a certain level of IT expenses can enable some organizations to enjoy opportunities of obtaining the application platform of higher depth (the management level supported by IT) and breadth (business activities supported by IT) as well as best infrastructure (more and more users can easily enter into the system and share resources and information, and at the same time, IT staff have certain level of technological and business knowledge). The process through which an organization transforms IT investment into IT asset is called "IT management." Markus and Soh presented, when studying IT transformation efficiency, four basic IT management priorities (Christina Soh et al., 1995): developing an IT strategy; choosing a proper organizational structure for the implementation of the IT strategy; selecting an appropriate IT project; and effectively managing the IT project.

Influence of enterprise informatization on enterprise performance

Kathuria observed that simply introducing information technology will not give a competitive edge. In the production environment, the application of information technology must be consistent with the competitive edge and the process structure of an organization and must closely link with corporate strategy so that the enterprise can achieve a competitive edge. Xie Kang et al. (2005) believed that the influence of information technology on enterprise performance is mainly imposed through four paths, namely indirectly through corporate strategy, organizational structure, business process and organizational learning (see Fig. 6.2).

Fig. 6.2.   Analytical model of IT influence on enterprise performance

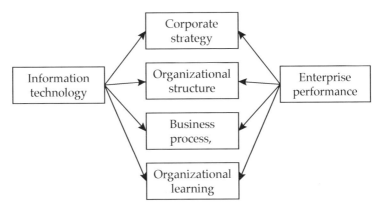

Source: Xie Kang et al., "Studies on the Influential Mechanism of Information Technology on Business Performance," *Value Engineering* 3 (2005).

1. IT influences business strategy through supporting business goals and further influences enterprise performance (Mahmood M. and Soon S. K., 1991). But according to the characteristics of the enterprise itself and the features of the industry it belongs to, IT has different degrees of strategic impacts on different types of enterprises. Specifically, the strategic influences of IT on different enterprises can be broken into four types : strategic, transformative, factory and support (Xie Kang, 2005). The strategic type refers to the fact that the role of IT becomes more and more prominent in both the present and future strategic processes; the transformative type refers to the fact that curently enterprises do not totally rely on IT to realize their operation targets and so the influence of IT on the strategy is comparatively small, but that in the future theywill increasingly rely on the new IT application to realize their strategic targets; the factory type refers to the fact that a smooth business operation depends on a safe, reliable and efficient IT support at a low cost At present, IT application is just inits development stage thus cannot form a core business competitiveness. The support type refersto the fact that IT does not have big influence on operating and furture strategy because the organizational status of an IT sector in these enterprises is rather low, and these enterprises do not link IT with their business planning activities. For example, most retail companies defined IT as

the "support type" two decades ago, but the emergence of new technologies has allowed a small company such asWal-Mart to grow at an enormous rate. Wal-Mart has gained enormous advantages with its extremely low costs and has thoroughly changed the competitive landscape largely because it successfully shifted the position of its IT from a support type to a strategic type.

2. IT influences enterprise performance through promoting changes in its organizational structure (Jooh Lee and Utpal Bose, 2002). First and foremost, IT can increasingly promote the flattening of business organizational structures and improve enterprise performance by increasing the speed and accuracy at which a business can react to environmental changes; secondly, IT facilitates the networking among an enterprise's organizational structures. After studying the the supply chains of 6 Finnish industries, Kemppainen and Vepsalainen (2003) observed that informatization is one of the three elements for the realization of organization networking (the other two elements are specialization and outsourcing). Networking in organizations has not only weakened the interactive and sequential dependence between many positions, it has also decreased the costs of coordination between individuals and groups. At the same time, a networked organization can also strengthen its core business so as to meet the goal of enhancing its core competitiveness and performance.

As early as 1985 in the classic article, "Management in the 1980s," Leavitt and Whisler predicted the shrinking of middle management. They argued that introducing information technology into an organization would reduce the size of middle management because information technology requires that the upper level management cross the middle management to deal directly with both the upstream and the downstream. This, then replaces many of the tasks performed by middle management. Much of the decision making can be automated and information can be easily and broadly accessed within an organization thus replacing many of the information and decision-making functions of middle management. However, theoretical and empirical studies conducted in the three decades following this 1985 paper have demonstrated that the influence of information technology is not universal. To this end, Pinsonneault concluded that it is only when organization and computer decision making are all centralized and when upper level management has more control over the use of computers, that information technology can be used to reduce the size of the middle management.

3. IT influences enterprise performance through influencing enterprise process (Mohsen Attaran, 2004). The spread and infiltration of information technology within an enterprise have constituted the crisscross information network inside the enterprise. Database, network and communication

technology can break through the constraints of division of labor. Information systems like MIS and ERP have enabled one person to do the work formerly done by many people, without any decrease or increase in efficiency. Being able to immediately share and flow of information has greatly reduced the barriers and time delay between different links, and has promoted the flattening of organizational structures and the optimization and restructuring of business processes. These changes in organizational structures and the restructuring of business processes have in turn deepened informatization and provided leverage to the potential for informatization.

4. IT can improve enterprise performance through boosting organizational learning (Michael J. et al., 2003). That is, IT promotes organizational learning through accelerating the acquisition of information, the dissemination of information and the interpretation of same. By doing so, it assists in the development of organizational memory and indirectly influences enterprise performance. A modernized and free flowing information environment is now a basic requirement for enterprise which wishes to be innovative and vibrant.

Theoretical Hypothesis

Our review of the above literature indicates that informatization can create benefits including in direct operation performance and can even result in strategic advantages. Bowersox and Daugherty (1995) observe, when considering supply chains and logistics, that informatization will create strategic advantages for businesses through integrated strategic plans and day-to-day integrated operations. Information will influence an enterprise's financial performance (Sangho Lee and Soung Hie Kim, 2006; Barnet J., 1997; Kohli and Debaraj, 2004; Poter and Millar, 1985). As an enterprise's financial performance can be easily quantified and measured, we will now consider the effect of informatization on the financial performance of existing Chinese medicine distribution enterprises.

Enterprise informatization and sales and operations planning

With the development of modern supply chain logistics management and information technology in the early 1990s, sales and operations planning (S&OP) emerged as a new type of management mode and platform. The purpose of S&OP in the supply chain is to promote a deeper level of cooperation among the various parts of the business and to integrate sales forecasting into the process of business operation. This can better balance supply and demand. Wing and

Perry asserted that S&OP is the core of business planning in the pharmaceutical industry and that it is a working process composed of a series of functional activities, including strategic planning, sales planning, order undertaking, production planning, transportation management and supplier planning.

The typical effect of informatization in the supply chain is to reduce friction among members in the supply chain through low-cost information flow. However, a more important role of informatization is believed to be the realization of information sharing and integration among members in the supply chain and the promotion of a supply-demand balance. For example, Lee and Whang regarded informatization as the key factor for alleviating the "bullwhip effect" in a supply chain. In addition, informatization can provide support to decision making. The information sharing and integration between members in the supply chain will effectively alleviate the "information island" phenomenon common in medicine distribution supply chains, reduce the uncertainties of supply and demand in the supply chain. It also allows members in the supply chain to better understand the requirements of other members and thereby promotes better co-ordination among these members in the supply chain. All this will allow more efficient ordering The result is effective management and coordination of the distribution channels, an improvement in the coordination of supply and demand, and, ultimately, better enterprise performance. In this connection, this study concludes that:

H1a: The capacity of information acquisition and information mining exerts a positive influence on S&OP;

H1b: The capacity of automatic information processing and information monitoring has a positive effect on S&OP;

H1c: The capacity of information sharing and integration has a positive effect on S&OP.

Enterprise informatization, S&OP and financial performance

Information, like human resources, fixed assets and business processes information, too, is an important business resource, and informatization will undoubtedly enhance the capacity of enterprises to access information resources. Markets are customer-oriented and obtaining, tracking and predicting customer wants and needs has become a key focus for businesses. Timely information enables businesses to promptly discover changing customer preferences, and make changes in a timely manner in order to address those changes. The gathering and discovering of business internal operating data helps enterprises learn more about their shortcomings, and provides information which can be used as the basis for continuous improvement. This, then, may allow cuts in operating costs and improvements in operational efficiency. The adoption of information technology and the improvement of productivity are significantly related.

Therefore, this study concludes:

H2a: Information acquisition and information mining exert a positive influence on financial performance.

Berger pointed out in his research report that Cisco Systems, Inc. restructured its internal processes, accomplished integration with its suppliers and clients and finally saved USD500 million in expenditures, all by leveraging on informatization and Internet technologies. In 2001, 90% of Cisco's sales were realized through the Internet. Informatization has enabled Cisco to use automated order processing to replace work previously done by hundreds of employees. The most profound change brought about by information technology to enterprises is on costs. Production costs, management costs and transaction costs are all reduced, including marginal. Cost management, especially that of inventory, is very important for medicine distribution enterprises. Information technologies relating to timely replenishment have improved the capacity of the enterprises in automated information processing and monitoring, and have exerted the substitution effect on inventory. Informatization has helped enterprises reduce inventory and thus decrease inventory management costs.

Therefore, this study concludes:

H2b: The capacity of automatic information processing and information monitoring has a positive effect on financial performance.

Grover and Malhotra et al. believe that information integration and sharing have enhanced the capacity of enterprises in information delivery and processing, and have made simultaneous decision-making possible. Simultaneous decision-making is the cornerstone of the supply chain management. Nada R. Sanders and Robert Premus demonstrated through empirical research that information sharing and integration technology can help a company realize internal integration and external alliances with other enterprises. Members in the distribution chain have realized smooth exchanges of business data and real time information sharing through the information sharing system. In particular, an information sharing system can objectively and truly reflect the demand of customers in a timely manner and thus allow businesses to better respond to customer demand. Through information sharing, members can, coordinate their operations, build mutual trust among them and form strategic alliances, thus creating a highly efficient distribution chain. Studies indicate that information integration and sharing among members of the medicine supply chain will enhance the capacity of medicine distribution enterprises in chain coverage and management This will then improve the performance of medicine distribution enterprises.

Therefore, this study concludes:

H2c: Information sharing and integration have positive effects on financial performance.

S&OP is basic to the financial performance of distribution enterprises. Optimizing S&OP will encourage various divisions within an enterprise, (such as sales, marketing, financial and operations) to develop an annual forecast or plan which can best meet the business' needs.

Therefore, this study concludes:

H2d: S&OP has positive effects on financial performance.

Data Gathering and Measurement

Samples and their features

This study was carried out using data obtained through handed out and returned between July 2005 and March 2006. These questionnaires had two categories respondents: firstly, the medicine circulation enterprises participating in various bidding platforms; and, secondly, the distributors in all kinds of medicine MBA classes. Altogether 666 copies of the questionnaires were handed out, 225 were returned with 119 effective responses. As for the respondents, 30 were small enterprises with annual sales of less than RMB30 million, (25.2% of respondents); 52 were medium-sized enterprises with annual sales between RMB30 million and RMB300 million, (about 43.7% of respondents); 37 were large enterprises with an annual sales of more than RMB300 million, (about 31.1% of respondents).

Measurement of variables

All questions in this questionnaire used Likert's 5-point scale from "very good" to "very bad," requiring respondents to demonstrate their recognition of each question. On the basis of having acquired scales and data, we employed SPSS software to make factor analysis and multiple regression analysis according to the theoretical model constructed. Our aim was to investigate the influence the three elements of informatization (information acquisition and mining, automatic information processing and information monitoring, information sharing and integration) have financial performance and potential competitiveness of the enterprise. When it comes to the validity of the scale, we employed the orthogonal rotation method in factor analysis to make an exploratory testing of the model construct, and the result shows that the indices designed in the questionnaires have better reflected the construct of the research model. As regards the reliability of the scale, in terms of the information acquisition and mining capacity index (IF1), we measured three relevant indices, and the questions included "the capability to acquire supply and demand information (the serial number is is1)," "the accuracy of acquired information (is2)" and "the capability to analyze and discover information (is3)," the Cronbach's α value is 0.818 0. When it comes to the information sharing and integration capacity index (IF2), we measured five indices: which are "the capacity to place an order according to demand (ic1)," "the capacity to control distribution and delivery (ic2)," "the degree of information sharing within an enterprise (ic3)," "the degree of information sharing between upstream and downstream enterprises (ic4)" and "the capacity to provide useful information to upstream and downstream enterprises (ic5)," and the Cronbach's α is 0.892 5. In terms of the automatic information processing and information monitoring capacity index (IF3), we measured four indices and the question items are "the automatic level of information process (it1)," "warehouse management system (WMS) construction (it2)," "the integrity of demand information structure (it3)" and "the capacity to control information (it4)," and the Cronbach's α value is 0.879 8. When it comes to the coordination between supply and demand index (S&OP), we measured four indices that are "the accuracy of market demand forecast (so1)," "the balance degree of purchase and distribution plans (so2)," "the rationality of the arrangement of human, material and financial resources (so3)" and "the accomplishment degree of distribution, supply and demand plans (so4)," and the Cronbach's α value is 0.887 7. In terms of the financial performance index (FP), we measured four indices that are "the terminal pure sales growth rate in the recent 3 to 5 years (sa1)," "the sales growth rate of business transfer and allocation in the recent 3 to 5 years (sa2)," "the net profits in the recent 3 to 5 years (np1)" and "the rate of return on assets in the recent 3 to 5 years (np2)," the Cronbach's α value is 0.766 0.

Results of Empirical Analysis

According to the data compiled from the questionnaires, this text uses path analysis to carry out model checking. Table 6.1 is the descriptive statistics and the correlation coefficient matrix of variables. Table 6.2 is the regression analysis of the influence of the three elements of informatization on the supply-demand coordination (Model 1). We can see from the test result of the equation that Model 1 explains 26.7% of the changes in the enterprise's S&OP.

Table 6.1. Data descriptive statistics and correlation coefficient (N = 97)

	Mean	SD	1	2	3	4	5
Information acquisition and mining capacity	10.680	1.670	1	—	—	—	—
Automatic information processing and information monitoring	17.966	3.162	0.501**	1	—	—	—
Information integration and sharing capacity	13.623	2.895	0.610**	0.606**	1	—	—
S&OP	14.641	2.520	0.530**	0.498**	0.490**	1	—
Financial performance	14.667	2.534	0.428**	0.200*	0.243*	0.581**	1

Note: $^*p < 0.1$, $^{**}p < 0.05$, $^{***}p < 0.01$.

Table 6.2. Multiple regression estimation of influence of informatization key factors IF1, IF2, IF3 on S&OP (N = 97)

Variable	Hypothesis	Reference model 1.1	Reference model 1.2	Reference model 1.3	Reference model 1.4
Control variable					
Information acquisition and mining (IF1)	H1a	—	0.488**	0.365**	0.310**
			(0.000)	(0.001)	(0.012)
Information integration and sharing (IF2)	H1b	—	—	0.265**	0.217*
				(0.016)	(0.069)
Automatic information processing and monitoring (IF3)	H1c	—	—	—	0.144
					(0.304)

(Cont'd)

Variable	Hypothesis	Reference model 1.1	Reference model 1.2	Reference model 1.3	Reference model 1.4
R square	—	0.128	0.327	0.373	0.381
Adjusted R square	—	0.003	0.221	0.266	0.267
ΔR square (Sig. F change)	—	—	0.199**	0.047**	0.008
			(0.000)	(0.016)	(0.304)
Sig. of Model	—	—	$p < 0.05$	$p < 0.05$	$p < 0.05$

Note: $^*p < 0.1$, $^{**}p < 0.05$.

(1) Supposing H1a is verified: In the Reference Model 1.2, the F test of the equation is significant under the condition that $p = 0.05$, and the tolerance of information acquisition and mining and the variance inflation factor (VIF) are 0.836 and 1.197 respectively, the equation does not have multi-collinearity. Information acquisition and mining have remarkable influence on S&OP of medicine distribution enterprises under the condition that $\alpha=0.05$, and the explanatory validity of information acquisition and mining towards the changes of S&OP has increased ($\Delta R^2 = 0.199$, $p < 0.05$). The effect of this model demonstrates: if other factors remain unchanged, information acquisition and mining have a strong linear relation with S&OP, and one percentage point increase of information acquisition and mining will cause 48.8% of rate of change of the supply-demand coordination.

(2) Supposing H1b is verified: In the Reference Model 1.3, the F test of the equation is significant under the condition that $p = 0.05$, and the tolerance of information sharing and integration variable with the information acquisition and mining variable is 0.662, the VIF is 1.512, the tolerance and VIF of the information acquisition and mining are 0.658 and 1.519, the tolerance and VIF of the automatic information processing and monitoring are 0.395 and 2.532, which indicate that the equation does not have multi-collinearity. Information sharing and integration have remarkable influence on S&OP under the condition that $\alpha=0.05$, and has increased its validity of explaining the changes in S&OP ($\Delta R^2 = 0.047$, $p < 0.05$). The effect demonstrates: if other factors remain unchanged, information sharing and integration has strong linear relation with S&OP, one percentage point increase of information sharing and integration means 26.5% of rate of change of S&OP.

(3) Supposing H1c is not supported by data: Under the condition that

$p = 0.1$, the regression test result is not significant (Reference Model 1.4, $p = 0.304$), we can conclude that automatic information processing and information monitoring have virtually no impact on S&OP.

Table 6.3 shows the regression analysis of the influence of the three elements of informatization and that of S&OP on financial performance (Model 2). The positive effect of information acquisition and mining (IF1) and S&OP on financial performance has been proven, but the data indicate that the effect of the information sharing and integration and that of the automatic information processing and monitoring on financial performance is not significant. The test results showcase that Model 2 has explained the 44.8% of changes of financial performance.

Table 6.3. **Multiple regression estimation of the influence of informatization key factors IF1, IF2, IF3 and S&OP on financial performance ($N = 94$)**

Variable	Hypothesis	Reference model 2.1	Reference model 2.2	Reference model 2.3	Reference model 2.4	Reference model 2.5
Control variable						
Information acquisition and discovery (IF1)	H2a	—	0.383** (0.000)	0.353** (0.002)	0.418** (0.001)	0.213* (0.063)
Information integration and sharing (IF2)	H2b	—	—	0.065 (0.552)	0.115 (0.328)	0.042 (0.675)
Information automatic processing and monitoring (IF3)	H2c	—	—	—	−0.164 (0.247)	−0.214* (0.079)
S&OP	H2d	—	—	—	—	0.534** (0.000)
R square	—	0.225	0.346	0.349	0.360	0.543
Adjusted R square	—	0.111	0.239	0.233	0.237	0.448
ΔR square (Sig. F change)	—	—	0.121** (0.000)	0.003 (0.542)	0.011 (0.247)	0.183** (0.000)
Sig. of model	—	—	$p < 0.05$	$p < 0.05$	$p < 0.05$	$p < 0.05$

Note: $*p < 0.1$, $**p < 0.05$.

(1) Supposing H2a is verified: In the Reference Model 2.2, the F test of the equation is significant under the condition that $p = 0.05$, and tolerance and VIF of information acquisition and mining with control variable are 0.827 and 1.212 respectively, the equation does not have multi-collinearity. The explanatory validity of information acquisition and mining to the changes of financial performance is high ($\Delta R^2 = 0.121$). The effect of this model demonstrates: if other factors remain unchanged, one percentage point increase of information acquisition and mining means the rate of change of the financial performance will be 38.3% ($p = 0.000$).

(2) Supposing H2b is not verified: In the Reference Model 2.3, the F test of the equation is significant under the condition that $p = 0.05$, and tolerance and VIF of information acquisition and mining are 0.656 and 1.524, and those of information sharing and integration are 0.688 and 1.453, indicating that the equation does not have multi-collinearity. But the increase of IF2 does not strengthen to a great extent the explanatory validity towards financial performance. ($\Delta R^2 = 0.003$). The effect demonstrates that if other conditions remain unchanged, information sharing and integration do not have an influence on the improvement of financial performance ($\beta = 6.5\%, p = 0.552$).

(3) Supposing H2c is not verified: In the Reference Model 2.4, the F test of the equation is significant under the condition that $p = 0.05$, and tolerance and VIF of information acquisition and mining are 0.524 and 1.908 respectively; those of information sharing and integration are 0.596 and 1.677; those of automatic information processing and information monitoring are 0.415 and 2.410 respectively. The equation does not have multi-collinearity. But automatic information processing and information monitoring has a weak explanatory validity to the changes of financial performance ($\Delta R^2 = 0.011$), and there is an insignificant β value in statistics ($\beta = -0.164, p = 0.247$); this can be understood as information automatic processing and monitoring having no significant influence on the improvement of financial performance.

(4) Supposing H2c is verified: In the Reference Model 2.5, the F test of the equation is significant under the condition that $p = 0.05$, and tolerance and VIF of information acquisition and mining are 0.468 and 2.138; those of information sharing and integration are 0.586 and 1.707; those of automatic information processing and information monitoring are 0.413 and 2.423; those of S&OP are 0.643 and 1.555 respectively, thus the equation does not have multi-collinearity, and S&OP has a strong explanatory validity to the changes of financial performance ($\Delta R^2 = 0.183$). The effect of this model demonstrates: if other conditions remain unchanged, S&OP has a strong correlation with business

finance, and one percentage point increase of S&OP means the rate of change of financial performance will be 53.4 % ($p = 0.000$).

Table 6.4 demonstrates that the path coefficient of information acquisition and mining to financial performance is 0.380, and that information acquisition and mining also plays a significant role directly or indirectly: When other conditions remain unchanged, a one percentage point increase in information acquisition and mining means 38% of the rate of change of financial performance.

The path coefficient of information sharing and integration to financial performance is 0.116, demonstrating that information sharing and integration have a positive effect on financial performance. When other conditions remain unchanged, one percentage point increase in information sharing and integration means the rate of change of financial performance will be 11.6%; but the direct effect of information sharing and integration on financial performance is not remarkable. In other words, information sharing and integration do not have any significant direct effect on financial performance from the perspective of the data analysis results. The reason could that a Chinese medicine distribution supply chain has yet been established, and if information sharing and integration is only limited within an enterprise, its influence on financial performance is limited.

The path coefficient of automatic information processing and information monitoring to financial performance is −0.214, demonstrating that automatic information processing and information monitoring has a direct negative influence on financial performance. When other conditions remain unchanged, a one percentage point increase in automatic information processing and information monitoring means the rate of change of financial performance will be −21.4%. At the same time, the indirect effect of S&OP on financial performance is not significant, indicating that the investments made by Chinese medicine distribution enterprises in automatic processing and monitoring, at the present, have not improved financial performance, on the contrary, they have decreased financial performance.

Table 6.4. Causality between the three elements of informatization and financial performance

Variable	Type of effect	Mark	Coefficient
X1 Information acquisition and mining	Direct effect	P_{51}	0.213
	Indirect effect	$P_{41} \times P_{54}$	0.167
	Total effect	$P_{51} + P_{41} \times P_{54}$	0.380

(Cont'd)

Variable	Type of effect	Mark	Coefficient
X2 Information sharing and integration	Direct effect	P_{52}	0.042 (insignificant)
	Indirect effect	$P_{42} \times P_{54}$	0.116
	Total effect	$P_{42} \times P_{54}$	0.116
X3 Automatic information processing and information monitoring	Direct effect	P_{53}	−0.214
	Indirect effect	$P_{43} \times P_{54}$	0.077 (insignificant)
	Total effect	P_{53}	−0.214

The path analysis of the concept model of this study is shown in Fig. 6.3.

Fig. 6.3.   Effect of informatization on financial performance of medicine distribution enterprises

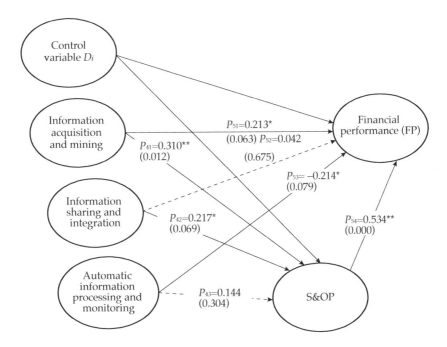

Note: Dotted line means that the path is not significant.

Conclusions and Suggestions for Further Studies

This research has tested the model constructed and the hypotheses through path analysis. The results indicate that our proposed model hypotheses are basically supported by the data. This research has tested, through empirical analysis, the causality between information acquisition and mining, information sharing and integration, automatic information processing and information monitoring on the one side and the financial performance of medicine circulation enterprise on the other. Based on the research results, we conclude: (1) Whether enterprise informatization can improve the financial performance does not depend on what information system has been chosen, which kind of information technology software has been employed and whether a modern medicine logistics center has been established. It depends on whether the enterprise's informatization program has truly integrated the functions of information acquisition and mining, information sharing and integration and automatic information processing and information monitoring so as to ensure the full play of these functions. These three aspects represent the three factors that need to be prioritized during the informatization process of Chinese medicine distribution enterprises. (2) The negative effect that automatic information processing and information monitoring has on financial performance (–21.4%) clearly shows that the huge investment made to date by China's medicine distribution field in automation has yet to yield corresponding returns. The employment of information technology and the improvement of production efficiency have no significant correlation. The research results in this regard are consistent with the research conclusions of Elliot. He believed that there is a need to take into consideration the existing business operation process, customization and other factors while implementing enterprise informatization. If these factors are overlooked, the effect on business performance might be not significant and might even be negative. At present, the building of a modern Chinese medicine logistics center lacks planning and there is only "inventory" but no "material flow." As a result, the employment of information technology has failed to improve the capacity of medicine distribution enterprises in automatic information processing and information monitoring in any real sense. The fundamental cause does not lie with the information technology itself rather, it lies in the fact that the informatization implementation group or department paid too much attention to the technology itself before first examining the business processes. This has weakened the effect of the information technology. Automatic information processing and information monitoring usually need substantial upfront financial investment for the purpose of establishing facilities such as an automatic stereoscopic warehouse and a modern logistics center, and developing corresponding information exchange platforms. But if there are no corresponding support of commercial distribution and logistics, the infrastructure of automatic information processing and information monitoring, after being established, will be a burden on medicine distribution enterprises. We can see from the questionnaire used in this research that 82 (69%) of the 119 enterprises have or are constructing medicine logistics center, which will undoubtedly harm the large-scale and intensive development of medicine logistics. However, medicine distribution enterprises believe that the establishment of modern logistics centers means the building of the capacity in automatic information processing and information monitoring. As a matter of fact, since commercial distribution is insufficient to support the operation of a logistics center, the modern logistics center with automatic information processing and information monitoring capacity is nothing but a noble "shackles." (3) Information sharing and integration is not yet a reality in the Chinese medicine distribution field. Information sharing among members of the supply chain can help improve S&OP, inventory management, channel management, market response and other key capacities as well as qualities of member enterprises. Information sharing and integration can facilitate alliances and coordination among members of the supply chain, and, finally, can improve the performance of member enterprises. But, the positive effect of information sharing and integration on the financial performance of Chinese medicine distribution enterprises is only about 11.6%, and so is statistically insignificant. The positive effect of informatization of medicine distribution enterprises is, therefore, yet to be realized.

In line with the empirical analyses and the current situation in the informatization and operation of Chinese medicine distribution enterprises, we believe: (1) Informatization development must be closely integrated with the optimization of internal processes and the improvements in management ideas. The supporters of process rebuilding observe that informatization must come with business process rebuilding if it is intended transform into business performance. Business performance is unlikely to be improved if all that is done is to employ information technology. This research also testified that the positive effect of the informatization of medicine distribution enterprises on financial performance is realized to a large extent through improving business processes, such as S&OP. So information development is not just restricted to utilizing information technology and constructing modern logistics centers. It should also involve updating management philosophy and continuously optimizing business processes. (2) There is a need to keep strengthening the implementation and improvement of quality management measures, such as GSP and ISO9000, and attach importance to the gathering and utilization of the data base of enterprises. Empirical research showcases that the information acquisition and mining of Chinese medicine distribution enterprises plays a significant role in improving business financial performance, which is closely related to the mandatory quality management systems like GSP and ISO9000 practiced in China in recent years. Both GSP and ISO-9000 stress the process and operation quality management. In particular, the ISO9000 quality management system requires the records of business operations and this will consequently make enterprises put more value on the gathering and mining of fundamental operation and management information. (3) There is a need to strengthen alliances and M&A in the pharmaceuticals industry, and facilitate the information integration and sharing between members of the supply chain. The contents of information integration and sharing must be wide-ranging and sufficient, and should not be restricted to items like order and confirmation sheets. Information integration and sharing must extend to information relating to operations, logistics and even strategic plans. (4) Building logistics standardization within the Chinese pharmaceuticals industry is required. A low degree of industry standardization is an underlying factor constraining the capacity of medicine distribution enterprises in automatic information processing and information monitoring. Only by realizing the standardization of medicine logistics can the database of a logistics center become a real-time and dynamic data acquisition system, and be used to improve efficiencies and the quality of logistics in any real sense.

Two aspects of this research still need to be addressed. We have been unable to fully deal with them, given the limitations of this text. It should be stressed that the demonstration of positive effects of informatization generally takes some time, but the research materials are mainly based on the middle and short-term, while China, especially its pharmaceutical industry, is a late comer to informatization. As a result, follow-up studies on the performance brought about by informatization should be carried out. In addition, the influence of informatization on financial performance can be realized not only through optimizing S&OP; other influencing paths need to be further explored in further studies.

7 Chapter

The Marketing Channel Innovation of Chinese Enterprises

Huang Jiangming, Dai Yingqiong, and Zhang Junhai

Introduction

Research background

Since 2003, with the rapid development of the retail market, the dominant position of Chinese large-scale retailers in distribution channels has been fully exposed. In this case, the subsequent conflicts between retailers and suppliers occur frequently. Particularly within the household appliance industry, the new and prosperous appliance chain enterprises keep increasing the entrance fees, collecting various management fees, violating sales agreements to cut down prices unilaterally, demanding the increase of sales rebates, and so on, by leveraging their dominant positions in the retail market. These behaviors have substantially cut down the profit space of manufacturers and dealt a heavy below to their traditional marketing system and market order. One of the worst cases was between Gome (a powerful household appliance chain), and Gree (an air-conditioner giant) in March 2004. Problems in the sales channel between the two broke out and evolved into a full-blown conflict. To date, Gree air-conditioner has yet to formally enter into Gome for marketing.

At the moment, the dominant position of retailers can rarely be changed, and retailers typically make use of their channel power to influence the transaction behaviors of manufacturers to gain profits. The Chinese government is trying to constrain the transaction behaviors between retailers and suppliers through legislations in a bid to reduce such channel conflicts. Academia is also actively exploring the root cause and influencing factors for the conflicts between retailers and suppliers to come up with suitable resolutions. The Ministry of Commerce of China has implemented the *Administrative Measures for Fair Transactions between Retailers and Suppliers* in October 2006 in an effort to regulate the transaction behaviors between retailers and suppliers, but it failed to realize the expected results. On July 28, 2007, the *Anti-monopoly Law of the People's Republic of China* was formally enacted but it failed to regulate market transaction behaviors in due to the lack of implementing regulations and law enforcement agencies.

For manufacturers, this means a weakening voice and squeezed profit. We can see from Fig. 7.1 and Fig. 7.2 that the main business of the four largest Chinese household appliance enterprises demonstrated a general rising trend between 1996 and 2006, with an average annual growth rate of more than 10%. Comparatively, the total net profits of the big four household appliance enterprises witnessed a standstill during the same period. As of 2004, the four household appliance leaders, except Gree, saw a decline, to different degrees, in their net profits. As a matter of fact, most listed companies of the household appliance industry witnessed a decline in net profits as a whole.

For retailers, this means a strengthening voice in sales channels and the expanding room for profit. Terminal retailers have better access to customer information, as they have direct contact with the customers. The intensifying competition resulting from product homogenization has increased the power of retailers and completely changed the previous channel power and profit distribution model. Table 7.1 tells us that nearly 25% of Gome's gross profits on sales were contributed by the fees charged from manufacturers. By 2006, the fees charged from manufacturers by Gome had totalled RMB929 million—nearly the same as the whole-year net profits (RMB942 million).

Such being the case, the sales channel power has directly determined the total profit and profit structure of manufacturers in the Chinese market where relevant national laws and regulations are in shortage. Therefore, the huge pressure exerted on manufacturers is clear. However, this is also the fundamental driving force for channel innovation. The research objectives of this chapter include: (1) through empirical studies on the status quo of the sales channel of the Chinese household appliance industry, revealing the mode and features of the sales channel of this industry under the environment that retailers dominate the market; (2) through the analysis of the Gree case in building its sales channel, summarizing the cause for the continuous rising of profits from the perspective of channel innovation; (3) abstracting the characteristics of and conditions for the sales channel innovation in the Chinese household appliance industry based on a case study.

Fig. 7.1. Main business income of household appliance manufacturers between 1996 and 2006

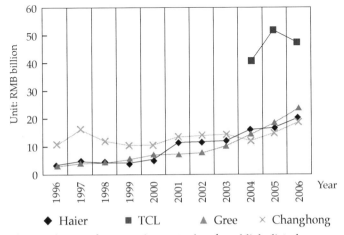

Sources: Prepared according to the annual report of each publicly listed company between 1996 and 2006.

Fig. 7.2. **Net profit of household appliance manufacturers between 1996 and 2006**

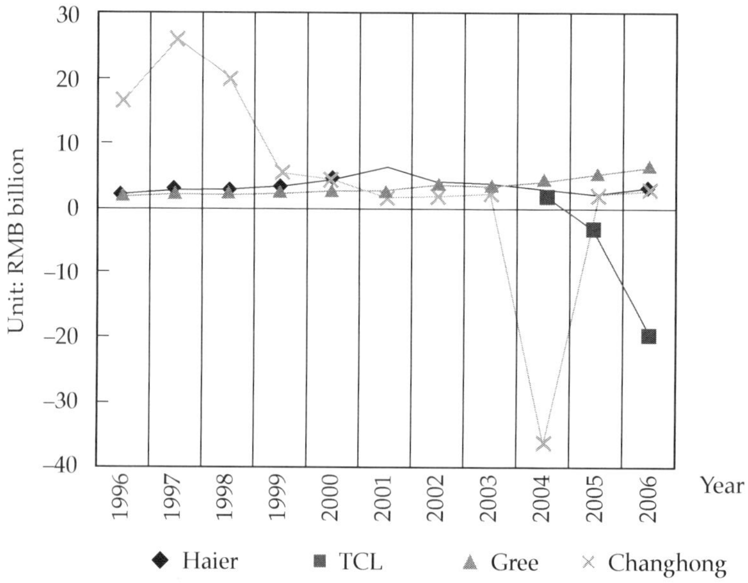

Sources: Same as those of Fig. 7.1.

Table 7.1. **Business profit statement of Gome**

Unit: RMB

Year	2003	2004	2005	2006
Sales Revenue	10,233,767,880	9,715,903,000	17,959,258,000	24,729,192,000
Cost of Sales	9,406,107,160	8,762,730,000	16,307,478,000	22,369,445,000
Gross Profit	827,660,720	953,173,000	1,651,780,000	2,359,747,000
Rate of Gross Profit	0.08	0.10	0.09	0.10
Other Business Profit	353,415,660	385,965,000	740,535,000	1,251,780,000
Other Business Profit from Manufacturers	276,811,580	247,959,000	492,707,000	928,879,000
Operating Profit	466,247,360	602,954,000	798,495,000	913,953,000
Net Profit	258,767,200	580,594,000	777,199,000	942,624,000

Sources: Prepared based on the annual reports of Gome.

Research method and objective

This chapter plans to adopt a research method combining both quantitative empirical research and the case study.

First, macroscopically analyzing the innovation environment for Chinese sales channels based on quantitative empirical studies on the manufacturer's sample survey results.

Secondly, adopting the case study method and taking Gree as the research objective to analyze the innovation process and the success story of the sales channel innovation of Gree from the strategic perspective. Through field surveys of the Gree sales company of Jiangxi province and its lower-level retailers, we have obtained a great deal of first-hand data about the structure and function of the sales channel mode and the distributor management of Gree, and have objectively analyzed and evaluated the applicability and achievements of the channel innovation of Gree.

We chose Gree as our research objective for three main reasons:

1. As shown in Fig. 7.1 and Fig. 7.2, among the four leading Chinese household appliance enterprises, Gree is not the largest in market size but stands out in terms of profit. In the air-conditioner field, Gree has surpassed Haier, the market leader in the Chinese traditional household appliance industry, to become the national number one in terms of market share and profit.

2. Different from other household appliance leaders who have adopted the strategy of giving up the controlling power of the sales channel, Gree has adopted the strategy of autonomously controlling the sales channel. Gree has created the "joint-stock regional sales company" mode. This channel mode has gradually enabled Gree to make its sales channel internalized and organized, and to effectively control it through effective integration of the marketing resources of external distributors, and by leveraging share-holding and franchise management. At the same time, this channel mode has added to the bargaining chips of Gree in the confrontation with retail giants like Gome and Suning, effectively eliminating the price war among distributors and halting the short-sighted action of distributors in brand operation.

3. In 2004, Gree and Gome engaged in a fierce conflict in the sales channel. This event came as a typical epitome of Chinese sales channel conflicts, and has reflected the weak points of Chinese household appliance manufacturers in channel innovation.

Literature Review of Sales Channel Innovation

Sales channel power and channel satisfaction

Marketing channel power is a control power or influence of one channel member on another at a different level in the same channel. The meaning of "channel power" has two dimensions: the first is that one channel member has rights over other channel members; the second is that these rights are powers that will promote other members to do something that they otherwise would not do. The core idea of the channel power theory is that members in any sales channel, such as manufacturers, agents, and retailers, have different statuses. Some channel members retain a certain influence on other members by leveraging on the resources under their control.

Academia has formed two schools of thought regarding the sources of channel power: Power-dependency Theory and Power-based Theory. But, Zhuang Guijun (2004) believed that these two theories unite into one at a more extensive level involving valuable resources. The interdependence between channel members is nothing but the demand for the abundant resources of one another (Power-dependency Theory); the resources enjoyed by one channel member can be assembled into different power bases (Power-based Theory). Generally speaking, one member's resources that are urgently needed and cannot be easily replaced are more likely to form a strong power over its counterpart.

As shown in Table 7.2 that channel power falls under six categories, including reward power, forced power, legal power, referent power, expert power and information power. Channel power can also be classified into different new categories, such as mandatory power and non-mandatory power, profitability power and non-profitability power, arbitration power and non-arbitration power.

Table 7.2.   Classification of channel power

Classification of channel power	Forced power	Reward power	Expert power	Legal power	Referent power
Characteristics of power	Mandatory	Non-mandatory (Mandatory)	Non-mandatory	Non-mandatory (Mandatory)	Non-mandatory
	Profitability	Profitability	Non-profitability	Profitaåbility	Non-profitability
	Arbitration power	Arbitration power	Non-arbitration power	Arbitration power	Non-arbitration power

Channel satisfaction refers to the contribution of one channel member to an enterprise, or the contribution of an enterprise to a channel member in the amount of sales and profit, and the assessment of each other in capacity, applicability, customer satisfaction and loyalty, and so on (Zhuang Guijun, 2004). Channel satisfaction can be divided into economic satisfaction and social satisfaction, and it is a general assessment of the difference between one channel member and other channel members in economic benefit, emotional recognition, expected economic benefit and emotional recognition obtained during their cooperative cooperation.

Channel satisfaction is not only the key factor for the establishment of channel relationship but also a major factor for the long-term cooperation between channel members. The main way for the channel leader to influence channel members' behaviors is to use channel power. In so doing, the influence of channel power on channel satisfaction is particularly important. When the theory circle studies the influence of channel power on channel satisfaction, the channel power is classified as mandatory power and non-mandatory power in most cases, and the channel satisfaction is classified as economic satisfaction and social satisfaction.

When a channel member uses mandatory power to influence another channel member, the affected will adopt certain response measures to counteract this kind of mandatory power and these response measures will consume certain costs of the one being influenced (Anderson and Narus, 1990). As such, the total economic benefits of the affected will shrink, which will decrease economic satisfaction. The social satisfaction will also drop as the confidence of the affected in the channel transaction behaviors suffers a blow.

When a channel member uses non-mandatory power to influence another channel member, the overall economic benefits of the one being influenced will increase as it gains extra support, which will, in turn, increase the economic satisfaction of the one being influenced. At the same time, as the use of the non-mandatory power has strengthened the confidence of the affected channel members, they should work together to overcome channel conflict as it occurs, and the social satisfaction of the affected will increase (Frazier and Summers, 1984).

With regard to the influence of legal power on channel satisfaction, the research results of Soumava and Robert (1998) indicate that in countries where the position of manufacturers and distributors is seriously asymmetric, relevant regulations, laws and contracts are urgently needed to safeguard their respective interests. In this case, legal power plays a significant role in improving the satisfaction of channel members. It is clear that channel power, channel satisfaction, and their relationship will also be impacted by the legal environment for the channel.

China's sales channel satisfaction and channel innovation

Traditionally, research of Chinese academia on the sales channel mainly focus on channel structure, channel function, channel mode, and channel management, but the research on channel conflict and cooperation and channel innovation only started in late 1990s. The research results on the channel innovation of the manufacturing industry are rarely seen under the background that retailers dominate the market and even abuse their dominant positions.

In the buyers' market, as retailers are at the forefront of the channel and are able to approach customers in the target market, and they can choose their commodity supplier, the dependence of manufacturers on retailers has increased, and the power within the distribution channel system has shifted from manufacturers to retailers. At the same time, distributors with a strong capacity in sales and service are well-positioned to boost the sales of commodities, and their role in this benefit distribution among channel members has become increasingly important. In this connection, the power and benefit mechanisms for members in the sales channel in China have undergone changes (Liu Qionghui et al., 1999).

Jia Xin (2006) observed that the relationship between members in Chinese sales channels has transformed from transaction-based to relation-based, which is an important hallmark of marketing innovation. The traditional relationship between channel members was purely transaction-based, and each member of the channel was independent and engaged in short-term cooperation or fierce competition with other members for profit maximization. The relation-based sales channel highlights the coordination in its strategic process, bidirectional communication in information, and the mutual benefit of marketing activities, requires mutual understanding in respective interests, and seeks common ground.

In addition, household appliance enterprises are diversifying and have placed higher expectations for the radiometric force and the control force of the channel (Wang Zhengxuan et al., 2006). Under a market environment with cutthroat competition in the channel shares, and in order to avoid excessively high channel costs and retailers' malicious control of the channel, some manufacturers choose to build a permanent vertical cooperative relationship with retailers, or choose to establish self-owned sales channel terminals. These strategies help increase the business internationalization degree of the channel, such as the establishment of their own wholesale departments, marketing companies, outlets and chain stores so as to maximally control the channel (Han Kun et al., 2006).

The study of Zhuang Guijun and Zhou Xiaolian (2004) on the relationship between Xi'an Department Store and manufacturers has also proven that the use of mandatory power will add to channel conflicts and hinder cooperation. In contrast, the use of non-mandatory power will decrease channel conflicts and improve cooperation.

Chinese retailers are in a dominant position and their "abuse of dominant power" has severely affected the market competition structure. But the *Anti-Unfair Competition Law of the People's Republic of China* does not include relevant articles to regulate and restrain the power in this regard. This has certain influences on the channel power behaviors of retailers and on the channel satisfaction of manufacturers (Wu Xiaoding, 2004).

Given the current situation, Gree's sales channel of "joint-stock regional sales company" mode can be regarded as the most distinguished and successful sales channel innovation of the Chinese household appliance industry. It demonstrates distinctive features of its time. As previously stated, Chinese retailers are now enjoying a dominant market role in the channel and they primarily use their channel power, especially mandatory power, to affect the transaction behaviors of manufacturers and obtain profits — even monopoly profits. In this special market and legal environment, Gree's channel innovation has been a wise choice. This innovation lasted for more than a decade and it has not only expanded the market share but also shaken off its dependence on household appliance hypermarkets like Gome. At present, 70%–80% of Gree's total sales are contributed by its outlets, while only a small amount of its sales come from household appliance hypermarkets (Zhang Shijun, 2007).

Theoretical framework

Although scholars have achieved great progress in their research on channel power and channel satisfaction, this research is still inadequate:

First, most research fails to take into account the influence of environmental factors for channel relations and seldom add environment variables into models for analysis. The precondition for the studies of most scholars is to suppose that the market is in a state of perfect competition, and that the channel power is obtained fully through self-strength (Power-based Theory). In fact, the obtaining and implementation of channel power is largely influenced by external environmental factors, particularly legal environment. The channel is impossible to exit in vacuum and thus its evolution also needs to take into consideration the stimulation of general macro-environmental factors, including economy, society, law, and policy, among others (Zhou Quanxiong, 2007). These factors can make the model more close to reality, but few people have researched the influence of the legal environment on channel power and on channel satisfaction and its influence mechanism.

Second, there is little empirical research on the relations between channel power and channel satisfaction in China at the moment; most are surveys on distributors based on the traditional sales channel structure (manufacturer-distributor). Virtually all research has neglected the basic precondition that retailers dominate the market.

Third, research on the channel relations at present generally lay stress on customer (retail terminals like Gome) satisfaction but turn a blind eye to the satisfaction of manufacturers — in particular small and medium-sized manufacturers.

Fourth, research conclusions more often than not stress a long-term strategic cooperation philosophy and seldom touch upon the independent innovation of the sales channels of manufacturing enterprises.

Our research concludes that the studies on the sales channel innovation (dependent variable) also need to regard channel satisfaction as the intermediate variable, apart from the commonly-used channel power (independent variable), and there is a necessity to include China's legal environment into the list of independent variables.

Regarding the legal factor as a variable to make a quantitative analysis of its influence on the retail channel relations, and even channel innovation, is still a thorny issue in theoretical and empirical research, and there is still no unified proxy variable. From the macro perspective, the legal status of a country within a certain period is at a fixed level, which has made it difficult to use this proxy variable to make a quantitative analysis.

However, we found through surveys on manufacturers in different areas that some local regulations exist in China on the management of the retail channel behaviors. These local regulations vary from place to place and these local governments are different from each other in their roles to coordinate channel relations. At the same time, China is still fledgling in building its legal system, and different regions are also different in understanding, quoting and judging national laws while handling the conflicts between retailers and suppliers. Consequently, we can assume that different regions are at different levels in regulating and constraining channel relations, which has transformed the proxy variable of the legal factor from fixed into unfixed. We can find the degree and process of the influence of legal environment on various variables of channel relations through analyzing the impact of local regulations and constraints on channel relations.

Therefore, our research starts from local regulations and restrictions of various regions among which a certain degree of differences exist, and propose to use "regulation deficiency degree" as a proxy variable for the legal environment based on surveys of nationwide manufacturers. Regulation deficiency degree mainly refers to: (1) the deficiency degree of relevant regulations of different regions on the transaction behaviors between retailers and suppliers and the differences between regions in their understanding of relevant national laws; (2) the deficiency degree of relevant institutions of different regions for managing the transaction behaviors between retailers and suppliers and the differences between different regions in the implementation of relevant regulations.

This proxy variable cannot represent the overall legal environment for China's retail field macroscopically and has certain limitations in both a theoretical and empirical sense, but it is still significant in research of the influence of legal factors and state actions on channel relations.

In conclusion, the basic premise of our research is: regulation deficiency—abuse of channel power—decline of channel power—channel innovation. The specific research path and method are shown in Fig. 7.3.

Fig. 7.3.   Theoretical framework for sales Channel innovation

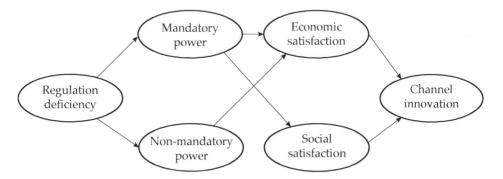

Environment for Sales Channel Innovation in China

Research objective and research design

The study of this section is built on a questionnaire which includes a total of 61 questions and falls under four parts.

The first part is to investigate the channel satisfaction of manufacturers. The investigation is divided into two aspects: the investigation in economic satisfaction and the investigation in social satisfaction, with a total of 16 questions. The second part is to investigate the channel power used by retailers, comprised of two parts: namely the investigation in mandatory power and the investigation in non-mandatory power, with a total of 17 questions. The third part is to investigate the regulation deficiency degree of the Chinese retail industry, mainly in the deficiency of existing regulations and the lack of system feasibility, with a total of 8 questions. The fourth part is specific information about the transactions between manufacturers and retailers, including an investigation in commercial bribery and channel conflicts, with a total of 20 questions. The first three parts adopt Likert's 5-point scale; the fourth adopts multiple-choice and fill-in-the-blank questions to carry out the test. The first two parts of this questionnaire are the adapted questionnaires based on those of domestic and foreign scholars;[1] the latter two parts were independently composed by the author through literature research.

The objects of this study are the manufactures in the retail industry and the questionnaire was handed out through the branches of the Professional Committee of Retail Suppliers of China General Chamber of Commerce. A total of 200 questionnaires were handed out, and 166 were completed and returned. The specific information about the objective of this study is shown in Table 7.3.

The statistical software SPSS 11.0 for Windows and the software of LISREL Version 8.51 were adopted to analyze the collected data, and the major statistical analysis methods adopted are as follows:

(1) Reliability analysis;

(2) Descriptive statistics;

(3) One-way ANOVA;

(4) Regression analysis;

(5) LISREL structure equation analysis.

Table 7.3. Basic information of the research objectives

	Classification	Quantity	Proportion
Business type	Production-based enterprises	62	37%
	Trade-based enterprises	104	63%

(Cont'd)

Classification		Quantity	Proportion
Main products	Wine and beverage	46	28%
	Reconstituted food	12	7%
	Snack food	30	18%
	Daily food	6	4%
	Fresh foods	24	14%
	Cleaning products	20	12%
	Home supplies	11	7%
	Culture and sports supplies	9	5%
	Household appliance	8	5%
Number of employees	100	124	75%
	100–300	32	19%
	300–500	6	4%
	More than 500	4	2%
Business type	State-owned enterprises	15	9%
	Private enterprises	139	84%
	Collectively-owned enterprises	12	7%
Annual sales amount	Less than RMB10 million	89	54%
	RMB10 million–30 million	57	34%
	RMB30 million–50 million	8	5%
	More than RMB50 million	12	7%
Annual net profit	Less than RMB3 million	142	86%
	RMB3 million–8 million	18	11%
	RMB 8 million–15 million	6	4%
Region	Yunnan	21	13%
	Anhui	20	12%
	Jilin	22	13%
	Shanxi	29	17%
	Hebei	27	16%
	Heilongjiang	24	14%
	Jiangsu	23	14%

Reliability analysis

This study uses Cronbach's α coefficient to test the internal consistency of relevant research variables. See Table 7.4 for the Cronbach's α coefficient of each variable. The Cronbach's α value of 5 factors in the first three parts is larger than 0.7, and the reliability ratio is higher.[2]

Table 7.4. Test of the reliability of scales

Name of scale	Factor	Number of questions	Cronbach's α Value
Scale of channel satisfaction (Part 1)	Economic satisfaction	8	0.756 2
	Social satisfaction	8	0.867 1
Scale of channel power (Part 2)	Mandatory power	8	0.828 4
	Non-mandatory power	9	0.861 4
Scale of regulation deficiency degree (Part 3)	Regulation deficiency degree	8	0.773 4

Analysis of the relations among regulation deficiency, channel power and channel satisfaction

Due to limited space, this chapter will not give a detailed account of the factor analysis of regulation deficiency, channel power and channel satisfaction. But, the explained variances of the three have all exceeded the required 50%, and thus the questionnaires have better construct validity. The fully standardized burden rate between factors and indices is 0.52–0.95, both of which are higher than the 0.50 standard, and the T value is larger than 2, having fulfilled the requirement. Most of the fix indices of the research model are near the recommended value and the model fitting degree meets the requirement of the research.

The calculation of the path coefficient of the model through LISREL software is shown in Table 7.5.

Table 7.5. Model path coefficient

Path	Fully standardized path coefficient	T value
Regulation deficiency degree ⟶ Using mandatory power	0.38	4.10
Regulation deficiency degree ⟶ Using non-mandatory power	−0.33	−3.23
Regulation deficiency degree ⟶ Economic satisfaction	0.14	1.38

(Cont'd)

Path	Fully standardized path coefficient	T value
Regulation deficiency degree ⟶ Social satisfaction	–0.16	–2.02
Using non-mandatory power ⟶ Economic satisfaction	–0.64	–3.75
Using non-mandatory power ⟶ Social satisfaction	–0.50	–4.33
Using non-mandatory power ⟶ Economic satisfaction	0.32	2.62
Using non-mandatory power ⟶ Social satisfaction	0.87	5.50
Economic satisfaction ⟶ Social satisfaction	–0.39	–2.80

Table 7.5 showcases that the "regulation deficiency degree ⟶ Economic satisfaction" path coefficient has a T value of 1.38 and an absolute value of less than 2, and the statistical result is not significant. Based on the above result, we can make the following evaluation of the theoretical assumption: see Table 7.6.

Table 7.6.   Test result of assumptions

Assumption	Contents of assumption	Test results
H1a	Regulation Deficiency Degree ⟶ Using mandatory power (+)	Accepted
H1b	Regulation deficiency degree ⟶ Using non-mandatory power (–)	Accepted
H2a	Regulation deficiency degree ⟶ Economic satisfaction (–)	Not Significant
H2b	Regulation deficiency degree ⟶ Social satisfaction (–)	Accepted
H3a	Using mandatory power ⟶ Economic satisfaction (–)	Accepted
H3b	Using mandatory power ⟶ Social satisfaction (–)	Accepted
H4a	Using non-mandatory power ⟶ Economic satisfaction (+)	Accepted
H4b	Using non-mandatory power ⟶ Social satisfaction (+)	Accepted
H5	Economic satisfaction ⟶ Social satisfaction (+)	Refused

Descriptive statistics analysis

The analysis of the status quo of manufacturers' channel satisfaction

In the economic satisfaction research, the satisfaction degree of manufacturers in each index is shown in Fig. 7.4. In the Fig. 7.4., the ordinate value 5 is the highest, representing "very satisfied," and value 1 represents "very dissatisfied." The indices that satisfy manufacturers are sales revenue (ES1), price (ES2), training support (ES8), promotion support (ES7), and profit (ES3), in proper sequence. The indices that dissatisfy manufacturers are rebate amount (ES5), total expenses (ES4), and payment days (ES6), in proper sequence.

The analysis of the indices of manufacturers' social satisfaction is shown in Fig. 7.5. The indices with which manufacturers are rather satisfied are the atmosphere of cooperation (SS5), worth doing business (SS7), regular communication (SS4), friendly cooperation (SS1), candid approach (SS3), and optimistic relations (SS6), in proper sequence. Indices that manufacturers are not satisfied with are mutual respect (SS2) and the loyalty of retailers (SS8), in proper sequence.

In this connection, the satisfaction degree in expenses is a sensitive question. Given that, this study makes a specific survey on the issues of expenses and discounts and has listed 14 items of common expenses and discounts; the proportion of the expenses and discounts in the sales revenue of manufacturers are demonstrated in Fig. 7.6 and Fig. 7.7 (the ordinate represents the proportion of expenses and discounts in the sales volume).

It is clear that the fees for product slotting, renewing contract, and new listing are the main part of the expenses of manufacturers; festival promotion fees are basically at the same level with the Spring Festival promotion fee remaining at the top; the proportion of fees for data query and business promotion are rather small. In addition to these expenses and discounts, manufactuers are also asked to reach the average minimum promotion times each year.

Fig. 7.4. **Satisfaction degree of manufacturers in economic satisfaction index**

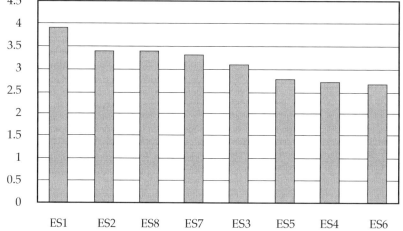

Satisfaction mean value

Fig. 7.5.   Satisfaction degree of manufacturers in social satisfaction index

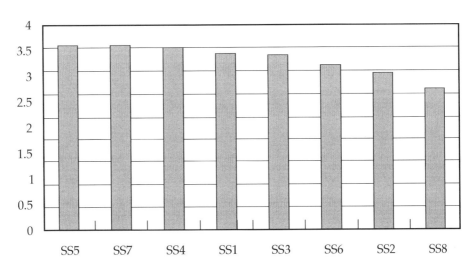

Satisfaction mean value

Fig. 7.6.   The proportion of the expenses and discounts paid by manufacturers (I)

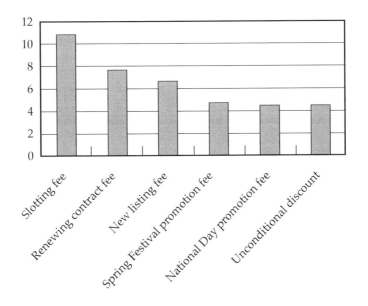

Proportion of expenses and discounts

Fig. 7.7.   The proportion of the expenses and discounts paid by manufacturers (II)

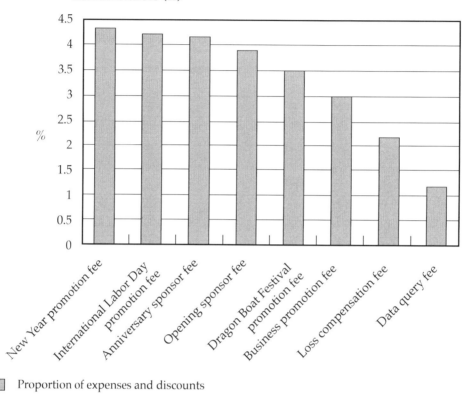

☐ Proportion of expenses and discounts

Analysis of the status quo of retailers' use of channel power

Fig. 7.8 and Fig. 7.9 respectively list the use frequency of different mandatory powers and non-mandatory powers by retailers, among which vertical axis value 5 represents "always use," and value 1 represents "never use."

We can see from Fig. 7.8 that the mandatory powers frequently used by retailers include prolonging payment days (CP4), demanding the cut of product prices (CP2), refusing to order goods (CP6), threatening to reduce orders (CP5), postponing ordering (CP6), demanding to give up immediate interests (CP8), and cancelling previous arrangements (CP7); the mandatory power sometimes used is the threat of litigation (CP3). However, compared with the frequency of mandatory power use, the use frequency of non-mandatory power by retailers is rather low (see Fig. 7.9). The powers occasionally used include the offering special promotion activities (UCP5), increasing exhibition stands (UCP6), assisting product exhibition (UCP1), providing suggestions on promotion (UCP7), providing market information and marketing information (UCP2), and offering suggestions on product improvement (UCP3); the powers rarely used include handling affairs as requirements (UCP8), assisting manufacturers to improve the operation mode with experiences (UCP9), and providing information about competitors (UCP4). Generally speaking, retailers tend to use mandatory powers.

Fig. 7.8.   The use frequency of mandatory powers by retailers

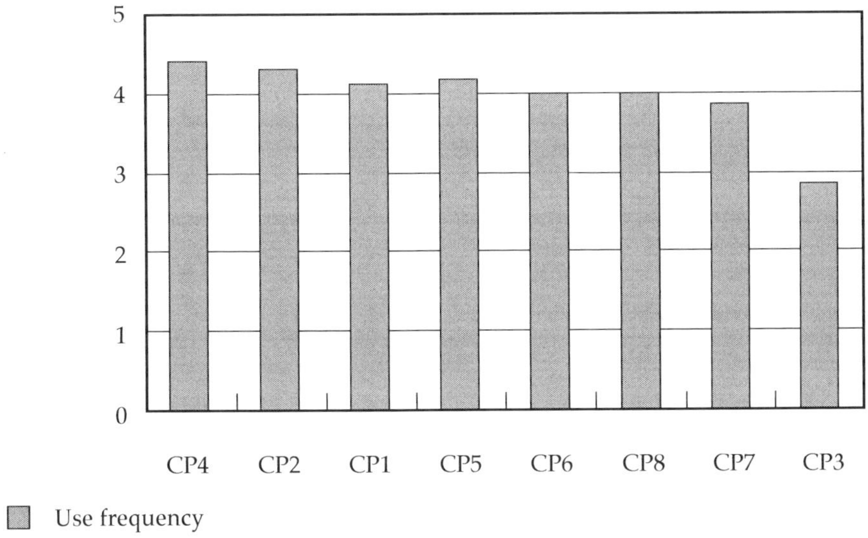

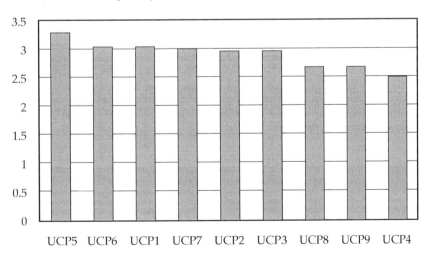

Fig. 7.9.   The use frequency of non-mandatory powers by retailers

Payment days are items which both retailers and manufacturers have paid close attention to. Accordingly, this research makes a comparative analysis of contract regulated payment days, actual payment days, and manufacturers' expected payment days (see Fig. 7.10). In Fig. 7.10, the ordinate is a month referring to the length of the payment days. We found that the actual payment period is generally 1.5–2 months longer than the contract regulated period, while the manufacturers' expected payment period is generally 0.5–1 month longer than the contract regulated payment days. There are great differences between the three.

Fig. 7.10. **Competitive analysis of average payment days**

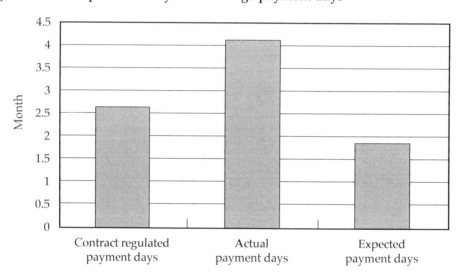

☐ Average payment days

Analysis of the status quo of regulation deficiency

In order to fully understand the regulation deficiency, it is necessary to make a statistical analysis of the indeces of different regulation measures' deficiencies. In Fig. 7.11, the ordinate represents the deficiency degree of regulation measures; value 5 represents "extreme deficiency" and value 1 represents "no deficiency."

We can learn from Fig. 7.11 that regulation deficiency is serious in the following aspects: the deficiency of laws and regulations on the conficts between retailers and suppliers (LAW3), the deficiency of the *Anti-unfair Competition Law* and the *Contract Law* on the abuse of dominant position (LAW2), the deficiency of local regulations on unfair trading behaviors (LAW6), the deficiency of local governments on commercial monopoly (LAW7), the lack of specialized arbitration institutions in handling conflicts between retailers and suppliers (LAW4), the deficiency of the *Anti-unfair Competition Law* and the *Contract Law* on commercial bribery (LAW1) and the deficiency of governmetns and associations in protecting the lawful rights of manufacturers (LAW5). The large majority of manufacturers are prudent yet optimistic about the constraints of the *Regulatios on the Stock Trade between Retailers and Suppliers* on trading behaviors (LAW8).

Fig. 7.11.   Statistical results of regulation deficiency degree

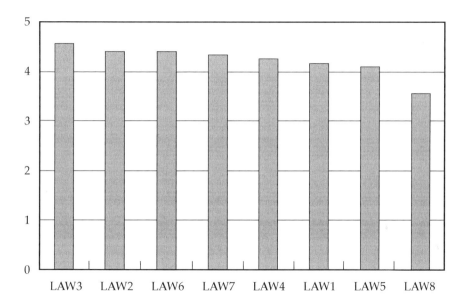

■ Regulation deficiency degree

In addition, this study also makes an investigation of commercial bribery and the results are as follows:

(1) 76% of manufacturers believe that the slotting fee charged by retailers can be deemed as taking bribes.

(2) 45% of manufacturers believe that they have offered bribes to retailers in various disguised forms, such as actively providing a promotion display fee, increasing promotion fee, and offering discounted prices and free shopping opportunities.

(3) 64% of manufacturers believe that their competitors offer bribes to retailers through active payment or preferential policies.

As commercial bribery is a sensitive issue, the research data might have some errors, but it should have reflected the reality of the general trend, and highlighted certain referential and guiding values for practices.

Empirical result analysis

This research found that regulation deficiency can positively influence the use of mandatory power by Chinese retailers but negatively influence their use of non-mandatory power. This is mainly because when the deficiency degree of laws and regulations is high, retailers can easily generate excess profits through mandatory powers, such as prolonging payment days, bringing down prices, and collecting numerous fees. Compared with making profits through normal operations, retailers prefer this "easy way of making profits."

When researching the influence of regulation deficiency on channel satisfaction, we found that regulation deficiency has a negative influence on social satisfaction, but not an obvious influence on economic satisfaction. This is due to the fact that social satisfaction is mainly measured from the perspective of psychological feelings. When the regulation deficiency degree is comparatively higher, manufacturers tend to lose their confidence in the trading environment, causing their relationship with retailers to lack the environmental factors guarantee, and the two cannot easily form a close relationship, leading to a lower level of satisfaction. It is easier to experience a higher degree of social satisfaction under the guarantee of an enabling environment. Economic satisfaction typically measures the factors in economic benefits, and thus the external influence on it is obviously weak.

When researching the influence of channel power on channel satisfaction, we found that the use of mandatory power by retailers can negatively influence the economic satisfaction and social satisfaction of manufacturers, but the use of non-mandatory power by retailers can positively influence the economic satisfaction and social satisfaction of manufacturers. This is consistent with the general research hypotheses. When retailers use mandatory power to influence manufacturers, manufacturers would take certain response measures to counteract this kind of mandatory power; these response measures cause certain problems for manufacturers and result in the decline of their economic benefits, thereby lowering their economic satisfaction. Mandatory measures also cause decreased confidence for manufacturers in transactions between one another, and their social satisfaction degree will decrease. When retailers use non-mandatory power to influence manufacturers, the overall economic benefits of manufacturers will increase as they have gained extra support, increasing their economic satisfaction; however, their social satisfaction will also rise, as the use of non-mandatory power will increase the trust of manufacturers in retailers.

When studying the relationship between economic satisfaction and social satisfaction, we found that a higher economic satisfaction degree of manufacturers means a lower social satisfaction degree — this is the opposite of our hypothesis. The cause for this is that China's retail industry market is still immature. Retailers are in a dominant position, they usually default on payment for goods, and charge various kinds of distribution fees; therefore, manufacturers have to provide retailers with extra expenses, discounts and preferential policies, among other "commercial briberies." This has increased the economic satisfaction to a certain extent, but manufacturers are not very confident in normal transactions with retailers, since commercial briberies have a corrosive role in normal communication relations.

To sum up, we can make the following conclusions through empirical research and literature research: in China's retail market, regulation deficiency can positively influence the use of mandatory power by retailers while negatively influencing their use of non-mandatory power; regulation deficiency plays a negative role in the social satisfaction of manufacturers; retailers can negatively influence both the economic satisfaction and social satisfaction of manufacturers by using mandatory power, but positively influence the economic satisfaction and social satisfaction of manufacturers by using non-mandatory power; under the current retail market environment in China, a higher degree of economic satisfaction means a lower degree of social satisfaction.

Sales Channel Innovation in China: Taking Gree as an Example

According to the empirical analysis of the Chinese sales channel environment, this study concludes that the cause and driving force for the sales channel innovation of Chinese enterprises are comprised of three aspects.

First, some large-scale retailers abuse their dominant position to a certain extent. Under the current retail market environment in China, retailers have a dominant role in the market, tend to use more mandatory power but less non-mandatory power, and the major demonstrations of the use of mandatory power include prolonging payment days, bringing down prices and charging various fees.

Second, the degree of satisfaction in the sales channel terminal is low. The three focal points of their dissatisfaction are: higher rebate amounts, diversified charges, and the excessively long payment periods.

Third is the regulation deficiency. At the moment, the legal system of Chinese competition laws and relevant law enforcement agencies have yet to be established, the *Anti-monopoly Law* has yet to be enacted, the existing *Anti-unfair Competition Law* and the *Contract Law* together with local and industrial regulations cannot fundamentally address channel conflicts in the retail market.

The above factors have resulted in the fact that many manufacturers lose the rights to control or have a voice in the market channel, business profits in manufacturing have been taken over by large-scale retailers, and their room for making profits and potential development has been squeezed.

To conquer these difficulties, many enterprises have started to explore new channel modes, and as a result, the sales channel innovation of Chinese enterprises has recently emerged. Gree, as a pioneer in channel innovation, is one of these enterprises,

The background for the emergence of the sales channel mode of Gree

The history of Gree Electric Appliances can be traced back to 1989; it was formally renamed "Gree" in 1992. When Gree was initially founded, it adopted the "encircling the cities from the rural areas" strategy due to its weak strength, focused on the development of the regions where the then-famous "Chunlan" and "Huabao" air conditioners had little influence, built up its brand image in provinces including Guangdong, Zhejiang, Jiangxi, Hunan, Guangxi, Henan and Hebei, and set up and consolidated its market base there. During the implementation of this strategy, the leading channel employed was focused on the operation of outlets and guaranteed the benefits of its clients through strong after-sale services.

Between 1992 and 1994, the company's business expanded rapidly. In 1994, its production and sales volume of air conditioners ranked second nationwide. But, during this period, the sales channel of Gree seemed in disarray. As the competition of the air conditioner market became fierce, the retailers, particularly the dealers of the same brand, often cut prices, sold beyond agreed areas and engaged in cutthroat competition, seeking profits. Similarly, Gree also faced these issues.

In 1997, Gree had four large dealers who had a strong business performance. That said, in the air-conditioner price war that same year, these four dealers started to engage in similar destructive competitions to cut down prices. After several rounds of competition, the market price system of Gree was shocked, which gravely influenced the normal market system of Gree and ultimately resulted in the "lose-lose" situation of retailers and manufacturers.

On December 20, 1997, having paid a heavy price, those four dealers voluntarily worked together to establish Hubei Gree Electronic Appliance Sales Co. with assets as the bond, and with the vice-general manager of Gree, Dong Mingzhu, acting as a go-between. The four dealers merged and became shareholders of the company. A general manager was elected and Dong Mingzhu became the chairman of the board. The four dealers took share capital as the basis to share risks and interests, united their networks, and wholesaled their products at a uniform price. After the founding of the new company, they only sold Gree's products, excluding all other brands, and formed a comparatively exclusive sales channel. Thus, manufacturers finally reached a consensus on their target and realized a basic consistency in interests. They have not only controlled prices but also diverted all their strength from internal friction to jointly expanding their market. Thus, the unique marketing mode of "joint-stock regional sales company" emerged.

The establishment of Hubei Gree Electronic Appliance Sales Co. has greatly regulated Gree's air-conditioning market in Hubei and enabled them to become a strong secondary administrative institution in the local region. In time, the company brought the sales to a new level in only its second year, recording a growth rate of 45% and a sales amount of over RMB500 million. With this success, Gree stepped up its pace and extended it to the rest of nation. It has set up regional sales companies in 32 provinces or cities, including Chongqing, Hunan and Hebei, which has become the "trump card" of Gree air conditioner amid the fierce market competition.

The characteristics and functions of the Gree sales channel

The characteristics of the Gree sales channel

The specific practices of Gree in the introduction of the "joint-stock regional sales company" mode targeting large dealers include: Gree joins hands with several large dealers in the same region, establishing a joint-stock sales company, with Gree holding shares and the dealers serving as the agencies for the sale of all Gree air conditioners. In other words, Gree joins the once-separate marketing and service networks and sells products at a uniform price (see Fig. 7.12). At present, Gree has established joint-stock regional sales companies in various provinces, municipalities and autonomous regions, and realized a more effective control through an increased share holding.

Fig. 7.12.   Structure of Gree distribution channel

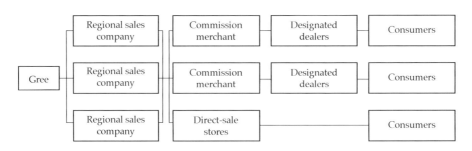

Source: Guangzhou Ark Marketing Research and Consulting, *The Three Major Marketing Modes of Air Conditioner* (2002).

The marketing companies in each province are not the organizations dispatched by Gree (subsidiaries or offices); rather, they are the consortium of manufacturers founded jointly by Gree and some competitive household appliance dealers from each province. The marketing company is equivalent to the sole agent of Gree in the province and practices an exclusive dealership system. In the secondary market of each province, Gree adopts selective distribution.

At each prefectural level, Gree chooses several stronger household appliance dealers as the wholesalers (known as the commission merchants) and some distributors (known as designated distributors) to serve as retail terminals. The marketing company strictly divides the wholesale region or wholesale target (designated dealer) of each commission merchant. Even within a same region, there might be more than one wholesaler, but dealers within this region are designated to pick up goods from a certain wholesaler. Dealers are affiliated to different wholesalers and thus are called designated dealers.

The headquarters of Gree is equivalent to a manufacturer and a nationwide wholesaler, and the marketing company is equivalent to the general agency at the provincial-level market. Gree and marketing companies are linked to each other by capital and are all shareholders of marketing companies. Gree becomes a shareholder of marketing companies with intangible assets, like the brand, while other shareholders buy shares with their monetary assets. Gree has the right to control marketing companies: Gree practices cash on delivery and does not sell products on credit to marketing companies; the chairman of the board of a marketing company is assumed by Gree, who also has the right to appoint or remove the general manager of a marketing company; Gree can control the marketing companies through its brand and product and can also support new marketing partners if necessary (changing shareholders or commission merchants).

A sales company is the primary wholesaler of the province while a commission merchant is equivalent to the secondary wholesaler. A sales company has no transit warehouse in the secondary market, thus commission merchants play the role as a transit warehouse. A commission merchant is a nominal wholesaler, they do not earn any profit in wholesale and they only earn a 2.5% agency fee and a 3% transportation subsidy. A commission merchant is the only place for designated dealers to pick up goods. They then sell the goods in a designated region. The payment for goods of designated dealers is directly remitted to the account of the sales company, who will issue a notice to the commission merchant regarding the order. At that point, the designated dealer may go to the commission merchant to pick up goods with a delivery notice. Wholesalers also monitor the behavior of the designated dealers, as the amount of goods picked up by a designated dealer are counted as the wholesale volume of the commission merchant.

The designated dealers directly remit money to the sales company, who keep a current account for each designated dealer. All sales channel policies are developed by the sales company, directly targeting designated dealers, and the sales company carries out relevant policies in accordance with the remittance amount, time and the delivery volume. The networks of designated dealers are mostly controlled by the sales company rather than by manufacturers. The sales companies, commission merchants and designated dealers identify their respective market responsibilities and interests by signing a three-party agreement.

Through this type of functional positioning and division of labor in the channel, Gree's sales companies have weakened the control of the secondary wholesalers over the distributors' network (retailers' network). The weakening of the functions of commission merchants has reduced the dependency of the sales companies on channel members, especially on wholesalers. But the full use and integration of retailers' resources has also cut down the channel cost of the sales company (mainly the fees for establishing transit warehouse and storage management fee) and increased the efficiency of the use of channel resources.

Functions of Gree sales channel

Gree's sales channel has all the functions that a distribution channel should have, including product flow, capital flow, information gathering and transfer, ownership transfer, promotion, service, production positioning and corporate culture expression; of these, product flow and capital flow are the most outstanding functions.

1. Product flow function

Gree has perfectly accomplished the product flow from manufacturers to dealers and then to consumers by leveraging on regional sales companies. An air conditioner is different from other household appliances as it must be installed by professionals. . This characteristic has determined that it is unfeasible for manufacturers to have the capability and resources to carry out all direct marketing; they need to capitalize on the resources and professional advantages of dealers to realize the marketing of products.

Commodities of Gree in the distribution channel mainly flow to the market through commission merchants who are the center of product flow in the channel. Through controlling the direction and range of product flow in the channel, Gree has not only saved the cost of building the transfer station but has also better controlled the direction of products' flow, creating a solid foundation for restraining sales beyond agreed areas and stabilizing the market price. Gree has better controlled the whole market order from the source of product flow, which has paved the way for a series of management activities.

2. Capital flow function

Another important function of Gree's distribution channel is that it has realized the flow of capital in the channel and alleviated the financial stress.

(1) Payment: the payment of designated dealers is made directly to the sales company. Gree practices the policy of making the payment first and then delivering air conditioners. The payment can be made by cash, spot exchange and bank acceptance.

(2) Financing: as the peak season and off season for the sales of air conditioners are rather obvious, there may be insufficient capital for production in off seasons, which may result in insufficient products in hot seasons. Manufacturers would be unable to organize production if capital is unavailable. However, the manufacturers of Gree have gathered capital of dealers as production funds through preferential policies for the participants in the channel, which has addressed these financial difficulties. The financing function is one of the most important functions of the channel, without which the capital chain of air conditioners would collapse. Therefore, Gree marketing companies have rolled out an array of preferential policies on the advance payment in the channel in order to meet the off-season capital demand.

The marketing company is the capital gathering center of the entire channel. Designated dealers are the downstream companies of commission merchants, but they make the payment directly to sales companies instead of first paying commission merchants who would then transfer the payment to the sales companies. This has guaranteed the fund specificity and centricity, strengthened the control of sales companies over the secondary merchant (designated dealers or retailers) network, weakened the control of commission merchants over the secondary merchant network, and reduced the dependency of sales companies on wholesalers (commission merchants).

Gree's incentive policies on the sales channel

Gree controls the market through holding the resources of dealers. In order to increase the enthusiasm of dealers, Gree has implemented policy incentives and benefit incentives for dealers, including the off-season advance payment policy and Gree's vague rebate policy.

Policy incentives

Identify distribution scope

The main purpose of controlling sales is to prevent conflicts between networks, protect the interests of dealers, and guarantee network efficiency. In the annual sales agreement, Gree sales companies clearly regulate the scope and objectives of good delivery for each commission merchant, and strictly prohibit commission merchants to deliver goods to non-designated or disqualified dealers. At the same time, the purchasing channel of designated dealers is also strictly regulated. Bar codes are used on the merchandise delivered to commission merchants and the bar codes delivered by commission merchants to designated dealers are all recorded. Additionally, the bar codes of all merchandise sold must be scanned, and dealers must send the scanning records to the sales company as part of the demonstration for sales and as proof for the settlement of installation fees. Because the merchandise is traceable via bar codes, it helps the company monitor the final flow direction of the merchandise. Therefore, it will be easy to trace the source of the goods if sales beyond agreed areas occur. This type of monitoring could guarantee the interests of regional dealers and the stability of the network. As for dealers who tear up bar codes or conduct other behaviors to violate regulations in wholesale, once detected, they are fined RMB3,000 the first time and disqualified as a dealer the second time, enjoying no year-end rewards.[3]

Price policy

Price confusion is bound to trigger dealers within the network to compete by cutting down prices, which would disrupt the sales order and result in conflicts of interests among dealers. As a result, if the dealers find it unprofitable, they would give up selling Gree air conditioners and the entire channel would totally break down. One of the keys in Gree's distribution channel is to strictly control the terminal retail price, build a more stable price platform, and regulate the market order. Before the peak season starts, or during a certain period in the sales process, Gree will publish a recommended retail price to dealers, demanding dealers to abide by the price strictly and not to sell products lower than the specified price. The product price takes the supply price of the provincial sales company as the standard, and dealers are required to follow the standard to sell goods. Dealers must also sell products above the recommended retail price; the losses resulting from underselling goods shall be absorbed by dealers themselves. If they are found underselling, they will be fined, goods delivery to them will cease, and they will be disqualified as dealers. Cross-regional hype of goods is prohibited, and if detected, dealers will be fined RMB2,000–3,000, their sales will be haltedfor two to four weeks, and all year-end awards will be cancelled. Cross-provincial sales for any reason is completely prohibited, and if detected, Gree Headquarters and provincial companies will fine dealers RMB10,000–30,000. At the same time, if it is discovered that the cross-regionally (including cross-county) sold installation card is not paid, the dealers will be fined the amount of 2–10 times more than the installation fees. Dealers have both the right and obligation to report the any party which has sold products at lower prices or violated regulations in wholesale. If there is sufficient evidence, the reporting dealer will be rewarded with 80% of the fines from the violated party.[4]

Year-end vague rebate policy

Gree's year-end vague rebate policy started in 1996 when Gree established itself as a leading market player nationwide. In the same year, the air conditioner market was in fierce competition, and dealers tried all means to cut down prices but suffered heavy losses. Gree, after internal discussions, diverted RMB100 million to subsidize its dealers, which was a rebate to dealers without any conditions. Through this rebate, Gree sent out a message to all dealers that they have the same interests and it established a win-win relationship with dealers.

In 1997, Gree created the vague rebate policy in the air conditioner field. This policy entails that a manufacturer will study the yearly sales and the profits of its dealers, and if fierce market competition results in lower profits for its dealers than dealers of other brands, or shrinking profits compared with the previous year, the manufacturer will transfer part of its whole-year profits to its dealers according to the rebate policy, in order to protect the enthusiasm of its dealers. The rebate policy has great uncertainties, mainly that it depends on the profits made by manufacturers and market competition situation, and the policy is kept secret from dealers until the end of the year. Before the policy is unveiled, the rebate proportion is unspecified, so dealers are unaware of know their specific rebate amount. Because of this, they dare not boost their sales volume by means of cutting down the price in advance, reducing the possibilities of selling beyond the agreed region and disrupting the market price. In the meantime, the year-end rebate policy can facilitate dealers to focus on performing well in the market, assisting the manufacturer in enthusiastic promotion mobilizing.

The year-end rebate policy of Gree is also a sales mode with distinctive Chinese features. The rebate policy has gained popularity among dealers because they believe that choosing Gree air conditioners means entering into a safe box, having no risk of losses. Although many manufacturers are now following this practice, the rebate policy cannot attract more dealers and mobilize their enthusiasm because there are no contracts signed in advance and most manufacturers do not have the market credibility Gree does.

Brand support policy

Gree attaches great importance to brand building and invests a huge amount of capital in its image promotion in the central government-supported media each year. The brand image of "Good Air Conditioner, Gree Made" has been very popular among consumers (see Fig. 7.13).

Fig. 7.13.   **Investments in advertisement of air conditioner brands on newspapers between January and June 2003**

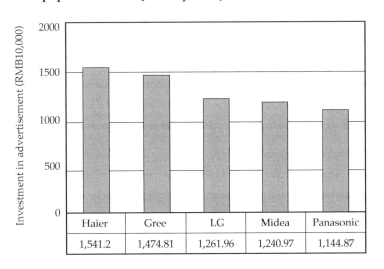

	Haier	Gree	LG	Midea	Panasonic
	1,541.2	1,474.81	1,261.96	1,240.97	1,144.87

Source: China Market Monitor Co., Ltd. (CMM), *Investigation Report on the Air Conditioner Market between January and June 2003* (2003).

Gree also invested a lot in retail terminals for the promotion of its brand image. The investment of Gree in exhibition counters is the highest, compared with other air conditioner brands, and the investment per sqaure meter is almost double that of other air conditioner brands. Gree has compiled a manual of the standardization of terminal image, and the contents include:

(1) uniform regulation on the material, color, logo of exhibition booths and overdoors;

(2) uniform regulation on the placement order of advertising leaflets and the display of samples;

(3) uniform regulation on the placement, use and maintenance of counter, light box and poster stand and on the hanging of POP.

Gree has made standarized requirments and management for each detail. Taking Jiangxi province as an example, dealers of Gree who have achieved an annual sales of RMB1.5 million (one million for those in secondary regions) and at the same time have more than 12 square meters of Gree shop signs and light boxes can apply for RMB2,000 reimbursement (RMB1,500 for those in secondary regions) through photo, contract and advertisement invoices. Gree also gives priority to support exclusive stores, offers county-level exclusive stores various forms of advertisement support, reimburses RMB12,000–14,000 per year to newly established exclusive stores in locally administrated cities (with an annual sales of more than RMB2.2 million) for renovation costs; county-level exclusive stores (with annual sales of more than RMB1.2 million) can apply for RMB7,000–9,000 per year reimbursement for renovation. If a store is founded in the peak season, it could get reimbursement of RMB7,000–9,000 per year (RMB5,000–7,000 per year for county-level exclusive stores). All exclusive stores (including existing ones) could enjoy a subsidy of RMB1,000 (RMB600 for county-level ones) per month in the off season.[5]

Interests reward

Off season advance payment policy

Air conditioner sales have an obvious seasonal characteristic, especially when air conditioners were still a luxury goods for most families a few years ago. Generally speaking, the off season is from September to April, during which the sales volume is quite low; between April and August, the products may be out of stock. For air conditioner manufacturers, production in off seasons is a big problem. The halt of production means that there may be insufficient stock for the sales in the next season; but, continuous production in off seasons will result in stockpiling, adding to inventory and costs. Previously, all air conditioner manufacturers would implement various kinds of preferential sales policies before the coming of a peak season in order to compete for dealers. The pressures on production and competition for dealers had made many manufacturers carry a heavy burden.

In 1995, Gree created the "Gree off season sales policy," i.e. "off season payment and interests subsidies policy," well known in China. The dealers who make payment to Gree before the coming of a peak season can enjoy subsidies, and the specific subsidy proportion is determined by the time of payment. As the subsidy is larger than the deposit interest rate of banks, this policy has gained great response and support from dealers. This policy has not only solved the financial difficulties faced by manufacturers in off seasons but also alleviated their pressure of supplying in peak seasons. At the same time, this policy has attracted dealers, taken an earlier step to tie up the capital of dealers, reduced the possibility of changing brands by dealers before the coming of a peak season, and squeezed out competitors in advance (as air conditioning manufacturers will roll out many sales policies before the coming of a hot season; many dealers will wait and see and might change their interest in brands).

Dealers actively respond to the off-season advance payment policy and almost all dealers would try to give more discounts to customers in off seasons. One of the reasons behind this is that dealers value the Gree brand as it is reliable and profitable. Another reason is that the interest rate subsidy for off-season advance payment is much higher than a bank's interest rate, and payment during off seasons means depositing money into a safe box with a high interest rate. Many air conditioner manufacturers have also replicated the off-season advance payment policy practiced by Gree, but the dealers' responses are not as enthusiastic, because these manufacturers have neither the same brand appeal as Gree nor the same credibility among dealers as Gree has accumulated over the years.

Gree stipulates that the off season is the period between August 20 and March 25, and the peak season is the period between March 26 and August 19. Taking the off-season advance payment policy in 2004 as an example, all dealers who had made the payment for purchasing goods between August 20, 2003 and March 25, 2004 enjoyed off season interest rate subsidies and rebates from Gree (see Table 7.7 and Table 7.8).

Table 7.7.   Off Season interest rate subsidy for payment by acceptance bill

Payment time	Month	9		10		11		12		1		2		3	
	Day	10	25	10	25	10	25	10	25	10	25	10	25	10	25
Interest rate subsidy (%)		6.1	5.6	5.1	4.6	4.1	3.6	3.1	2.6	2.1	1.6	1.1	0.6	0.1	0.1

Source: Annual sales policy of Jiangxi Gree sales company in 2004.

Table 7.8.   Off season interest rate subsidy for payment by cash

Payment time	Month	9		10		11		12		1		2		3	
	Day	10	25	10	25	10	25	10	25	10	25	10	25	10	25
Interest rate subsidy (%)		7	6.5	6	5.5	5	4.5	4	3.5	3	2.5	2	1.5	1	0.5

Notes: 1. Interest rate subsidy and rebate is carried out by means of lowering normal supply price or offsetting the payments for sold goods on the 25th of each month;
2. The payment time for the interest rate subsidy and rebate is calculated by the time of receiving the bill of exchange;
3. Dealers who make the payment through telegraphic transfer or mail transfer cannot pick up goods until the payment is received.

Source: Same as Table 7.7.

At the same time, Gree also encourages dealers to purchase goods in off seasons in order to bring down inventory costs, and it practices price subsidy and returned purchase policy on purchasing goods in off seasons.

The Circular No. 2004 (No. 002) of Jiangxi Gree Electronic Appliance Company stipulates that when the total amount of goods purchased (except bargain goods) before October 31, 2003 accounted for 20% of the total of 2003, there would be a subsidy equivalent to 1% of the total sales (except bargain goods) of 2003; if the total amount of goods purchased (except bargain goods) before March 31, 2004 reaches 60% of the total of 2003 and the returned money accounts for 65% of the normal amount of cash collection of 2003, the subsidy would be 2.5% of the total sales (except bargain goods) of 2003. The subsidy would be delivered by means of purchasing goods or offsetting the payments for sold goods. All Gree air conditioners purchased in off seasons, if not unpacked, could be returned between June 20 and 30, 2004.

Related cost subsidy

In order to keep improving the popularity, attracting consumers and increasing the market share, Gree has invested a lot in advertisement in different regions.

Gree encourages each dealer to engage in active promotion, invest more in advertisement and organze various promotion activities. Dealers can enjoy subsidies from the sales company with photos of promotion banners, outdoor advertising pictures or TV advertising invoices.

Other awards

Taking Jiangxi province as an example, for the top 10 best sellers in the region in 2004, each was awarded one quota of people to travel abroad and the top 3 also enjoyed certain extra awards. The top 10 best sellers in secondary regions also enjoyed the same awards. Exclusive stores and dealers realized 70% of their total sales would be awarded a Gree Loyality Prize.

Gree's channel support

Brand support

The fundamental reason for the survival of Gree sales companies is that Gree brand helps them make money. Since 1996, Gree has been ranking first in terms of the production, sales volume and the market share in the entire industry, making it the well-deserved leader and the number one brand in the air conditioning circle. Gree enjoys a strong popularity and recognition across the nation. This number one position has become a powerful driving force for the channel sales. For manufacturers, the brand and products of Gree is a market resource; for dealers, selling Gree air conditioners is undoubtedly a profitable choice. Since profit is the most important issue for dealers who do not want to lose the huge profits and long-term guarantee appeal of Gree brand, they are willing to offer discounts in off seasons and strictly abide by Gree's price and regional policies. In this connection, it can be said that it is the brand of Gree that has made its channel policy, channel management and channel regulations effective.

Technical and product support

As the most professional enterprise in China's air conditioner industry, Gree has been regarding technological innovation as its lifeline. At the end of 2003, Gree accomplished the heavily-invested construction of the world largest air conditioner research and development center, covering a land area of nearly 40,000 square meters. This center has 170-plus labs and has built a technology R&D team of more than 1,000. In December 2004, Gree's lab was accredited by the Underwriters Laboratories' Client Test Data Program (UL-CTDP) of America and became the first Chinese lab accredited with such recognition. In January 2005, Gree's lab became the nation's first batch of labs for data comparison and confirmation from air conditioner testing devices, which gained the national authoritative certificate (No.1). The huge investment in technology made by Gree has produced great results. Since 1991, Gree's 350-plus patents have been authorized and it ranks top among its national peers in terms of the number of different specifications and series of its air conditioners.

Gree's market management support

A good market order is one of the most important parts of maintaining the distribution channel. If the market is out of order and the business environment for dealers changes dramatically, then sales companies would bear heavy losses and their confidence in Gree products would suffer setbacks. Confidence is one of the important assets of sales companies to maintain Gree's sales channel mode. Gree headquarters and its sales companies put a high premium on the management of the market order in sales, aiming to build an enabling business environment for dealers, safeguard the interests of dealers, prevent conflicts between networks, maintain a virtuous development of networks and make sure this marketing mode plays its role steadily. The maintenance of the market management by Gree headquarters and its sales companies can be demonstrated in the following four aspects:

First, controlling the flow direction of merchandise and capital. As detailed earlier in this chapter, sales companies practice a bar code management on all goods from the warehouse. All goods delivered to each region are put on record and dealers must scan the bar code of products being sold, then send the results to the company as evidence for the settlement of installation fees. The enthusiasm of commission merchants and designated dealers to protect their own interests has been fully mobilized by using the interest leverage, which displays their role in assisting the sales companies to control the market.

Due to the direct control over various dealers' payment for goods, sales companies can also make quick responses to any verified violation, make decisions on punishment, and implement the decision within a short period of time. This takes a hard line against and deters the unfair competition practices of some dealers.

Second, dividing sales territory. Gree divides the sales area of each distributor based on the market consumption ability and the marketing capacity of each distributor, and strictly controls the marketing territory of distributors through a series of monitoring means. For instance, designated distributors could only purchase goods from designated commission merchants, while the commission merchants could only supply the designated distributors; if cross-regional sales or supplies are discovered, Gree's sales companies will impose a fine and even disqualify them as distributors for serious cases.

Third, managing retail price. A rational price platform is a key support for Gree's distribution channel. If the price platform is out of control, distributors will suffer profit losses and the original business process of commission merchants and designated distributors will lose effectiveness accordingly. Gree made a recommended retail price and requires all distributors to sell products based on this price. If violations occur, Gree will fine and even disqualify the violated distributors for serious cases.

Fourth, offering after-sale service support. Air conditioners are known as "semi-product" and depend more on installation and maintenance, as along with other after-sale services, than any other household appliance. With the increase of China's overall consumption level and the decrease of the price of the entire air conditioner industry over the past two years, price is no longer a last straw for seizing market share; instead, the quality and services have become an effective means to get an upper hand. The warranty period of Gree's air conditioners has been formally adjusted to 6 years for the complete unit (including compressor, fan motor, master control board and all other parts and components of the air conditioners) since January 1, 2005.6 Before that, the state had standardized a 1-year mandatory warranty of the complete unit and a 3-year warranty for main fittings and spare parts of air conditioners. Presently, 3 years is the longest warranty period for a complete air conditioner unit provided by other domestic air conditioner brands. That means Gree's warranty period has surpassed not only the national standard but also the standards of all other air conditioner brands nationwide, making Gree the air conditioner enterprise with the highest standard of after-sale services. The general lifetime of air conditioners is 8–10 years, so the 6-year after-sale warranty period offered by Gree almost equals to free repair for the life of the product.

Channel conflict with Gome

In 2004, a conflict between Gree and Gome, one of China's largest electrical appliances retailers, broke out, attracting much attention. After the outbreak of the conflict, Gree actively adjusted its sales channel layout, which increased its sales volume. Against the backdrop of the expanding household appliance retail chain and the conflict between the traditional and the emerging sales channels, adjusting the channel mode by Gree became a focus of the public.

Cooperation between Gome and Gree

In 1992, Gome started to directly cooperate with domestic household appliance manufacturers to build a new mode of supply and sales – that is break away from middle man, contact manufacturers directly, and practice exclusive sales. As such, Gome could enjoy more price discounts from manufacturers due to its large buying demand.

But Gree did not enter into hypermarkets, including Gome and Suning, until the second half of 2001. Different from other household appliance enterprises that totally depend on the sales channels of hypermarkets, Gree only took these hypermarkets as its general distribution outlets and treated them as any other distributor; the price of its supply to Gome was the same as that to other distributors. This is the mode that had been expanded by Gree across the nation, which had guaranteed the interests of distributors at all levels.

In 2003, Gree recorded sales revenue of RMB300 million, with only 10% realized through hypermarkets like Gome and Suning. Gome, as one of the distributors, took up even fewer shares. Gree only took up 3–4% of the air conditioner sales of Gome, while the sales revenue of Gree's products in Gome was only about 5% of Gree's total sales revenue. There are 194 kinds of air conditioner brands selling in Beijing, including Panasonic, Sumsung, Electrolux, LG, Hitachi, Haier, Midea, TCL, AUX, Aucma, Changhong, Chunlan, Gree, Hisense, Hualing, Keolon, Mitsubishi, Mitsubishi Electric, Mitsubishi Heavy Industries, Sharp, Little Swan, Shinco, and Chigo. Among these, 10 are produced by Gree (about 5.2% of the total), 18 are produced by Haier (accounting for 9.3%), and 17 are produced by Chunlan (taking up 8.8%). In 2001, Haier sold 33,000 units of air conditioners in Gome, 45% of which were sold in Beijing. By contrast, the shares of Gome in the sales of Gree air conditioners were less.

Conflicts between Gree and Gome

The rancor between Gree and Gome has long existed. As early as in 2003, Gree was dissatisfied with the "Buy Willy, Get Gree Free" promotion activities of Gome.

On February 21, 2004, Chengdu Gome Electronic Appliance cut down the price of a 1P wall hanging air conditioner from RMB1,680 to RMB1,000 and the price of a 2P floor standing air conditioner from RMB3,650 to RMB2,650. Gree was extremely angry about being passively involved in this price war, and argued that Gome cut down the retail price without its consent, demanding that Gome stop cutting down the price immediately.

However, Gome did not answer the request of Gree and kept the promotion prices of two kinds of Gree air conditioners. Gree wrote again to Gome, demanding the stop of the low price sales and formally ceased its product supply to Gome.

On March 9, the same year, the domestic household appliance giant Gome issued "An emergency notice about selling all remaining stocks of Gree air conditioners" to its branches nationwide, requiring them to sell all their stock of Gree air conditioners and clear all related business. Gree headquarters recriminated: "Gree will eliminate Gome from its marketing system if Gome fails to abide by the game rules of Gree."

Gome argued that the major problem of the fight with Gree was that the sales channel mode and retail price of Gree could not meet the market operation requirements of Gome. Gome had been committed to the direct supply mode—to cooperate with manufacturers directly in order to eliminate dealers' profits, lower the terminal sales price and ultimately bring benefits to consumers. But Gree had been providing supplies to Gome through sales companies in different regions, and the price had included the year-end rebate and various promotion rebates. As such, its recommended retail price could not meet the zero purchase price demanded by Gome, thus Gome was unable to realize its proposed principle of "small profits but quick turnover."

He Yangqing, the deputy general manager of the sales center of Gome headquarters, noted: "March is the time for the start-up of the air conditioner market. Other air conditioner manufacturers provide supplies to Gome directly, but Gree still provides the supplies through commission agents and refuses to make a concession in price, which violates the business idea of Gome."

An Mi, news director of the sales center of Gome Electrical Appliance Co., Ltd, said that Gree's agent system had decided that Gome must negotiate one by one with distributors in different regions if it wanted to sell Gree electrical appliances across the nation. This would not only waste a lot of time, manpower, and materials, but also would bring difficulties to Gome's management; in addition, Gree at that moment had focused on the secondary and tertiary markets and neglected the primary market; there were signs that Gree would stop its line of top brands, as their products were rarely seen in the mainstream markets. Additionally, the sales revenue of Gree products in Gome hypermarkets was not large, so the temporary termination of cooperation with Gree would make little difference to Gome.

Dong Mingzhu, general manager of Gree, was firm in his attitude: "I cannot say for sure how future relations between Gree and Gome will be. But, what is clear is that Gree has been ranking the first in sales volume for nine years in a row nationwide. We believe that there are only three principles for our cooperation: fairness, justice and sincerity. Otherwise, nothing is negotiable."

Gree's spokesperson, Huang Fanghua, said in the interview that Gree treated all distributors equally and would not treat Gome differently, as it would be unfair to other distributors.[7]

According to Huang Fanghua, Gree insisted that the supplies must be provided by local sales companies to distributors, including not only Gome but also Suning and Yongle. Furthermore, Gree did not exclude the chain operation mode, but the precondition was that distributors must accept the agent sales mode of Gree. He said that there are no "big clients" and all distributors are equally treated.[8]

The fight between Gree and Gome is not only the conflict between the two enterprises but the conflict between the traditional regional agent sales mode, represented by Gree, and the chain store sales mode, represented by Gome.

Adjustment of the sales channel of Gree

After the fight between Gree and Gome, many marketing experts pointed out that Gree would lose the entire market after losing Gome, and thus would lose the future. But, the sales performance of Gree in 2004 strongly refuted these experts.

The 2004 annual report of Gree indicated that Gree electrical appliances realized a sales revenue of RMB13.832 billion in 2004, up by 37.74% compared with the previous year; the net profits reached RMB420 million, up by 22.74% compared with the previous year, and the net return on assets stood at 17.24%, representing a sound economic benefit and continuously maintaining a leading role in the air conditioning industry. In the report "Successful Marketing—Survey on the Most Competitive Brand of the New Generation in China," the composite competitiveness index of Gree air conditioner ranked top in the industry and its market share index reached 13.96%, followed by Haier and Midea.

The sales growth in 2004 is closely related to Gree's adjustment of the entire sales channel after its conflict with Gome. The adjustment principle was to reduce the dependency on one or two distributors through strengthening the relations with most distributors, including cooperating with chain household appliance enterprises like Suning and enhancing the partnership with traditional distributors.

Boosting cooperation with retail chain giants like Suning and Dazhong

The high quality and strong brand of Gree have enabled all distributors, including Suning and Dazhong, to reap high profits even if they have to purchase Gree's products through local sales companies. Therefore, these big distributors choose to maintain a close cooperation with Gree.

After Gome issued the notice of "selling out all of the stocks of Gree's products," Gree began negotiations with Suning about the increase of supplies. Suning boosted the promotion of Gree air conditioners either in its product layout or shopping guide.

On March 20, 2004, Dazhong and Gree signed an annual agreement on the exclusive sales of air conditioners valued at RMB180 million with expected sales volume of 80,000 units, while the sales figure was only RMB10 million in Dazhong in 2005. Dazhong said that they leveraged on the consumption guide power of Gree distributors to increase the market capacity of Gree air conditioners.

Dazhong has 42 chain stores in Beijing, ranking first in terms of the number of hypermarkets in Beijing. The unexpected involvement of Dazhong broke up the rumor that Gree's intrinsic sales mode is difficult to be accepted by emerging large-scale household appliance chain distributors at the moment. The collaboration between Dazhong and Gree is a win-win situation—Gree had found a desirable chain hypermarket while Dazhong eliminated a competitor in mainstream air conditioners.

Not long before signing of the agreement with Dazhong, Gree inked a similar regional cooperation agreement with Yongle Electrical Appliance, the largest household appliance retailer in Shanghai. After Gree stopped cooperation with Gome in Beijing, the sales volume of its products in Dazhong, Suning and Yongle increased significantly.

Gome failed to dominate the sales channel. Gree still had many other choices besides cooperation with Gome, and its sales volume was not really affected.

Strengthening the building of exclusive stores

After the conflict with Gome in early 2004, Gree claimed that its sales volume in Guangzhou increased instead of decreasing, and the costs on marketing expenses had been saved by a large margin. According to surveys, this can be largely attributed to the success of Gree's self-established sales channel.

Sales companies encouraged distributors to open exclusive shops to sell Gree's products. Up to that point, many distributors expressed willingness to open Gree exclusive stores. In the marketing mode of Gree, distributors are capable of generating more profits, which has encouraged distributors to work harder to sell Gree air conditioners.

In Gree's sales channel, both exclusive stores for Gree products and stores specializing in selling air conditioners are mainstream—exclusive stores in particular. Across China, there are now more than 2,000 Gree exclusive stores which remain the direction of Gree's future development.

As chain stores sell more than 100 varieties of products, and there are a dozen brands of air conditioners alone, chain stores would neither invest much energy in one brand nor focus on promoting one brand. On the contrary, exclusive stores have more incentive in selling air conditioners than chain stores, since air conditioners depend more on installation and maintenance as well as other after-sale services than other household appliances.

Safeguarding the interests of original distributors

The relationship with distributors has always been at the core of Gree's channel mode; the "representing the interests of distributors" is an active representing rather than a passive maintaining. Gree has an attractive cohesiveness for distributors.

After ten years of hard work, Gree has developed more than 12,000 loyal distributors; theseare the most critical and precious resources of Greer to succeed in the market. These distributors can be found in all levels of the market across China, and they take control of all levels of terminals. The growth of household appliance chain stores is gaining momentum, but there is still a long way to go before they set foot in the secondary and tertiary markets. In addition, the unique characteristics of air conditioners have also determined that Gree should not immediately give up the traditional channel and distribution team. The joint-stock regional sales companies of Gree have played a significant role in regulating and stabilizing market competition, protecting the interests of distributors and consumers, and safeguarding a healthy and orderly development of the industry. Under the current, insufficiently-regulated market competition environment, this mode can guard against the practice of selling beyond agreed areas, control the retail price, guarantee the benefits for distributors, increase the enthusiasm of distributors, and thus capitalize on the strength of distributors and use the minimal possible investment to explore the market. Gree sales mode will continue to play its part for a long time to come.

With regard to the conflict between Gree and Gome, many marketing experts and entrepreneurs observed that Gree must adjust its channel mode as soon as possible in order to adapt to the emerging household appliance chain channel led by Gome. Otherwise, Gree may lose their entire market due to the loss of chain retailers. In the meantime, citing the case of Haier's channel mode adjustment, those experts and entrepreneurs try to prove that Gree must adjust its channel mode. Before joining the chain retailer like Gome, Haier originally adopted the channel mode of Haier Industry Trade Company, characterized by direct contact with terminal retailers without relying on the strength of wholesalers; as with numerous retailers of Haier, Gome is also a terminal that has direct contact with consumers. As a result, the channel changes of Haier, which is to meet the requirements of Gome, would not harm the interests of other distributors, and thus would not affect the entire sales channel. The adjustment of Haier is nothing but the adjustment of some functions of the Haier Industry Trade Company and would not harm the entire sales system.

But the mode of Gree differs vastly from that of Haier; Gree's sales depend largely on the operation of distributors. Since the personnel allocation and capital investment of Gree headquarters are insufficient in each regional market, Gree must operate by leveraging on the staff and capital of its distributors. If Gree adjusts its original channel mode to directly supply Gome, it is bound to trigger conflicts between the traditional channel and the emerging channel, harm the fundamental interests of traditional distributors, lead to serious turbulence of the entire distribution channel, and affect Gree's national market layout. In 1997, Chunlan Group changed its "controlled dealership system" into a "terminal system." Due to the coexistence of these two channels, the original "dealer giants" were dissatisfied and lost the enthusiasm to promote Chunlan air conditioners, which harmed the popularity of Chunlan in the market. The story of Chunlan is a good lesson for Gree: it must stick to its original sales channel mode and safeguard the interests of its original distributors in the short run.

At the same time, amid the challenges brought by household chain stores and the changes of consumer demand, Gree needs to adjust the market positioning of distributors, improve the service quality of exclusive stores and air conditioner shops, level up the services and make up for the disadvantages in product variety and scale through services.

Moderately adjusting the mode of joint-stock regional sales companies

From a long-term perspective, the general trend is that of household appliance chain stores. Therefore, Gree has to adjust its original channel policy and value emerging household appliance chain stores if it seeks to maintain its leading position in the market. That said, the changes of the channel cannot be easily accomplished and must be a slow process of adjustment and reform. Dramatic changes of the channel will likely lead to the failure of the entire market.

Short term tasks: attaching importance to the household appliance chain enterprises that have the willingness to cooperate, and actively trying to collaborate with them under the precondition of not disrupting the original distribution channel. As the No. 1 Chinese air conditioner brand, Gree has a strong brand appeal and certain brand premium which could be used together with large profit space to attract chain enterprises. In order to safeguard market order and stabilize the market retail price, considering the low price policy of household chain enterprises, Gree should provide several types of custom-made air conditioners with a bargain price (other distributors do not have) to household chain enterprises.

Medium- and long-term tasks: the primary and secondary markets are the strong markets for household appliance chain marketing enterprises, and take up more than 60% of the market shares in large cities. In order to ensure the large sales volume in the primary and secondary markets, the sales policy in large and medium-sized cities must be adjusted. Household appliance chain enterprises tend to directly purchase goods from manufacturers rather than through distributors in order to realize low sales prices. At the same time, there is also a need to eliminate distributors to realize channel flattening, as the profit in air conditioners keeps shrinking. Gree should adjust the functional positioning of provincial sales companies in the sales channel and transform the profit center into the market management center. That is to say, a provincial sales company, after being adjusted would no longer be a distributor, and would not obtain differential proceeds from the selling of products. Instead, the sales company would become the market management department, formulating market policies and managing distributors, and its proceeds would be service charges rather than the profits from products. Gree needs to slowly adjust the share proportion of provincial sales companies, invest more capital and human resources, and have more say in the sales companies in order to transform its future services. After the transformation, the sales companies would be equivalent to a sales department of Gree, to serve the emerging distributors.

In the tertiary and quaternary markets, Gree should strengthen its relationship with terminal distributors, actively communicate policies with distributors, stabilize their confidence, and directly control the terminal and assist distributors to develop the market intensively.

Evaluation on Gree sales channel innovation

Gree has been the national champion in the sales volume of air conditioners for 14 consecutive years and its market share also ranks first. Gree has also seen increasing gross profit rates while Haier and Koelon as well as other top brands have witnessed decreasing gross profit rates. This is closely related to the innovation of Gree sales channel. The core of Gree sales channel innovation is to bind large distributors together by means of a joint-stock system and make manufacturers and distributors become a community of interest. This mode has played a significant role in regulating and stabilizing market competition, protecting the interests of distributors and consumers, and safeguarding the healthy and orderly development of the industry. Gree has so far extended this channel mode to more than 30 provinces and cities, making it an effective means for the company to succeed. The channel mode of joint-stock regional sales companies has been proven to be an effective weapon to fight against price wars and cutthroat competition, even after several years of operation.

The benefit brought about by this innovative relationship between manufacturers and distributors is reflected in the following:

(1) Substantial capital has been saved compared with the self-constructed channel network. Gree binds manufacturers and distributors together, which has helped it save a huge expenses on building the network by itself, cut down the sales costs, and spread the risks.

(2) The price war among distributors has been eliminated. Distributors have become shareholders and their profits will come from the year-end dividends of the sales companies, thus it is unnecessary for them to keep fighting for sales areas and low prices. Even though there are problems, they can be solved at internal meetings.

(3) The short-sighted behaviors of distributors in brand operations have been eradicated. Previously, distributors often did something that might damage brand value, because they chose to pursue the current profit maximization due to the worry that manufacturers would change the policy. As the confidence of Gree's distributors in manufacturers improves through capital cooperation, they should make a long-term plan and work hard in the sales of Gree products.

(4) The cooperation between influential manufacturers and distributors is conducive to increasing the market share of Gree. Gree could promptly gain local market advantages and quickly occupy the broadest range of market shares by joining forces with local influential distributors and leveraging on their capacity in market distribution and coverage. This is also the fundamental cause for Gree to rid itself of the control of large retailers like Gome.

(5) The channel for goods delivery has been effectively controlled and the market order has been stabilized. Only one company delivers goods in one region, which has prevented the price confusion brought about by multi-channel delivery, stabilized the retail price and guaranteed the healthy and orderly development of the market.

On top of that, Gree's operation mode has saved a considerable amount of expenses for salespeople, including salaries, compensations, travel expenses, and communication costs, as well as substantial product marketing costs including advertisement expenses and promotion costs. For distributors who have participated in the Gree sales channel, the supply and price of goods have been guaranteed and, with the year-end rebate and dividends, their profits are secured and their operation risks are reduced.

Conclusions and Suggestions for Future Research

This chapter, adopting both case study and quantitative empirical study, has made clear the status quo and characteristics of the sales channel of Chinese household appliance enterprises under the environment that retailers dominate the market. Through analyzing the case of Gree in the building of its sales channel, this study probed into the causes for its continuous increase of profits from the perspective of channel innovation.

Research conclusions

The environment for the sales channel innovation

(1) Retailers enjoy the market dominant position in the channel and abuse their position, which is the environment for the sales channel innovation of Chinese enterprises. Large retailers typically make use of their channel power, especially mandatory power, to influence the transaction of manufacturers and reap benefits or monopolistic profits. The use of mandatory powers by retailers would negatively influence the economic satisfaction and social satisfaction of manufacturers, but the use of non-mandatory powers by retailers can positively influence the economic satisfaction and social satisfaction of manufacturers.

(2) Regulation deficiency is a key factor influencing the sales channel innovation of Chinese enterprises. Regulation deficiency degree has a positive influence on the use of mandatory powers and a negative influence on the use of non-mandatory powers by retailers. Regulation deficiency can negatively influence the social satisfaction of manufacturers. Therefore, when the economic satisfaction of manufacturers improves, their social satisfaction will drop.

(3) The satisfaction of manufacturers in the sales channel relations and effects is at a lower level. Based on the recognition of the special market and legal environment in China, it can be said that the sales channel innovation of Chinese

enterprises is not merely a passive choice, but an active and wise one.

(4) The pursuit of channel power and channel profits is the basic driving force for the sales channel innovation of Chinese enterprises, which could not only help the enterprises to gain more say in the channel, but also expand the space for market profits.

Sales channel innovation of Chinese enterprises

(1) Channel mode innovation. Gree's regional sales company mode has played a vital role in regulating and stabilizing market competition, protecting the interests of distributors and consumers, safeguarding the healthy and orderly development of the industry, especially in regulating the market, preventing selling beyond agreed areas and stabilizing market price. Regulating the market and retail price means that the interests of distributors are fundamentally guaranteed, and the stable development of the entire channel is secured. Gree's sound market performance for 12 consecutive years is the strongest evidence.

(2) Channel policy innovation. The off-season advance payment policy and the vague rebate policy created by Gree have been replicated by many household appliance enterprises, but only Gree has been successful with these policies, which also reflects the confidence of distributors in Gree and the strong influence of Gree on distributors. Gree has developed a host of loyal and influential distributors that are not only the largest wealth and resources for Gree but also the strongest competitive advantages of Gree, compared to other air conditioner manufacturers.

(3) Channel management innovation. Channel mode is undoubtedly important, but its implementation is more important. Gree values not only the mode itself but also more the specific operation, management and implementation of the mode. When it comes to the regulation of the market, other enterprises have also adopted the policies and regulations that are similar to that of Gree, but many of them failed to abide by those policies and regulations. Specifically, many enterprises failed to perform in accordance with regulations on the management and punishment for cross-regional sales. The mechanism for discovering the cross-regional sales is still not infallible, mainly referring to the bar code management mechanism and the surveillance and reporting mechanism. On the other hand, the brand does not have strong influence and deterrent force toward distributors so that distributors would not care about being disqualified for their violations. Therefore, other manufacturers need to learn not only Gree's channel mode but also the specific operation and implementation of the channel mode.

(4) Channel resource integration mode innovation. Different from other well-known enterprises like Haier and TCL, Gree's channel members mostly operate within its sales system. Major retail terminal resources (such as Gome and Suning) all operate outside the sales systems of Haier and TCL, and belong to uncontrollable channel resources which are difficult to integrate. But, the over 10,000 franchised stores and exclusive shops of Gree all operate within the sales system of Gree, although they are independent in terms of property rights and operation. At the same time, different from some enterprises that invest heavily in building their own sales channel, Gree has accomplished the integration of these external sales resources with the lowest costs and become a community of interest. Arguably, Gree has put in place the mode of making external sales resources operate internally, in an organized manner, with the lowest possible fund — the core of the innovation of Gree sales channel.

In conclusion, the overall environment of Chinese sales channels is harmful to manufacturers. Against this backdrop, we can divide the channel strategies of manufacturers into two categories: passive adaptation and active innovation. The former category is represented by Haier and TCL. They cooperate with the large-scale public and open sales channels like Gome and Sunning, and advocate the service philosophy of "helping customers (retailers) make profits." In addition, the category lacks a controllable autonomous sales channel and has basically lost its say in the sales channel and the channel profit allocation rules. Channel power is controlled by others, which is a leading cause for the modest profits in the sales margin of this kind of enterprise.

Gree, as a representative of the actively innovative enterprises, has gradually organized its autonomously controlled and close-end exclusive store channel system since 1997 while cooperating with large retail organizations like Gome and Suning, and thus truly realized the "walking on two legs" strategic vision of the sales channel. Given this, Gree enjoyed more channel resources than Haier and thus had more say in the market when competing with Gome to seize the power in market domination, pricing, and profit distribution. The internalization of organization and the control of the sales channel turned out to be the major causes for its higher profits in sales than Haier and TCL over the years, despite the fact that Gree is smaller than its competitors.

However, Gree sales channel mode also has its limitations:

(1) Large difficulties in monitoring. In Gree sales channel system, as the secondary distributors (cost centers) are stripped of their real power, it may result in insufficient market management and service of secondary distributors.

(2) The market in the primary large cities is weaker than the secondary and tertiary markets. The strong point of Gree's sales channel system is in the

secondary and tertiary markets, where large household appliance enterprises like Gome have rarely entered. But, once these large household appliance giants, including Gome and Suning, start to become intensively involved in these markets, Gree's sales channel system would be inevitably challenged and affected.

(3) It is difficult to coordinate with large-scale household appliance chain enterprises. At the moment, Gree has established the partnerships with Suning and Dazhong, among others, but these are not strategic alliances based on long-term strategic interests; it has an obvious short-term and tactical feature. Moreover, one of the key issues that Gree has to face for a long time is still its relationship with Gome, the largest Chinese retailer.

However, we cannot deny the achievements of Gree mode because of these shortcomings. The market is the only standard to judge a mode. The market data of 10 consecutive years indicate that Gree's sales channel mode is workable and practical under the current market environment in China.

Future research subjects

This chapter has obtained a great deal of first-hand data about Gree's sales channel mode and its specific operation and management through field surveys on Gree's sales companies and their lower-level distributors. Much of these findings are published for the first time. Through substantial first-hand data and in line with the distribution channel theory and the distributors' management theory, we can make a more thorough and objective analysis and assessment of Gree's sales channel mode and provide a more significant reference and lesson for theoretical research and the practical operation of the sales channel.

Due to various reasons, there are still some difficulties to be solved in this study, including:

(1) Under the requirement of keeping internal data in confidentiality, Gree's internal data in 2006 and 2007 were not used.

(2) The number of samples for empirical research is insufficient.

(3) Theoretical research still has some problems in its breadth and depth.

(4) The concept of channel innovation covers a broad range and includes the channel electronization, networking innovation, channel process rebuilding, and a series of other important subjects.

These issues as the future subjects need to be further investigated and discussed.

Appendix: Questionnaires on Retail Channel Relationship

Section one: survey on the relationship between your company and the sales channel of major retailer

Please grade each question according to your company's satisfaction degree with your main retailer (the retailer with the largest sales volume of your products) in the following economic relationships ("1" represents extremely dissatisfied, "2" dissatisfied, "3" indifferent, "4" satisfied, "5" extremely satisfied).

1	Your company's satisfaction with the sales volume of your major retailer	1	2	3	4	5
2	The price of your company's products sold to your major retailer	1	2	3	4	5
3	The profits brought by your major retailer	1	2	3	4	5
4	The total expenses of your company caused by your major retailer	1	2	3	4	5
5	The amount of various rebates demanded by your major retailer	1	2	3	4	5
6	The length of the payment period of your major retailer	1	2	3	4	5
7	The material and manpower support provided by your major retailer for the promotion of your company's products	1	2	3	4	5
8	The training provided by your major retailer to your company's salesmen	1	2	3	4	5

Please grade your company's recognition of your main retailer in the following social relationship ("1" represents extremely disagree, "2" disagree, "3" indifferent, "4" agree, "5" extremely agree).

1	Your major retailer and your company have a friendly cooperation	1	2	3	4	5
2	Your major retailer often respects your company's opinions	1	2	3	4	5
3	Your major retailer does not conceal something that your company deserves to know	1	2	3	4	5
4	Your major retailer and your company can exchange views regularly	1	2	3	4	5
5	Your company gets along well with your major retailer	1	2	3	4	5
6	The relationship between your company and your major retailer is optimistic	1	2	3	4	5

(Cont'd)

7	Your major retailer is a partner worth doing business with	1 2 3 4 5
8	Your major retailer is loyal to your company	1 2 3 4 5

Section two: survey on your company's view on the commercial activities of your major retailer

Please give a score to the frequency of the negative commercial activities conducted by your compan y's major retailer during the cooperation ("1" represents never conducted, "2" conducted seldom, "3" conducted sometimes, "4" conducted often, "5" always conducted).

1	Refuse to order goods from your company, or threaten to cancel an order	1 2 3 4 5
2	Demand your company to cut down product price	1 2 3 4 5
3	Threaten to file a lawsuit against your company	1 2 3 4 5
4	Prolong payment period	1 2 3 4 5
5	Threaten to reduce the order quantity as a punishment for not fulfilling its requirements	1 2 3 4 5
6	Postpone an order as a punishment	1 2 3 4 5
7	Revoke some original arrangements (such as withdraw in-store advertising and shelving position, and so on)	1 2 3 4 5
8	Demand your company to surrender some immediate interests for a long-term interests	1 2 3 4 5

Please give a score to the degree of the following positive commercial activities conducted by your company's major retailer during the cooperation ("1" represents never conducted, "2" conducted seldom, "3" conducted sometimes, "4" conducted often, "5" always conducted).

1	Actively assist your company in in-store exhibition of your products	1 2 3 4 5
2	Provide your company with market information or sales information	1 2 3 4 5
3	Offer suggestions to your company on improving your products	1 2 3 4 5
4	Report to your company the situation of your competitors	1 2 3 4 5
5	Conduct special promotion activities for your company's products	1 2 3 4 5

(Cont'd)

6	Increase the in-store exhibition booth for your company's products	1	2	3	4	5
7	Provide your company with helpful promotion suggestions	1	2	3	4	5
8	Follow through your company's requirements	1	2	3	4	5
9	Use its experience in retail to persuade your company to improve the operation mode	1	2	3	4	5

Section three: survey on your company's view on the legal environment of Chinese retail industry

Please give a score to your company's view on the legal environment of the Chinese retail industry ("1" represents extremely disagree, "2" disagree, "3" indifferent, "4" agree, "5" extremely agree).

1	The *Anti-unfair Competition Law* and the *Contract Law* in China lack restrictions on taking bribes, offering bribes and other commercial bribes.	1	2	3	4	5
2	The *Anti-unfair Competition Law* and the *Contract Law* in China lack restrictions on large retailers' abuse of their dominant position.	1	2	3	4	5
3	China is now lacking laws and regulations on settling the conflicts between retailers and suppliers.	1	2	3	4	5
4	China is now lacking specialized arbitration organizations to solve the conflicts between retailers and suppliers.	1	2	3	4	5
5	At the moment, local and industrial associations fail to better protect the legitimate interests of suppliers.	1	2	3	4	5
6	At the moment, local governments are lacking relevant laws and regulations to restrain unfair transactions.	1	2	3	4	5
7	At the moment, local governments are lacking relevant laws and regulations to restrain monopolistic behaviors.	1	2	3	4	5
8	The *Administrative Measures on Transactions between Retailers and Suppliers* published recently by the Ministry of Commerce of China cannot solve the fundamental conflict between retailers and suppliers.	1	2	3	4	5

Section four: survey on the basic transaction situation between your company and your major retailer

Please use "√" to mark the options meeting your company's situation. Please write down in _____ if needed.

1. Your company is a • production-based enterprise • trade-based enterprise

2. Your company's main products are • liquor and beverage • reconstituted food • snack food • grain and edible oil • daily food • raw and fresh food • cleaning products • family products • cultural and sports supplies • children's products • knitting supplies • household appliance • garment • others

3. Your company's major retailer (the retailer with the largest retail volume of your products) is_____.

4. How many employees does your company have? • fewer than 100 • 100–300 • 300–500 • more than 500

5. Your company is • state-owned enterprise • collectively-owned enterprise • private enterprise • foreign-funded enterprise • enterprises with three types of foreign investment

6. Your company's annual sales revenue is • less than RMB10 million • RMB10–30 million • RMB30~50 million • more than RMB50 million

7. Your company's annual net profit is • less than RMB3 million • RMB3–8 million • RMB8–15 million • more than RMB15 million

8. In the contract between your major retailer and your company, the stipulated payment period is _____months, while the real payment period is _____months. How many months do you think is proper for the payment period? _____

9. The cooperation between your company and your major retailer started in the year of _____.

10. Your company's promotion costs account for _____% of your sales revenue, and the sales costs account for _____% of the sales revenue.

11. Your major retailer and your company meet and communicate _____times each month, and communicate _____times through phones each month.

12. Whether your company's major retailer charged the following fees or discount from your company (including the fees stipulated in the contract and temporarily charged fees). Please mark "√" before the fees or discounts and fill in the proportion, and write down "contractual" or "contemporary" in the blank to indicate whether it is contractual or temporarily.

• Slotting fee, accounting for _____% of the sales revenue • Unconditional discount, accounting for _____% of the sales revenue

• Renew contract fees, accounting for _____% of the sales revenue • Data query fees, accounting for _____% of the sales revenue

• Listing fees for new products, accounting for _____% of the sales revenue • Opening sponsorship, accounting for _____% of the sales revenue

• Sponsorship for anniversary, accounting for _____% of the sales revenue • Loss compensation, accounting for _____% of the sales revenue

(Cont'd)

- The number of promotion is at least _____times/year • Business promotion discount, accounting for _____% of the sales revenue

- Promotion fees for New Year Day, accounting for _____% of the sales revenue

- Promotion fees for the Spring Festival, accounting for _____% of the sales revenue

- Promotion fees for May Day, accounting for _____% of the sales revenue

- Promotion fees for Dragon Boat Festival, accounting for _____% of the sales revenue

- Promotion fees for National Day, accounting for _____% of the sales revenue

- others_____

13. Can the fees provided in the contract be regarded as the commercial briberies taken by your major retailer from your company? • Yes • No

14. Can the above fees charged temporarily be regarded as the commercial briberies taken by the major retailer from your company? • Yes • No

15. Has your company actively paid some fees or provided some preferential policies to your major retailer in order to make them better sell your company's products, or reject the products of your competitors?   • Yes • No

    If it is "Yes," then these costs or preferential policies include_____.

16. Have your company's competitors actively paid some fees or provided some preferential policies to your company's major retailer to compete with your company? • Yes • No

    If it is "Yes," do you think these costs or preferential policies belong to commercial briberies?

    • Yes • No

17. Over the past year, have any of the following conflicts happened between your company and your major retailer?

    • The number of promotion is at least _____times/year • Business promotion discount, accounting for _____% of the sales revenue

File a lawsuit to the court • Withdraw from the marketplace • Fierce quarrel in negotiation • Recover payments leading to the worsening of relationship • Others

18. If your company has conflicts with your major retailer, which of the following would you like to have happen?

    • File a lawsuit to the court • Submit to arbitration authority • Settle privately • Other means

19. Do you think the existing *Contract Law* of China can settle the conflicts between retailers and suppliers?

    • Yes • No

(Cont'd)

20. (Not Compulsory)

The name of your company is_____

Your company's registration address is _____city _____province

Your company's telephone number is _____; the fax is_____

E-mail _____Website_____

The contact person of your company is_____; his or her title is_____

Your title is_____

8
Chapter

Proprietary Brand Innovation and the Performance of Chinese Enterprises

Liu Fengjun, Wang Liuying, Jiang Tao, and Ou Dan

Introduction

Brand is the focus and even the determinant for enterprises engaging in market competition. Some brands witnessed transient glories while some others were able to preserve their reputation. Building a proprietary brand, effectively operating the brand and continuously making explorations and innovations amid market competition have become the common goal of many enterprises.

Proprietary brand is a symbol for national industries, the source for a series of business opportunities, and more importantly the critical condition for national industries to obtain long-term interests during globalization. At present, 10% of global famous brands take up 60% of the market share. Among them, the most famous global brands are less than 3%, but take up as high as 40% of the market share and over 50% of the sales revenue with several even surpassing 90%. In 2004, the U.S. *Business Week* published the world's top 100 most valuable brands, 58 of which were from the United States, 34 from Europe and 8 from Asia. The brands from Asia include Japan's Toyota, Honda, Sony, Canon, Nintendo, Panasonic and Nissan, and South Korea's Samsung. Chinese brands have yet to be ranked among the world top 100 most valuable brands. In 2005, among the world top 100 most valuable brands published by the *Business Week*, 52 were brands from the United States, 9 from France, 8 from Germany, 7 from Japan, 5 from Switzerland, 4 from Britain, 4 from Italy and 3 from South Korea, but none of Chinese proprietary brands were ranked on the list even though China had become the third largest country in GDP and the fourth largest manufacturer in the world. Therefore, the fact that Chinese brands failed to be ranked for eight consecutive years, between 1998 and 2005, in the ranking of the "Global Top 100 Most Valuable Brands" released by the *Business Week*, has reflected the grim reality that Chinese economy is large in scale but not powerful. Furthermore, 170-plus varieties of Chinese products are world leaders in output, but only 4 Chinese brands were listed among the 2005 "World Top 500 Brands" published by the World Brand Lab. Among these Top 500 world brands, 249 were from the United States, 46 from France and 45 from Japan. In this sense, China is a typical "large manufacturer with few proprietary brands." In 2005, China's total export reached USD750 billion, but only 20% of the export enterprises had their proprietary brands, and the export value of these proprietary brands only accounted for 1% of the national total.

In recent years, as China has become a member of the WTO, many Chinese industries are undergoing profound changes. Taking the beer industry as an example, the influx of foreign capital, the increasingly fierce competition, the expanding industry scale, the rising number of enterprises, and the accelerating integration of industries have further improved the industrial concentration. It is worth mentioning that the increasing brand awareness of consumers has exerted more and more influences on the competition of the beer market in China. Brand recognition and reputation have a direct effect on people's buying behavior and on the competitiveness of like products in the market and accordingly on the speed and scale of the development of the beer enterprises. China Resources Snow Breweries (CRSB) is undoubtedly an eye-catching dark horse among Chinese beer brands. Since its founding in 1993, CRSB has carried out a host of mergers and acquisitions (M&A) and become one of the largest Chinese beer production groups. In 2006, the group's sales volume reached 5.3 million tons and one of its products— Snow Beer—alone recorded 3 million tons, both ranking first nationwide. At the same time, CRSB has shaped a brand image recognized and favored by consumers within a few years through successful brand positioning, effective brand promotion and innovative brand management. In 2006, CRSB realized a brand value of RMB11.185 billion and became the fastest growing Chinese beer brand in value. How did they increase brand value during the brand innovation? How did they form a competitive edge over other enterprises through effective brand operation and management? We believe that the experiences of CRSB in brand innovation and brand operation can serve as a reference and inspiration for other Chinese enterprises. This chapter adopts the research method of case study and takes CRSB as an example to analyze and evaluate its industrial environment, brand positioning and brand operation strategy. In addition, this chapter also employs the method of field surveys to assess the brand image of CRSB, and has generalized the valuable experiences of CRSB in innovation, revealed the problems in the practices of business brand innovation, discussed the insufficiency of existing research on brand innovation, and provided future research orientation.

Literature Review

Researches on Chinese proprietary brands have typical Chinese features, each determined by its unique economic situation. China has now become a world trading power, but the problems of low export product quality and weak brand competition edge have remained salient, which has resulted in the passive and weak position of Chinese enterprises in international economic competition. In international exchanges, most Chinese products depend on selling cheap resources and labor force to generate meager trade gains, which is bound to attract the attention of different categories of people to the proprietary brand.

Characteristics of proprietary brand

What is a proprietary brand? Generally speaking, it includes two aspects: first, the independently developed product (the intellectual property rights are self-owned); second, the product is not self-developed, the leading (dominant) rights can still be owned based on the capital nature, which means a brand that a company has the right to use (produce) can also be called a proprietary brand. Some scholars also believe that it is more meaningful to interpret proprietary brand from the perspective of the control rights. They observe that from the perspective of economic benefits, a proprietary brand emphasizes the controllable brand that is to share, control, and dominate the economic benefits brought about by the brand. The control rights over a proprietary brand include two aspects: first, it is the control rights over the intellectual property rights of the brand. At the moment, many companies have the right to use the brand operated by them while the ownership and the right of disposition are controlled by their counterparts, thus this cannot be deemed as a proprietary brand; second, it is the control rights over the brand enterprises, which refers to the rights to make decisions in business asset disposal, business operation and development strategies of the enterprises owning the brand.

One of the key issues in the definition of "proprietary brand" is to lay emphasis on the "proprietary intellectual property rights" owned by enterprises. In a broad sense, an enterprise buys out another's patent or other intellectual property rights, and has the rights to freely dominate and dispose the intellectual results it purchased without intervention by the original patent holder. Hence, the resulting new products can be called the products with proprietary intellectual property rights. In conclusion, no matter whether it is a local independently innovated brand or a brand obtained through M&A, as long as the ownership belongs to domestic enterprises, the brand can be called the proprietary brand of a country. A proprietary brand takes root domestically and is nourished with proprietary intellectual property rights. It is an integrated reflection of business capital, product, quality, technology, market, credibility, after-sale services and many other aspects, and also the demonstration of the comprehensive strength of enterprises and the state. Under the conditions of marketization and globalization, the proprietary brand has important roles; that it is a kind of intangible asset and can exert historical and social influences within a nation, territory, and country.

Studies on the status quo of Chinese proprietary brands

Existing research generally believes that Chinese proprietary brands grow slowly and their major obstacles include: first, Chinese enterprises prefer to cooperate with multinational corporations, giving insufficient attention to the building of proprietary brands; second, Chinese enterprises are satisfied with the identity of being at the low-end production chain in the international division of labor, lacking confidence in building proprietary brands; third, Chinese enterprises are still unfamiliar with the theory of brand building and also lack the related experiences.

China's weakness and shortage in proprietary brands has a significant impact on Chinese economy. First, some scholars believe that whether a country owns and how many proprietary brands it owns symbolize the country's international competitiveness from a national perspective. If Chinese enterprises intend to occupy the high end of the industrial chain in international competition, they must focus on independent innovation and brands with proprietary intellectual property rights. As such, enterprises would have a strong competitive edge and could also improve China's international competitiveness. The lack of internationally competitive proprietary brands will not only lead Chinese enterprises to a disadvantageous position in the international market competition and reduce their export gains, but also influence the upgrading of industries and China's position in the international division of labor. This would fundamentally put a stop to China's leapfrog from a large exporter to a powerful exporter. Second, from the perspective of enterprises, the building of proprietary brands can promote sustainable business development, form a competitive edge and shape core competitiveness to obtain brand earnings, such as profit margin, sales volume and market share. In addition, Chinese enterprises can also develop toward the high end of the value chain through brand building and product-added value generation. In this connection, the essence of building Chinese proprietary brands is not just a matter of national sentiment, but also of the business profit-making model.

Studies on the building of a proprietary brand

How does one build proprietary brands of enterprises in facing of the harsh reality of China's weakness in proprietary brands? Existing research is mainly conducted from three perspectives: enterprises, consumers' involvement, and government guidance.

Studies from the perspective of enterprises

Some scholars have discussed issues about proprietary brand building of the automobile industry through theoretical analyses. They observed that when enterprises integrate brands and the source of technology, there are three models of proprietary brand development: (1) create an original brand by integrating independent research and development (R&D) with the introduction of some foreign technologies; (2) establish an original brand via obtaining technologies through merging and acquiring foreign automobile design and R&D companies; (3) merging and acquiring foreign finished automobile manufacturers to obtain technologies and using acquired brands to introduce new products or original brands.

There are also some scholars who use western brand theories and models to study issues surrounding building proprietary brands. They put forward that, by using consumer-based brand equity model (CBBE), building a strong brand needs four steps: building correct brand logo, creating appropriate brand connotation, guiding correct brand response, and building an appropriate consumer-brand relationship. At the same time, these four steps depend on the six dimensions of brand building: significance, performance, image, judgment, feeling, and resonance. Of these, significance applies to brand logo, performance and image to brand connotation, judgment and feeling to brand response, and resonance to brand relations.

Studies from the perspective of consumer involvement

Considering the latest development trend of western brand theories, consumer-involved brand building is a key means to build a strong brand image. According to Kevin Lane Keller (1993), one of the core scholars in international brand research, consumer involved brand building means that all dimensions of business brand building are closely linked with consumers and should leave positive brand images in the mind of consumers. Keller mainly defines the dimension of brand images from the perspective of brand association. He noted that brand image takes shape in the long-term contact between customers and the brand, reflects the recognition, attitude and emotion of customers toward the brand, and also heralds the future behavior tendency of customers or potential customers. Brand association has reflected the brand image as a whole and determined the position of a brand in the mind of customers. Considering the feasibility and effectiveness of the building of a brand image, the discussion of Alexander L. Biel (1992) about brand image can be taken as the theoretical basis. Specifically speaking, a strong proprietary brand image can be built from the following three aspects: (1) building a strong enterprise image; (2) building a strong product image; (3) building a strong user image.

Studies from the perspective of government guidance

Some scholars have focused on the government-guidance perspective to explore ways to effectively implement a proprietary brand strategy, from properly handling the relationship between government, enterprises and consumers, through theoretical analyses. They believe that enterprises represent the mainstay of proprietary brand building while consumers are the key for the success of proprietary brand building. They are all the results of autonomous selection, and the government needs to guide enterprises and consumers to meet the target of building proprietary brands through propriety brand strategy. The specific measures include: (1) encouraging enterprises to create and build proprietary brands; (2) creating an enabling external environment for proprietary brand building; (3) properly guiding the demand of consumers.

Other scholars have analyzed through empirical research on endogenous choice mode the proprietary brand building of Chinese enterprises to realize the macroscopic issues of upgrading industries and developing a sustainable economy, which in a sense, has provided a reference to the formulation of governmental policies. They believe: (1) the government needs to give priority to durable products in industry selection while guiding enterprises to create international brands; (2) although China has become a member of the WTO, it is still necessary to properly protect Chinese enterprises in the domestic market and maintain their market share while conforming to the rules of the WTO; (3) great efforts need to be made to eliminate the domestic market segmentation and form a national, unified, big market; (4) protections need to be strengthened in terms of the business secrets of domestic enterprises, not only for intellectual property rights, but also for various implicit knowledge of enterprises. At the same time, the government has the obligation to encourage the general public to strengthen their involvement and understanding of domestic brands, and guide them to (a) choose national brands and (b) take the lead to purchase inexpensive and quality domestic brands.

The Current Development of the Chinese Beer Industry

The Chinese beer industry has developed steadily over the past few years. Toward the end of December 2006, there were 580 enterprises above the state-designated scale in the entire industry, 11 more than that of the same period in 2005. The growth rate slowed down compared with previous years, but it has remained the growth trend since 2002 (see Fig. 8.1 and Fig. 8.2).

At present, the per capita beer consumption amount in China is nearly 22 liters, but the figure is only about 10 liters in mid-western China and it is less than 5 liters in the rural areas, with a population of more than 800 million. In contrast, the world average level is presently 30 liters and the per capita annual

Fig. 8.1. **The number of beer enterprises nationwide between 2001 and 2006**

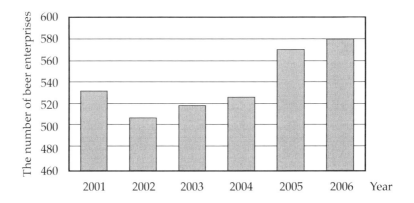

Source: Prepared based on the data from www.drcnet.com.cn.

Fig. 8.2. **Accumulative assets amount of the beer industry between 2001 and 2005**

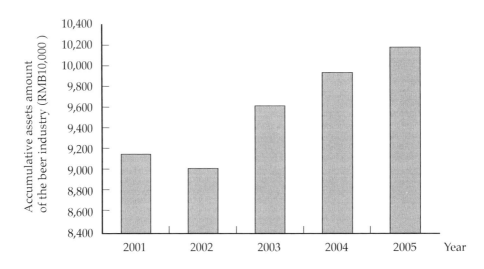

Source: Prepared based on the data from www.drcnet.com.cn.

consumption in Europe and the United States has surpassed 80 liters; Among them, it is more than 159 liters, 135 liters and 117 liters in the three largest beer-consuming countries of Czech, Ireland and Germany, respectively. The per capita consumption amount in China is only about one-eighth that of Czech, one-fourth that of the United States, and two-fifths that of Japan. If the per capita beer consumption in China reaches the world average level of 30 liters, it will bring about nearly 40 million tons of production and RMB100 billion in sales revenue. During this continuous increase of market capacity, enterprises have started to increase the production capacity. Fig. 8.3, Fig. 8.4 and Fig. 8.5 have demonstrated the growth of the output of the Chinese beer industry. In 2005, the Chinese beer industry realized an output of 30.61 million tons, which was 1.52 million tons more than during the previous year. In 2006, the Chinese beer industry witnessed a continuous growth and recorded an output of 35.15 million tons, up by 14.8% year-on-year, making China the world's largest beer producer and consumer and the fastest growing market for 4 years in a row.

The increase of the Chinese people's demand for beer has led brewers to make investments in fixed assets. The three leaders in the Chinese beer industry, Tsingtao Beer, Yanjing Beer and Snow Beer have expanded their production capacity. Fig. 8.6 showcases that the sales of Chinese beer industry witnessed a high growth rate, even in 2003 when China was in the grip of the "SARS epidemic." The annual sales revenue of beer grew by 8% in 2002 and 2003,

Fig. 8.3. Trend of Chinese beer sales volume in the recent six years

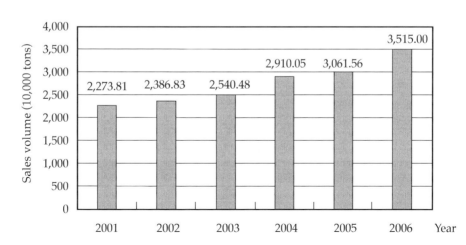

Source: Prepared based on the data from www.drcnet.com.cn

Fig. 8.4. **Growth rate of the beer sales volume in China between 2001–2006**

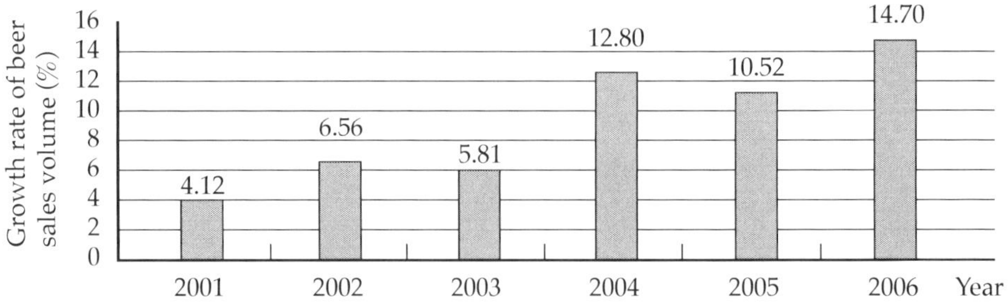

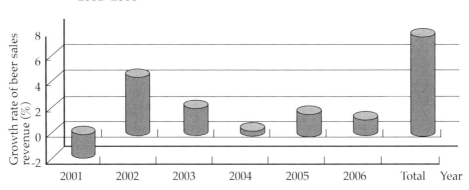

Source: "Re-analysis of Chinese Beer Industry in 2006," *Chinese Wine* 10 (2007).

Fig. 8.5. **Growth rate of the sales revenue of Chinese beer between 2001–2006**

Source: Same as Fig. 8.4.

and 16%, 17% and 15% in 2004, 2005 and 2006, respectively. This indicated that Chinese beer industry still had plenty of room for development as the macro-environment improved.

In 2004 and 2005, the revenue of listed beer companies grew faster than the industry average, but the growth rate of the profit was slower than that of the industry average. This was mainly influenced by the high cost and low price of the beer products (Fig. 8.6).

There are three factors for the rise of costs: first, due to the impact of raw materials and energy, Chinese beer enterprises have continued to grapple with stubbornly high costs and lower sales prices, thus their profits have grown slowly. Second, the expenses on management have picked up and those of listed

Fig. 8.6. Accumulative sales costs and expenses of Chinese beer industry between 2001–2005

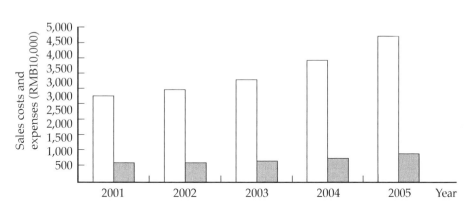

companies are more than those of unlisted companies. This is mainly because listed companies expanded their business on a large scale, after going public for financing, and both the good and bad acquired enterprises are intermingled, which have dragged down the performance of listed companies. Therefore, compared with listed companies, the expenses of unlisted ones are relatively lower. Finally, as companies have recognized the role of marketing in the competition of homogeneous products and the necessary investment for entering into new markets, the beer enterprises, especially national and cross-regional beer enterprises, have also seen higher increases in sales costs.

The beer industry is a capital-intensive industry featuring low profits. If scale advantage cannot be formed effectively, it will be difficult for them to survive in this fierce competition. The strategic advantage of using capital merger and acquisition to improve scale competitivenes s and then to enhance their business capacity to make profits has become increasingly obvious (see Fig. 8.7 and Fig. 8.8).

We can see from Fig. 8.7 and Fig. 8.8 that the proprotion of the total assets of the top 10 beer enterprises in the Chinese beer industry has risen from 23% in 2003 to 26% in 2005; the industrial concentration keeps rising. In addition, the profits of the top 10 beer enterprises have reached over 60% of the total industrial profits and the advantage in industrial concentration has been obviously demonstrated. During the industrial integration, the number of beer enterprises rose from 503 in 2002 to 580 in 2006. This figure keeps rising, which is a result of the industry's development, the increase of market capacity, and the expansion of the entire industry.

It can be said that the development of the beer industry is the integration of

Fig. 8.7. **The proportion of the total assets of top 10 enterprises to the industry's total**

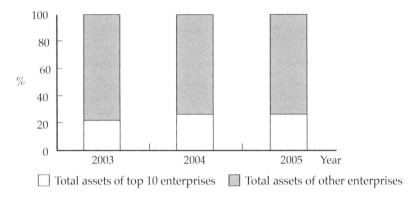

Fig. 8.8. **The proportion of the total profits of top 10 enterprises of the beer industry to the industry's total**

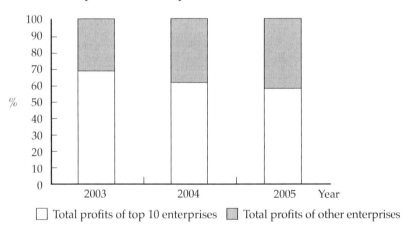

the beer industry. The beer industry of the United States took two centuries to advance from its initial stage to its mature stage. The number of American beer enterprises was once as many as 4,131, but this figure is now only 7. The number of Chinese beer enterprises was once as many as 1,000, but it took China 20 years to integrate the 1,000 beer enterprises into 500. The United States also spent 20 years to integrate 1,000 of its beer enterprises into 500.

The rapid integration and development of the Chinese beer industry is closely linked with the existence of the huge amount of foreign capital in the Chinese beer market. Foreign enterprises have brought about capital, technologies and marketing experiences, which have helped to accelerate its speed of integration and development.

The Competitive Landscape of the Chinese Beer Industry and the Market Standing of Snow Beer

During the integration of the beer industry, three groups have taken shape over time. The first group is the very large beer enterprises, led by "Tsingtao Beer," "Yanjing Beer" and "Snow Beer"; the second group is the large enterprises led by "Zhujiang Beer," "Harbin Beer," "Chongqing Beer," "Kingway Beer," and "Huiquan Beer"; the third groupis composed of other small, regional beer enterprises.

We can see the general tendency of the beer industry from the market share and the total profits. The market share of the three market leaders rose from 32% in 2003 to 36% in 2005. The market share of the top 10 reached 62.4% in 2006, up by 1.4% from 2005. The proportion of the profits of the top 10 enterprises recorded as high as 60% of the industry total. The oligopoly landscape of the Chinese beer industry has taken initial shape; however, compared with other consumer markets in China, the process of centralizing was obviously slower. This is mainly because of the increased difficulties after a large number of M&A, as well as the localized taste for beer of consumers, thus hindering the increase of market centralization.

From the territorial perspective, the three leaders have their core regions forming an industrial, competitive landscape in which Tsingtao Beer dominates the market of Shandong and Shaanxi, Yanjing Beer dominates the market of Beijing and North China, and Snow Beer dominates the market of northeast of China and Sichuan. In addition, these three groups are infiltrating each other's markets, creating an interpenetrative landscape. Yanjing Beer, Harbin Beer, China Resources Breweries and Tsingtao Beer all have their production bases in the northeast of China, but the strongest player is Snow Beer of China Resources Breweries in Shenyang. North China is basically the stronghold of Yanjing Beer, but after Tsingtao Beer acquired Beijing Five Star and Three Ring Beer, formerly controlled by Pan Asia Investment Group, the two entered into fierce competition in the Beijing market. Yanjing Beer taking Beijing and North China as the core markets puts forward the strategic plan of consolidating the market of Beijing, targeting North China and exploring the market nationwide.

The three leading beers have their distinct characteristics in development and compete fiercely. Tsingtao Beer has chosen to cooperate with strong foreign investment, realized the scale effect, and successfully accomplished the transformation from production-based enterprise to market-based enterprise by leveraging on the advantages of their brand and the capital and technology of foreign capital. Tsingtao Beer shared the middle and high-end beer market with Budweiser Beer by capitalizing on its brand advantage, and restructuring its product while capitalizing on the stable growth of this market. At the same time, Tsingtao Beer has also widened the gap with other competitors in China's middle and high-end market, and its profits continue to grow steadily.

Yanjing Beer has practiced the main-brand and sub-brand development strategy (Yanjing+Huiquan Beer, Yanjing+Li Quan Brewery, Yanjing+Snow Deer Beer) and has produced preliminary results. Yanjing also leveraged on the Olympics marketing opportunity to expand its brand influence in the hope of developing Yanjing Beer into a national brand. But, Yanjing Beer has shown a slow development trend.

China Resources Snow Beer is the product of the cooperation between China Resources Group and South African Breweries (SAB). In December 1993, the early days of its founding, the annual output of CRSB was less than 200,000 tons. After an array of large-scale M&A, CRSB has become one of the largest Chinese beer groups. From the mushroom strategy in the early days to the strategy of making its layout along river areas and costal areas, and then to the market decision of making Snow Beer the leading brand in 2002, CRSB has consistently committed to brand marketing. That means CRSB has taken sustainable development as the most important operation target but not just focused on the short-term strategy. In each region, CRSB has always been able to transfer regional brands it acquired to its main brand, Snow Beer, and followed the strategy of being the largest producer, to the largest sales volume, and ultimately developing snow beer into most famous brand in the local market. As of 2004, CRSB has been the beer-producing frontrunner in most Chinese provinces. In 2005, CRSB gained momentum and its sales volume neared that of Tsingtao Beer. In 2006, CRSB recorded a sales volume of more than 5 million tons nationwide (see Fig. 8.9) became the first to surpass 3 million tons, ranking top across the nation. From the perspective of market share, CRSB took up 15.1% of the total, surpassing Tsingtao Beer for the first time, and changed the landscape of the Chinese beer industry (see Fig. 8.10). In the same year, Snow Beer's brand value reached RMB11.185 billion, becoming the fastest-growing beer brand in China.

The top three beer enterprises in China have relied on acquisition to support the expansion of sales areas in recent years. What is different is that the immersion of CRSB after acquisition is more effective. Industry professionals believe that the strong point of CRSB is that it can integrate the acquired factories into its operations and put in place a unified strategy and management after it enters into a new region. It should be noted that CRSB has practiced

Fig. 8.9. **Growth trend of the sales volume of the three beer leaders between 2002 and 2006**

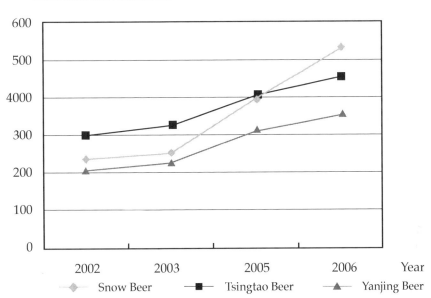

Source: Prepared based on the data of the Beer Branch of China Alcoholic Drinks Industry Association.

Fig. 8.10. **The market share of the three beer leaders in sales volume between 2003 and 2006**

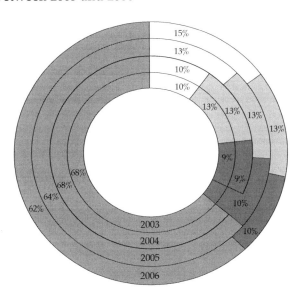

Snow Beer   Tsingtao Beer   Yanjing Beer   Others

unified mangement in purchases. At the moment, more than 95% of the raw materials of the 50-plus factories across the nation come from the group's unified purchases, which has guaranteed not only the price but also the quality and a stable supply of goods.

In early 2006, CRSB adjusted its strategy, extended beyond the central coastal cities, and adopted the strategy of a nationwide production layout. For instance, the group started to build or purchase factories in Inner Mongolia, Shanxi and Gansu, and realized a broader development as a result. They also focused on the integration of its acquired enterprises while engaging in expansions and made investments in reconstruction of acquired factories to improve product quality and strengthen professional skills. The 10 factories acquired by China Resources Snow Beer have been transformed from their previous states; many have witnessed expansion and reequipment several times, where investment has been heavy. Additionally, whenever entering into a region, the group will build its own newteam. CRSB will send general managers to regional companies and set up the management team based on individual capacities and qualities. At present, the management teams of its companies are composed of three parts: first, the management staff from the acquired factories; second, externally recruited employees; third, internally promoted employees. The group is good at assimilating excellent mangement methods and modes of production from the acquired enterprises. Therefore, the more factories the group has purchased, the more scientific and efficient the management becomes, which should result in a very strong team. In so doing, CRSB has equipped itself with a greater basis and capacity for development.

Brand Innovation and Growth of Snow Beer

Capital operation and brand operation under the mushroom strategy

Capital operation is the objective basis for brand operation. For China Resources Snow Beer, the mushroom strategy has become the effective mode for realizing an integration of capital operation and brand operation.

It is natural that the distribution costs will go up if the sales radius of beer is too large. If there are no factories within a given region, it will be difficult to take up a large share of market. In this case, there is a need to set up one's own factories within the region if it wants to enter into the market.

CRSB created a "mushroom strategy" after it straightened out the brands it acquired in different regions and realized the scale growth. The mushroom strategy is to set up a factory within a region and then to generate a large share of the market; when the "planting of a mushroom" succeeds within one region, this successful experience will be replicated in other regions, and different regions could realize mutual support. At an appropriate time, these mushrooms connect together form one big mushroom. The essence of the mushroom strategy is to transfer its sub-brands into famous local brands, then popularize one or two of these sub-brands nationwide. The mushroom strategy is the basic strategy of Snow Beer, which has closely linked the Snow Beer's target markets, sales channels and brand strategy together.

Since 2001, CRSB has been engaging in an astonishing large-scale expansion by leveraging on its capital and technology. It has established more than 50 breweries, including China Resources (Dalian), Lanjian China Resources (Sichuan), China Resources (Jilin), China Resources (Tianjin), China Resources (Anshan), China Resources (Anhui) and China Resources (Heilongjiang), owning more than 30 regional brands.

Product innovation and brand integration

Target market and brand positioning

In 2004, CRSB (China Resources Breweries at that time) cooperated with Kotler Marketing Group to set up a project team who identified the core of Snow Beer brand as "growth" after 4month-long surveys and analyses, and determined the target consumers as those between the ages of 20 and 35 who are are enterprising and spirited. The new Snow Beer tries to deliver a message that "Snow Beer is a partner to accompany and encourage those between 20 and 35."

The core aspiration of Snow Beer is to "enjoy growth"; that is also the main content of their brand messaging. The target customers of Snow Beer are passionate and enterprising young people who are between 20 and 35, dare to challenge themselves, and know how to enjoy their lives. At the same time, young people at this age are the driving force for the current social transformation and the future of the nation. As such, they are also facing realistic challenges and great pressures, and only those with innovative thinking and an unyielding spirit could become strong. Cellphones, the Internet, private cars, and private houses are developed under the drive of these young people. Therefore, Snow Beer is determined to attract this group of people to become its target customers, and promote and encourage them to choose their own way of life — working hard at their jobs, daring to face challenges, and actively consuming in their lives.

After defining the target market and clearly positioning the brand, various kinds of market promotion activities were carried out by Snow Beer. After entering into the national market, Snow Beer promoted the idea of "making beer for the young generation" and delivered this message to its target customers. The annual sweepstakes event of "Snow Beer, No Supers" is organized independently by China Resources Snow Beer. The event is mainly carried out by means of building direct relationships with its consumers and, after strict screening, the grand prize winners can take part in a global adventure together with scientists, a Discovery film crew, and outdoor explorers.

"1 + N" brand strategy

The Chinese beer industry has a distinctive regional characteristic, and regional brands experience high popularity among their region's abundant loyal consumers. In order to give equal consideration to regional and national brands and meet various requirements from different customers, CRSB has introduced the "1 + N" brand portfolio strategy that is the "national brand + regional brands" (regional strong brand and regional tactical brands) strategy. "1" refers to one main brand—Snow Beer, "N" refers to N regional brands, including the regional strong brands Lanjian, New Shanxing, Huadan and Taihu Lake. The "1 + N" brand portfolio strategy has guaranteed that Snow Beer can adopt different development strategies for different brands, maintained a clear-cut distinction between the national brand and regional brand, and promised to not allow its brands to engage in price wars. At present, CRSB has founded more than 50 breweries nationwide and has 30-plus regional brands. The brand strategy implemented by CRSB can help it consolidate the local market share of regional beer brands and at the same time drive the sales of Snow Beer across the nation.

Brand innovation

The product innovation of CRSB is mainly demonstrated in two aspects: product variety and product package. When it comes to the variety, Snow Beer has designed diversified varieties of products for consumers with different demands (see Fig. 8.11).

There is a growing story in the beer industry: if the trademarks of different brands are torn off, it would be difficult for even directors of these beer breweries to distinguish their own brands from others just by tasting the products. At present, the technological barriers in the beer industry are disappearing, the trend of homogenization is growing, and regional obstacles

Fig. 8.11.   Products of China Resources Snow Beer

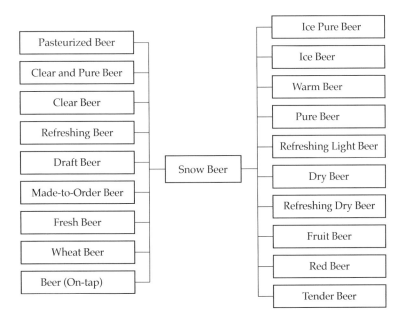

have increasingly become a big problem for merchants. Within these contexts, vivid packaging and logos are not only a visual enjoyment for consumers, but also a striking identification mark for a brand to stand out from its competitors. Modern product packaging communicates with consumers in brand spirit and has been increasingly valued by merchants.

Snow Beer has fought an orderly war for expansion by updating its packaging. Before New Year's Eve 2005, Snow Beer invested a lot of money to quickly change the packaging of more than one million tons of beer. Different from the previous trademark that separates the English words "SNOW" from the Chinese words "雪花啤酒 (Snow Beer)" on two sides, the new packaging integrated "SNOW" with "雪花啤酒" and added a dark green contrasting border outside of the new logo, strengthening the visual identity; furthermore, sharp edges and simple lines were also added to the design, which highlighted the key characteristics of Snow Beer—young and fashionable. The new packaging also made clever use of the orange and green of its standard colors, which have fully displayed the lively and enterprising personality of Snow Beer and its consumers. The brand-new design of the packaging of Snow Beer not only has a strong visual identity but also enhances its visual salience. This undoubtedly laid a foundation for Snow Beer to become a national brand.

Brand communication

Integrating the master brand into the company name

Generally, in media reports, the name of an enterprise is much more frequently mentioned than its brands. In order to effectively spread Snow Beer brand, China Resources Beer (China) Limited Company formally changed its name to China Resources Snow Beer (China) Limited Company on August 10, 2004.

China Resources Beer had been regarded for years a brand lacking brand influence by industry insiders, and although having acquired over 30 regional strong brands, it still lacked a nationwide influential brand. To this end, China Resources Beer announced in 2001 to commit itself to establish Shenyang Snow Beer, purchased in 1994, into a national brand. This renaming linked the Snow brand to its company name. This heralded the new starting point of the brand development of Snow Beer and created favorable conditions for further improvement of its brand image and brand value.

Non-Olympics marketing

Corporate sponsorship for sports events like the Olympics is a good opportunity not only to make a brand better known, but also to make profits. Samsung is a good example of a company that improved its brand image by leveraging on the Olympics.

The 2008 Beijing Olympic Games was an unavoidable topic for all brewers because "Olympics," "revelry" and "beer" seem to be closely linked with each other. The Olympics was the carnival for all; as a Chinese poem says, "singing while drinking during daytime for expressing joyful feelings," and beer can be quite effective for releasing these joyful feelings. Because Yanjing Beer and Tsingtao Beer foresaw this huge opportunity for sales growth in the beer market, they made a substantial investment in sponsoring the Olympics. However, a challenging question facing Snow Beer was how to present itself during the Olympics.

The Olympic Games are splendid, not only because of sports stars, gold medals or the sports themselves; the games are spectacular because of the enthusiastic involvement of the public. This is the fundamental basis for which manufacturers sponsor the Olympic Games. The marketing focus of general sponsors of the Olympics is to try to influence consumers indirectly through influencing the Olympic Games. However, consumers generally do not care about who sponsors the Games. What they care about is the benefits they can gain. That is to say, consumers would be more satisfied if beer brewers brought more direct benefits to them during the bidding and hosting period of the Olympic Games. Given this, Snow Beer chose to focus on and communicate with their consumers directly. Because the beer lovers who were benefitting from Snow Beer's direct marketing were also involved in the Olympic Games, Snow Beer, as "a formal partner of beer lovers" demonstrated its support for the Olympic Games. In a sense, the indirect Olympic marketing of Snow Beer, who was not a direct sponsor of the Games, demonstrated the spirit of the Olympic Games—"Nothing is more important than participation."

In this connection, the non-Olympic marketing of Snow Beer does not mean not participating in the Olympics, but is a term proposed by comparing it with Olympic marketing. If "Olympic marketing" was meant to cater to the Olympics, center on the Olympics, support the Olympics and influence consumers, then "non-Olympic marketing" is to target consumers, center around consumers, support consumers and leverage on the Olympics.

With the help of the China Central Television (CCTV) platform

One of the important components of the brand-building of Snow Beer was to extend its market nationwide with the help of the national authoritative media platform.

Undoubtedly, the FIFA World Cup in 2006 was one of the most influential social events of the year; it caught the eyes of the people across China in June and July that year, and became the most popular media event. Making use of sports events to conduct brand marketing can infuse a younger, healthier, and more enterprising spirit into the brand, effectively corresponding to public sentiment and also assisting other promotion activities for integrated brand communication.

CCTV has always been the first choice of major sports events for the Chinese mainland and also the core channel for Chinese viewers to get first-hand information about international sporting events. After gaining the exclusive broadcasting rights of the South Korea World Cup in 2002 and the Athens Olympic Games in 2004, CCTV purchased the exclusive rights to broadcast in China the 2006 FIFA World Cup. The prime time of CCTV 1, equipped with an unprecedented driving force for market and channel expansion, became the top choice for CRSB to promote its products. Snow Beer's popularity increased rapidly after capitalizing on its intensive investment in advertising and the 96% audience ratings of CCTV in China. In May 2006, Snow Beer started to purchase advertising during prime time on CCTV 1. Since then, CRSB has sponsored the

CCTV program *FIFA World Cup for Fans* during the World Cup days and offered Snow Beer as the prize. *FIFA World Cup for Fans* invited beer lovers to the studio to engage in a series of activities to demonstrate their knowledge and football skills, which doubled as a way for fans to participate in this festival, cheer for the World Cup, and demonstrate the value of fans during the World Cup season. Apart from sponsoring this program, CRSB also offered RMB1 million as a grand prize for fans who had participated in the program's activities, as well as for those viewers who had sent text messages during the event.

Integrating the brand spirit of "No Supers"

The efforts made by CRSB in brand operation can be demonstrated in its annual activities "No Supers." The significance of "No Supers" for CRSB is to bring about fundamental progress in its brand operation.

In 2005, CRSB launched the first "Snow Beer, No Supers" exploration, which was carried out in key regions—Liaoning, Jiangsu and Wuhan—and gained great popularity among consumers. It was reported that more than one million consumers actively responded to this activity. Through cooperation between the Discovery film crew and well-known scientists as well as explorers, the popularity and reputation of Snow Beer have been improved significantly.

In 2006, CRSB, with the assistance of the CCTV program—*Recovering The Yangtze River,* launched the second "Snow Beer, No Supers—Exploring the Source of the Yangtze River." During the 15 days of exploration, CRSB organized "Team mission, I Go, I Shoot Films" and other diversified team-based activities as well as a series of public benefit activities, such as taking an oath for environmental protection and holding a ceremony of water sampling at the source of the Yangtze River. As a result, "Snow Beer, No Supers" activities had attracted great attention across the nation and its popularity and influence had improved significantly.

It is worth mentioning that CRSB launched a product named "No Supers" to the national market by taking advantage of the popularity of this annual activity; because of this, the message delivered by "No Supers" beer had attracted customers who love challenges and have high aspirations.

Channel expansion

CRSB started to upgrade and improve its sales channel in 2002. They adopted the channel strategy of "deep distribution, terminal maintenance" based on the traditional channel mode.

Deep distribution adopts the "manufacturer–operator–distributor–terminal" mode—that is, the original wholesalers are transformed into operators mainly responsible for transportation. Manufacturers directly provide management and services in marketing, and the sales managers of Snow Beer go directly to retail shops and restaurants, among other terminals to market, serve, display and maintain the price system. This has been the most critical sales mode of CRSB in its main market, through which Snow Beer maintained its brand images and market order.

Based on the mode of deep distribution, CRSB also established the coordination-based exclusive distribution mode. This mode includes two aspects: operator exclusive distribution and distributor exclusive distribution. The manufacturer works together with distributors to set up a sales team, carry out joint management and integrated operation, and manage the terminals according to the mode designed by CRSB. The manufacturer, operators, and distributors have formed a strategic partnership, the sales channel was further consolidated, the relationship between enterprises and channel members became closer, the loyalty and trust for CRSB further improved, and the control power of the enterprise over the channel became even stronger.

CRSB choses the direct distribution mode for key retail shops, which meant the sales managers of CRSB directly received orders from them and took charge of the delivery and terminal display.

The most outstanding characteristic of these distribution modes of CRSB is the improvement of the control power over the terminal market and the firm control of the order of the entire system by the CRSB. But, compared with the traditional extensive sales channel mode, these modes of CRSB require the enterprise to have stronger executive force.

Comprehensive Assessment of the Brand Image of Snow Beer

In order to conduct an in-depth and comprehensive examination of the effects of Snow Beer's brand innovation and brand operation as well as brand management from the perspective of consumers, this section adopted the means of questionnaires and surveys to comprehensively assess the brand image of Snow Beer versus other well-established Chinese beer brands. The questionnaire and surveys were conducted in Beijing between November 2006 and March 2007. The respondents included MBA students, postgraduates, community residents, restaurant consumers and customers in supermarkets. The respondents were selected according to three requirements: (1) the respondents were over the age of 18; (2) the respondents and their family members were not working in the areas of marketing, sales, or advertising; (3) the respondents had not accepted any market surveys over the previous six months. During the survey, 280 copies of questionnaires were handed out and 206 questionnaires were collected; the response rate was 73.6%. 167 copies were deemed effective; the main reason for being ineffective is the questionnaire was not completely filled out.

The composition of consumers being surveyed among the effective questionnaires is shown in Table 8.1.

Before formal data processing, the reliability of the questionnaires needs to be tested by means of Cronbach's α coefficient. The test results are shown in Table 8.2. After calculation, the Cronbach's α coefficient is 0.952, the Standardized Cronbach's α coefficient is 0.953, much larger than 0.7, so the internal consistency of the questionnaire is good and the questionnaire can be effectively used to assess the brand image of the beer industry.

Table 8.1.   The composition of effective respondents

Demographic Variable		Percentage (%)	Demographic Variable		Percentage (%)
Sex	Male	77.2	Individual Annual Income	Less than RMB30,000	30.5
	Female	22.8		RMB30,000–50,000	22.2
Age	Below 25 years old	28.1		RMB50,000–100,000	37.7
	25–35 years old	65.9		RMB110,000–200,000yuan	9.0
	36–45 years old	3.0		More than RMB200,000	0.6
	46–60 years old	2.4			
	Above 60 years old	0.6			
Education Background	High school or below	1.8			
	Technical secondary school	1.8			
	Associate degree	4.2			
	College degree	29.3			
	Master degree or above	62.9			

Table 8.2.   Reliability test of the scale

Cronbach's α	Cronbach's α based standardized items	No. of items
0.952	0.953	17

According to the questionnaire's test of beer brand image, factor analysis is used to calculate the weight of different test indices and in turn the comprehensive score of the brand image of Snow Beer can be calculated. Firstly, KMO and Bartlett's test (see Table 8.3) on the data of Snow Beer is carried out and the test results indicate that the data can be used to calculate the weight of various indices through factor analysis.

Table 8.3.   KMO and Bartlett's test

Kaiser–Meyer–Olkin Measure of Sampling		
Adequacy		0.895
Bartlett's	Approx.ch–Square	4,307.829
Sphericity	df	861
	Sig.	0.000

In the factor analysis, factor loading matrix is obtained through factor rotation. The matrix points out the relations between factors and each test index. Table 8.4 is the factor loading matrix after factor rotation. After the factor rotation, the factor loading matrix of a test index, i.e. "I was deeply impressed by many advertising slogans of the brand," is less than 0.5, and thus it is eliminated.

Table 8.4.   Factor rotation matrix

Test index	Component		
	1	2	3
It will be a pity to not buy this brand (Snow Beer)	0.794	0.070	0.190
The production halt of this brand will make a great difference in my life (Snow Beer)	0.785	0.149	–0.012
This brand is one of my "loyal friends" (Snow Beer)	0.772	0.290	0.082
I have a strong loyalty to this brand (Snow Beer)	0.770	0.251	0.073
This brand is very important in my life (Snow Beer)	0.760	0.190	0.172
I will recommend this brand to other people (Snow Beer)	0.573	0.406	0.067
This brand has a high stability of quality (Snow Beer)	0.136	0.757	0.207
This enterprise upgrades its products rapidly (Snow Beer)	0.173	0.750	0.192
This enterprise has an abundant varieties of products (Snow Beer)	0.134	0.685	0.230
The products of this brand taste good (Snow Beer)	0.426	0.657	0.081

(Cont'd)

Test index	Component		
	1	2	3
I have been deeply impressed by many of the advertising messages of this brand (Snow Beer)	0.311	0.400	0.209
The cost performance of this brand is rather high (Snow Beer)	0.147	0.045	0.780
I can find the products of this brand in various shops quite often (Snow Beer)	−0.003	0.084	0.688
I feel the advertisements of this brand are reliable (Snow Beer)	0.034	0.213	0.681
I have a good feeling when drinking the products of this brand (Snow Beer)	0.173	0.304	0.645
This brand plays a leading role in its industry (Snow Beer)	0.144	0.315	0.589

Notes: 1. Extraction method: principal component analysis.
  2. Rotation method: varimax with kaiser normalization.
  3. A rotation converged in 5 iterations.

We can see from Table 8.5 that the first factor extracted from factor analysis explains the 37.44% of the variance of the brand image, the second factor explains 12.35% of the variance and the third factor explains 7.25% of the variance. According to the data acquired by the questionnaires, the final comprehensive score of Snow Beer's brand image is 17.27. Using the same method to calculate the comprehensive scores of the image of other brands like Tsingtao Beer, Yanjing Beer, Zhujiang Beer and Budweiser are 20.0, 17.13, 10.73, and 13.37 respectively. The results show that the comprehensive image of Snow Beer ranks top among consumers' assessments, which basically coincide with the assessment report on the value of beer brands made by Beijing Famous Brand Evaluation Co., Ltd in 2006.

Table 8.5. Factor explained variance

Component	Initial eigenvalues			Extraction sums of squared loadings		
	Total	% of variance	Cumulative %	Total	% of variance	Cumulative %
1	6.365	37.444	37.444	6.365	37.444	37.444
2	2.100	12.352	49.795	2.100	12.352	49.795
3	1.232	7.250	57.045	1.232	7.250	57.045
4	0.986	5.798	62.842	—	—	—
5	0.881	5.184	68.027	—	—	—
6	0.694	4.081	72.107	—	—	—
7	0.685	4.032	76.139	—	—	—

(Cont'd)

Component	Initial eigenvalues			Extraction sums of squared loadings		
	Total	% of variance	Cumulative %	Total	% of variance	Cumulative %
8	0.620	3.647	79.786	—	—	—
9	0.547	3.220	83.006	—	—	—
10	0.466	2.744	85.750	—	—	—
11	0.458	2.693	88.443	—	—	—
12	0.431	2.533	90.976	—	—	—
13	0.403	2.373	93.349	—	—	—
14	0.364	2.143	95.491	—	—	—
15	0.287	1.687	97.179	—	—	—
16	0.269	1.581	98.760	—	—	—
17	0.211	1.240	100.000	—	—	—

Note: Extraction method is principal component analysis.

Conclusions and Future Research Directions

Main conclusions

Based on the preceding analyses and referring to the practices of CRSB in brand innovation, this chapter concludes the following successful practices in brand innovation, brand operation, and brand management.

Taking quality as the basis for brand innovation

The rand takes the product as its carrier, and the product quality has a direct bearing on the effectiveness obtained by consumers in product consumption. The promise made by the enterprise to consumers must be realized on the quality and image of a brand, and thus determines whether a brand can survive in the market or not.

Product quality is the objective basis of a brand. No brand can achieve long-term success without quality products. In this sense, strengthening a brand image must start from the improvement of product innovation and product quality.

The phenomenon of homogenization of domestic beer products is salient, several beer products lack distinctiveness, and different brands are similar in product taste, thus it is difficult to differentiate between different brands. Therefore, beer makers should focus on product innovation, develop products with their own distinctive features, explore new breakthroughs, and keep improving product quality to meet the consumer demand. In the meantime, as the subordinate enterprises of large beer groups are decentralized, it is vital for each enterprise to conduct its own strict product monitoring, not allowing any arising problems to damage the image of their own brands, especially that of their main brands. Factories that do not have the production line for manufacturing the main brand products generally are not allowed to produce these products in order to avoid tarnishing the image of the main brand products.

We also need to make it clear that, based on the marketing management theory, the product, as an integrated product, refers to anything that can be provided to the market to meet certain desires and needs of the general public, including not only tangible goods but also intangible services. In this sense, product quality includes not only the quality of the intangible products but also the quality of the services that come with the tangible products. In practice, the pre-sales, in-sales and after-sales services have become the critical means for market competition.

Pooling promotion resources and integrating marketing communication

While trying to improve product quality, there is also a need to strengthen the effective promotion of products and brands, keep delivering knowledge about products to consumers, and increase consumers' recognition of product quality and brands. Enterprises need to regard excellent product quality as a unique resource, by which the brand impresses its consumers, and make it a key resource of brand innovation management.

Apart from leveraging on the self-promotion of a product, it is important to use strong media, represented here by CCTV, to improve brand popularity and effectively conduct integrated marketing communication regarding various publicity activities. If various promotion resources can be pooled together to deliver one image, one voice and one expression, then the positive effects of integrated marketing communication should be realized after a period of time.

Implementing effective channel management

While developing a channel strategy and carrying out channel management, an enterprise needs to build a guiding ideology for operations, with consumers at the center and the satisfaction of consumers as its major target.

First, enterprises need to carry out a classified management of the existing general distributors. Enterprises can classify general distributors into usable and unusable ones according to their attitude and capability, and then eliminate unusable distributors. Usable distributors can be subdivided into a group that must be trained and a group that must be reconstructed. The former group must accept training, since systematic and professional training represent some of the most important means to enhance the distribution channel capacity of enterprises. The latter group can be reassigned its business areas or market segments based on its operation capacity.

Second, enterprises need to re-determine the contents and function of distributors' records. The contents of distributors' records need to be extended from distributors' basic information, credit standing, sales landscape, price management, expenses and profits management, to regional competitors' basic information, consumers' feedback, downstream distributors' suggestions, and customer strategy. This will enable the enterprise to carry out professional and multifaceted management of the sales channel in an, and extend the business management to end customers.

Last, enterprises need to use modern information technologies to set up an information processing system. Enterprises must accomplish the gathering of information, realize the accumulation and sharing of clients' resources, and form valuable customer information and market information based on the gathered information.

Future research directions

Taking into consideration of the literature review of brand innovation at the beginning of this chapter and referring to the study of this chapter, we conclude that the existing brand innovation research still has some deficiencies in the following aspects:

First, there are insufficient comparative studies on the brand innovation management between Chinese and foreign enterprises. A large number of foreign enterprises have best practices in brand innovation, brand operation, and management and some of their success stories merit the recognition of Chinese enterprises. However, the existing research focuses on either introducing the successful experiences of foreign enterprises or case studies of Chinese enterprises; it is rare to find analyses comparing the Chinese and foreign enterprises. As a matter of fact, it is often easier to discover the gap and insufficiency of Chinese enterprises by means of comparative analyses, and would also be beneficial for the Chinese enterprises to learn others' strong points in order to improve their own weak points.

Second, most existing research is theoretical research and simple case studies, but there are few in-depth empirical research. Most scholars focus on business or government perspectives, which results in brand innovation research being mainly conducted from a theoretical or business internal perspective, instead of from consumers' perceptions. This chapter concludes that the final evaluation of the effectiveness of business brand innovation, brand operation, and brand management is determined by terminal consumers. The recognition and purchase power of consumers are the main source of brand value; it is quite necessary to conduct empirical research with statistical significance based on the brand perception and brand assessment by customers.

Third, data is limited. There is a huge demand for business operation data and assessment data of brand assets and brand value when studying an enterprise's brand, but gathering this data is difficult. On the one hand, business operation data (except the performance data published by listed companies) is often the business secret of an enterprise; on the other hand, various Chinese market evaluation and advisory institutions have not yet mastered the scientific method for evaluating brand assets and brand value, the evaluation of brand value has yet to form a system, which makes it difficult to access relevant industrial and business data. The data obtained by this research also has certain limitations, such as smaller amounts of samples and narrower sample regions.

Expanding and improving relevant research on brand innovation from three aspects would be helpful. First, choosing to conduct in-depth comparative research on Chinese and foreign brand innovation and brand management, and explore an effective channel to improve Chinese proprietary brands and strengthen brand innovation. Second, empirical research on consumers can be carried out in the field of brand innovation, and integrating empirical research with theoretical studies can continuously supplement and perfect the theoretical system of research on brand innovation. This chapter carried out empirical studies on the appraisal of brand image by consumers, but these studies are only limited to the beer industry. Similar studies can be extended to other industries in the future. Finally, there is a need to create conditions for absorbing more relevant data about business operations and brand values to support and strengthen the scientificity and reliability of the studies on brand innovation. Naturally, this calls for a long-term and friendly partnership between researchers and enterprises in order to conduct surveys inside enterprises and obtain first-hand data. It also requires Chinese market assessment and consulting institutions to modernize their evaluation systems for business brand value as soon as possible, and conduct long-term and broad business brand assessments to provide a basis and reference for theoretical studies on brand and enterprise brand operations.

9
Chapter

Developing Evaluation Indicators for the Continuous Technological Innovative Capability

*Wang Huacheng
and Guo Chunming*

Introduction

In the new division of labor of the global economic system, cost and quality are no longer the key factors underlying the competitive edge in the 21st century, but the basic conditions for entering the market. Continuous technological innovative capability has become the most important source for enterprises to develop their global competitive edge.

Since 1970s, the average life expectancy of enterprises worldwide has been shortened due to the intensifying market competition. According to statistics, the average life expectancy of European and Japanese enterprises is 12.5 years; in U.S. 500,000 new enterprises are founded in annually and 60% of them keep afloat after one year, but only 20% are left after 5 years, and a mere 4% survive after 10 years. When it comes to China, the technological innovation realized by most Chinese enterprises was either a single project or in a discrete state. After some Chinese enterprises implemented one or two technological innovation projects, if substantial copycats emerged or their competitors introduced more competitive innovation products, these Chinese enterprises found it difficult to keep afloat and they often struggle. According to statistics, since the reform and opening-up of China, and under the market economic system, the number of newly-established domestic enterprises with a life cycle and a brand life cycle of more than 5 years is minimal; still fewer are enterprises with a life cycle and a brand life cycle of more than 10 years. The average life expectancy of private enterprises is only 2.9 years. What has happened proves that the advantages of the new economy belong to technologically innovative enterprises that continuously and rapidly attract and develop new technologies and introduce new products, even after experiencing numerous hardships.

Continuous technological innovation is the organizational behavior of technological innovative enterprises under the stimulation of external environmental factors like technical progress and market demand, and is an essential requirement for technological innovation. Systematically studying the formation mechanism of a business' continuous technological innovative capability and building a scientific evaluation index system from the perspectives of the three dimensions (namely factor, space and time, and the influence of economic capital and intangible capital) are of great significance for guiding enterprises to foster and set up continuous technological innovation mechanisms and improving enterprises' core competitiveness.

Literature Review

Continuous technological innovation is synonymous with enterprises' continuous innovation. Schumpeter (1942) noted that business technological innovation features discontinuity, but he also clearly pointed out that enterprises are able to and are bound to keep carrying out technological innovation. In 1998, the European Union (EU) formally set up a project to study the strategies of continuous technological innovation in the 21st century. In China, Xia Baohua (2001) put forward the concept of the "continuous technological innovation of enterprises" and made more systematic research on its nature, driving force, structure, and management; Chen Zhongbo (2003) studied the characteristics of continuous technological innovation and the features of the continuous technological innovation system of high-tech enterprises.

The formational mechanism of the continuous technological innovative capability of enterprises is the foundation for studies on continuous technological innovation. Schumpeter (1942) observed that the driving force for innovation comes from the pursuit of excess profit and entrepreneurship; researchers of the technological innovation school put forward a series of technological innovation dynamic models, such as the technological trajectory theory and market-driven theory between the 1950s and 1980s. Li Yuan (1994) pointed out that under different environmental conditions, the specific innovation driving force of most enterprises is by no means the mix of a single or fixed combination of driving forces but a showcase of diversified combinations; Fu Jiajin (1998) noted that technological innovation is dependent on a path and the technological innovation of an industry or enterprise has certain self-reinforcing mechanisms. In terms of continuous technological innovation, Segerstrom et al. (1999) presumed that the existing leaders in technology have advantages in cost compared with external competitors, and they proved that these leaders would keep carrying out R&D if they had sufficient cost advantages in this regard. Schankerman et al. (2000) proposed the "selection effect" of market competition and believed that in a more competitive industry, the monopoly profits obtained by industrial leaders are much higher than that of other manufacturers, and R&D can be stimulated through increasing marginal profit from innovation and improving the competitiveness of their product in the market. Zhuang Ziyin (2005) presented that the core of entrepreneurship is the continuous technological innovation and imitation; entrepreneurs are the risk-takers and represent the microstructure mechanism of long-term economic growth. Ouyang Xinnian (2004) analyzed the effects of the internal and external driving force factors for enterprises' technological innovations.

At present, there are few studies about the evaluations on continuous technological innovative capability. Sun Xiaofeng (2007) established the evaluation indices system for measuring enterprise continuous technological innovation, including investment capacity, R&D capacity, input-output efficiency, enterprise management capability, and the macro environment for enterprise. But, the following research literature of technological innovative capability and the continuous innovative capability evaluation could provide beneficial references for studies of this chapter. Christensen (1995) made a classification of technological innovative capability according to the different effects and functions of the technological innovation process; Burgelman (1996) identified the structure of technological innovative capability from a strategic perspective; Organization for Economic Cooperation and Development (OECD) in 1996 promulgated the indices system for measuring innovative capability in *Oslo Manual*. Fu Jiaji (1988), Wu Guisheng (2000), Yang Delin (2004), Guan Jiancheng (2004), Zheng Chundong (1999), and many other Chinese scholars presented different dimensions for measuring enterprise technological innovative capability based on foreign research results and the technological innovation of the Chinese manufacturing industry. Zheng Qinpiao (2001) conducted a more systematic research on the structure and evaluation of continuous innovative capability and put forward a comprehensive capacity system, including input capacity, production capacity, marketing capacity, finance capacity, innovation potential, output capacity and environment adaptability (institutional factor); Xiang Gang (2006) defined the basic features of enterprise continuous innovative capability and built an evaluation indices system which included the capacity to seize opportunities for continuous innovation, the capacity to implement new combinations of continuous innovation, and the capacity to realize economic benefits of continuous innovation. Xu Jun et al. (2006) built the evaluation indices system for continuous technological innovation evaluation composed of an input sub-system, organization management sub-system, financing sub-system, and output sub-system.

At the moment, certain results have been produced regarding the studies on the building of this evaluation index, but this chapter has found that there are still insufficiencies, as follows: (1) most studies are conducted from internal perspectives of the business, and fail to pay attention to the important role of external stakeholders in the developing and building of enterprise continuous innovative capability; (2) there are many indices for measuring, but the design of these indices has certain subjectivity, redundancies, and fails to take into consideration the mutual effect of these indices; (3) a set of indices reflecting the new features of and dynamically and systematically assessing enterprise continuous technological innovation has not been established yet.

Continuous Technological Innovative Capability of Enterprises

Continuous technological innovation

Enterprise continuous technological innovation as an organization behavior can be interpreted as the support and boost of technological innovation at certain stages for its follow-up technological innovation activities, whose effects can be reflected in the upward spiral of the enterprise technological trajectory. For modern enterprises, continuous technological innovative capability is the most remarkable demonstration and important source of their core competence. Through continuous technological innovation, new products or processes keep emerging, and after rapid diffusion, become the core technology and the resulting core business of an enterprise before finally being recognized by the market. In this way, enterprises can gradually establish their new core competitiveness and competitive edge.

Research on continuous technological innovation aims to reveal the substantive characteristics of technological innovation. Therefore, compared with the research on general technological innovation, it pays more attention to the pioneering, systematic, institutional, and effective natures of technological innovation, and also are different in orientation, target, strategy, behavior model, and system structure. Compared with the general technological innovative capability, the continuous technological innovative capability of enterprises obviously has different components, thus many more aspects, including technological, economic and social development, and assignment of innovation works should be taken into consideration in order to carry out an open, dynamic and systematic assessment.

Soft financial theory

Soft financial theory believes that an enterprise is an open dynamic value creation system and its main body is its stakeholders. The links between tangible and intangible value exchange between stakeholders and the connections between different stakeholders have formed the business value network covering interpersonal, inter-organizational, and social networks, along with the organizational structure. Intangible assets are the deposited and accumulated knowledge or asset stocks in the above networks,[1] which can be specifically demonstrated in four aspects: human capital, organization capital, relation capital, and reputation capital. Human capital is the accumulation of the knowledge formed by an individual's investment in network relations, in the individual, and in the relations between individuals, including individual capital, group knowledge capital, overall staff competitiveness, and individual social capital. Organization capital is the accumulation of knowledge at organizational level, with two key elements: structure and process, and core knowledge and technology. Relation capital is the value creation capacity contained in the relationship between stakeholders in an organization and a network of organizations. Reputation capital means that an enterprise has gained a high social recognition or trust due to its actions and code of conduct, and thus has gained great support and better social status, allowing it access to the resources and opportunities it needed and the ability to fend off future uncertainties. The value exchange relations in the business value network has formed the connections between factors of intangible assets, and then facilitated knowledge flow and value transformation. Various intangible assets are mutually embedded and coupled, and thus form an inseparable whole (see Fig. 9.1).

Fig. 9.1.   **Coupling relationship of intangible assets based on the business value network**

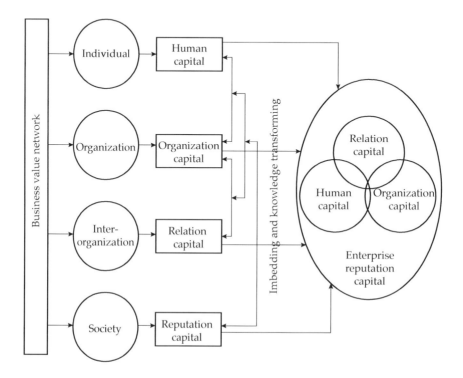

Fig. 9.2.   Capital transformation

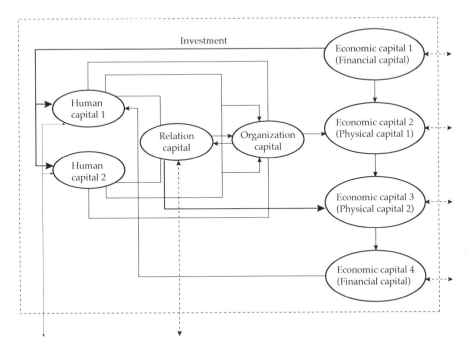

Note: The continuous thin line refers to the internal transformation between inangible capial and economic capital, the heavy line refers to the transformation between economic capital and intangible capital, and the dotted line refers to the transformation between internal and external environment.

For enterprises in the embedded value network, the input of various kinds of intangible resources have realized the accumuation of intangible capital. The mutual effects and transformation between intangible capital and economic capital (physical capital and financial capital) have constituted the enterprise capital network system. Fig. 9.2 describes the general capital transformation model of modern manufacturers.

Continuous technological innovative capability based on soft finance theory

The capability of enterprises in continuous technological innovation is, in essence, the capacity of dynamic knowledge integration embedded in a business value network. This knowledgeis not singularly technical knowledge, rather, it is systematic knowledge with certain hierarchical and relational natures, particularly the knowledge embedded in business behaviors and strategic behaviors. Fig. 9.3

Fig. 9.3. Enterprises' continuous technological innovative capability

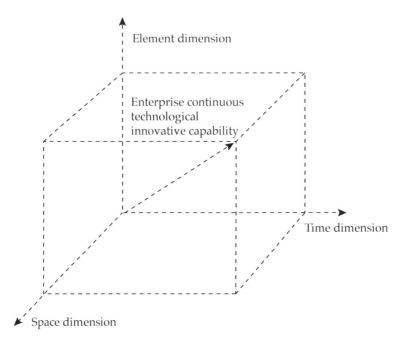

describes the continuous technological innovative capability of enterprises from three dimensions: space, time, and element. The space dimension reflects the knowledge involving the relationship between an enterprise and its stakeholders, and the relationship between the enterprise and other factors, i.e. its investors, creditors, clients, competitors, employees, community and environment. The time dimension reflects the knowledge involving the resource integration capacity and environment adaptability of enterprises during the consistent evolution of technological innovation. The element dimension reflects the knowledge involving a business' internal support system, including continuous economic capital, human capital, corporate culture, and the institutional framework of enterprises. The element dimension is the pre-condition and basis of enterprises' capability in technological innovation. The space dimension and the time dimension are the specific external and long-term reflections of a business' continuous technological innovative capability.

Element dimension

The element dimension of the continuous technological innovative capability of enterprises corresponds to part of the economic capital, human capital and organization capital of businesses.

1. Economic capital

Economic capital mainly includes fianncial capital and physical capital, among other elements. The formation of the capability of enterprises in continuous technological innovation requires investments and accumulation in both the enterprise system and the capital network system. It is demonstrated as a knowledge asset in this form. Because the basic characteristic of knowledge is that it is subject to the changes of technology and the environment, an enterprise must maintain the value of its knowledge assets through continuous investment; the competition in the economic market environment may change the position of an enterprise in the value network, and thus it needs to continue investing to eliminate the shocks induced by environmental changes. Enterprise value network, capital network, and enterprises themselves are all open and self-organizing systems; their losses and transfer may occur due to the intangible capital output during continuous technological innovation. Businesses enterprises need to realize a new positioning target in the value network through core resources investment.

2. Human capital

The human capital of the element dimension refers to group knowledge and the overall staff competitiveness. The group knowledge capital is mainly shown as the psychological contract among group members and the implied knowledge in the group. This can be measured by the indices of the knowledge complementarity of group members and the level of trust within the group. The overall staff competitiveness is the manifestation of the degree of human capital development and knowledge application, which can be measured by the overall competitiveness of the enterprise's staff and their degree of satisfaction.

3. Organization capital

Organization capital refers to the capability of an enterprise to transform various element inputs into a value. It will be maintained in the intangible assets of the organization even if group memers leave. Organization capital is demonstrated as the capacity of the knowledge structure. This includes production efficiency and the innovation-learning ability of asystem formed by the interaction of process, structure and information systems of an enterprise. The analysis of Atkeson et al. (2002) indicates that organization capital is an enterprise's learning accumulation, and its value is equivalent to two-thirds of the market value of tangible assets. Organization capital can be measured from the basic structure, core knowledge and continuous innovation of corporate culture. Basic structure measures the inter-organization coordination ability and the matching degree between technological innovation strategy and structure. Core knowledge examines the difficulty of imitation for its core technology and market potential. Corporate culture is the values acknowledged within an enterprise.

Space dimension

The space dimension of enterprise technological innovative capability generally reflects the relation capital in the enterprise value network. It must value R&D and the integration of internal and external knowledge resources, and emphasize the complementary knowledge resources and capital input required through cooprative networks. The relation capital is the value of the relationship between an organization and its. The stakeholders hereby refer to supplier, distributor, client, shareholder, partner, competitor, government, society, consulting agency and public research institute, among others. Supplier, distributor and client represent the continuous technological innovation source of an enterprise. According to some investigations, products such as scientific instruments, semiconductors, and printed circuit boards, 60% of the users have carried out the innovations, while many othersare performed by manufacturers and suppliers. The partners can be either research institutes, entities established through joint venture and holding shares, or financial institutions, all of which can assist enterprises to generate new growth opportunities and improve existing production structure. Additionally, for a business to realize continuous technological innovation, its industry positioning and selection cannot go against the government's industry policy and scientific policy.

Time dimension

The time dimension indicates that technological capability is continuous and lasting; the enterprise engages in short-term and long-term competition through actively shaping the innovation. This dimension contains three aspects:

(1) Enterprise continuous technological capability cannot exceed the carrying capacity of the enterprise's internal support system, which means that development and delivery need to align with the enterprise's capability in resource integration.

(2) Enterprises choose different technological innovation modes and have different technological innovative capacities at different development stages.

(3) Enterprise continuous technological innovative capability cannot exceed the carrying capacity of its external natural environment, which means they need to actively develop green technology, and reduce pollutant emission and wastes in production.

The main component of an enterprise's value creation system is its stakeholders. The links between tangible and intangible exchanges among stakeholders have formed the enterprise value network. The formation of continuous technological innovative capability in this system is affected by economic capital, human capital (individual capacity, group knowledge capital, etc), organization capital (process structure, core knowledge, etc), and relation capacity (customer relations, supplier relations, partner relations, etc). Those embedded, coupled factors have shaped the (see Fig. 9.4). Among many capital elements, the continuous innovation spirit of top enterprise managers is the key source for an enterprise to develop its continuous technological innovative capability; the coordination between R&D, production, and marketing is the fundamental guarantee for enterprise continuous innovation; rational and strategic human capital investment, organization capital investment (such as R&D investment, systematic informatization investment), relation capital investment (such as the cooperative investment between suppliers and enterprises) represent the influential factors for the realization of enterprise's continuous innovation.

Fig. 9.4.   The formation mechanism for enterprises' continuous technological innovation

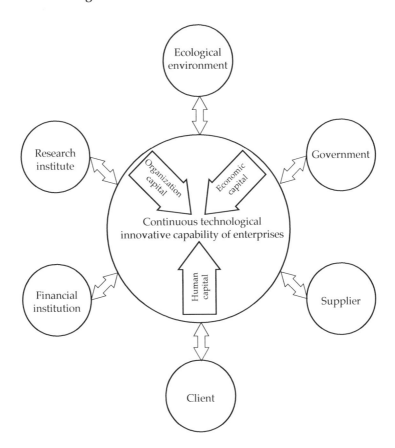

Indices System for Evaluating the Continuous Technological Innovative Capability

Initial building of the indices system

According to these analyses, the general structure of the evaluation indices system of enterprises' continuous technological innovative capability is shown in Fig. 9.5.

Fig. 9.5. **The framework of enterprises' continuous technological innovative capability evaluation**

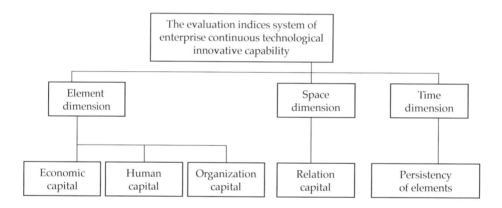

This chapter has taken listed companies as the objectives and has built the following evaluation indices system of enterprises' continuous technological innovative capability, with 34 indices (see Table 9.1).

Table 9.1. **Evaluation indices system of enterprises' continuous technology innovative capability**

Evaluation index system for enterprises' continuous technological innovative capability	Element dimension	Economic capital	Net asset of each share (p_1); profit rate of main business (p_2); profit margin on net assets (p_3); per share cash flow from operations (p_4); R&D investment intensity (p_5); non-R&D investment intensity (p_6); the proportion of human capital for R&D to sales revenue (p_7); contribution rate of R&D to profits (p_8);

(Cont'd)

Evaluation index system for enterprises' continuous technological innovative capability	Element dimension	Economic capital	Net asset of each share (p_1); profit rate of main business (p_2); profit margin on net assets (p_3); per share cash flow from operations (p_4); R&D investment intensity (p_5); non-R&D investment intensity (p_6); the proportion of human capital for R&D to sales revenue (p_7); contribution rate of R&D to profits (p_8);
		Human capital	Education background of core R&D staff (p_9); ability of core R&D staff in organization and coordination (p_{10}); sense of responsibility of R&D team members (p_{11}); knowledge complementarity of R&D team (p_{12}); stability of R&D team (p_{13}); satisfaction of R&D team within organization (p_{14}); continuous innovation spirit of top managers (p_{15});
		Organization capital	Preference degree of top manager training and appointment mechanism to continuous innovation (p_{16}); advancement of core knowledge (p_{17}); difficulty level of imitation of core technology (p_{18}); market potential of core technology (p_{19}); the number of patents per 100 people (p_{20}); degree of information (p_{21}); coordination ability between different departments (p_{22}); corporate culture of pursuing innovation (p_{23}); incentive system supporting innovations (p_{24});
	Space dimension	Relation capital	Adaptability to national macro policy (p_{25}); financial capacity (p_{26}); capacity of responding to customers' demand(p_{27}); capacity to cooperate with suppliers in technology (p_{28}); strategic cooperation with universities and research institutes (p_{29});
	Time dimension		Average growth rate of p_5 in recent 5 years (p_{30}); average growth rate of p_6 in recent 5 years (p_{31}); average growth rate of p_7 in recent 5 years (p_{32}); average growth rate of p_{20} in recent 5 years (p_{33}); environmental friendliness of innovation technology (p_{34}).

$$P_5 = \frac{\text{R\&D costs}}{\text{Sales revenue (physical investment is converted into monetary investment)}} \qquad (1)$$

$$P_6 = \frac{\text{Costs for introducing technology or improving technology}}{\text{Sales revenue (physical investment is converted into monetary investment)}} \qquad (2)$$

p_{21} = man-machine ratio;

p_{22} mainly refers to the coordination capability between R&D, manufacturing and marketing departments;

p_{25} mainly refers to the adaptability to national industrial policy, technology policy and environment policy, among others.

p_{30}–p_{33} adopt average growth rate index, and the calculation method of this index is as follows:

$$\text{Average growth rate} = \text{Average development speed} - 1 \ (100\%) \qquad (3)$$

The average development speed is the result of chain indices of development speed within a certain observation period, and its calculation methods include the geometric method and equation method. The geometric method is most commonly used:

$$\bar{x} = \sqrt[n]{\Pi x} \qquad (4)$$

Of which, $\bar{x}$ refers to the average development speed of index x; n refers to the observation period.

The average growth rate has both positive and negative differences, and respectively refers to the gradual average increasing degree and the average decreasing degree.

Explanations of some indices

Continuous innovation spirit of top managers (p_{15}), preference degree of top manager training, and appointment mechanism to continuous innovation (p_{16}), incentive system supporting innovation (p_{24})

In modern business organizations, it is the shareholders and managers who directly determine the enterprise's technological innovation strategy. Shareholders as consignors and managers as agents have not only different interest requirements but also different risk preferences, which determine their different tendencies in the selection of innovation strategies.

Conflicting interests between consignors and agents is the root cause of subjective agency issues. The settlement of subjective agency issues can effectively promote enterprise innovation, which involves the corporate governance structure. Governance structure is the company's basic level of organization, and determines its basic characteristics and value orientation. Effective governance structure can play an incentive and restrictive role in the investment, objective, decision making and implementation of enterprise innovation. Only when the continuous innovation objectives are formed under the restrictions and incentives of effective governance structure, can these objectives be acknowledged by all stakeholders.

In addition, the source of innovation is information and knowledge which are only useful when they are properly used the exchanges between working staff with necessary knowledge and skills. Human capital performs creates, applies and converts new knowledge during an enterprise's innovation process. Therefore, apart from encouraging managers to innovate, enterprises also need to further encourage working staff, especially the core staff, in innovation.

Knowledge complementarity of R&D team (p_{12})

In an enterprise, diverse knowledge exchange between individuals is necessary As Simon pointed out, the co-existence of different knowledge structures has promoted innovative learning and problem-solving. Effective exchanges and interaction between individuals with diversified and different knowledge structures will help organizations increase new knowledge connections and combinations—innovative capability.

Coordination ability among different departments (p_{22})

Enterprise technological innovation involves R&D, manufacturing, marketing and diversified activities of multiple departments. The high specialization within each division of labor will result in obvious differences among different departments in objectives, values and working orientation. Lawerence and Rche used to make investigations of the manufacturing, R&D, and marketing departments of 10 companies, and found that each department gradually shaped their own objectives and business characteristics, as they established connections with different external groups. According to the empirical investigations conducted by American technical management expert, Souder, when there were serious problems in R&D and marketing, 68% of R&D projects completely failed in the commercial sense, and 21% had difficulties. A related research in 1994 also indicates that when there were serious problems in R&D

and marketing sectors, about 40% of R&D projects technically failed, and 60% of the projects that succeed in technology failed to generate profits. Continuous technological innovations require guaranteed coordination between sectors with a relatively orderly organization structure; thus the organization structure should neither be extremely flexible nor extremely hierarchical. Instead it should be flexible and meet the requirements of different development stages of enterprises.

Indices system reduction based on rough set theory

In the above evaluation indices system, not all information is equally important for evaluating technological innovative capability, and some indices are redundant. In particular, when the data source is randomly chosen, the redundancy of data will be more universal. This chapter employs the attribute reduction method in the rough set (RS) theory and eliminates some indices under the preconditions of ensuring the classification capability of the indices system are unchanged, which has simplified the evaluation process and increased the system's operability.

RS theory is a new type of mathematical tool put forward by Professor Z. Pawlak, a Polish mathematician, for processing incomplete and uncertain issues. Its major role is to find tacit knowledge and reveal its potential laws based on people's knowledge of the data obtained, to effectively reduce the knowledge base and rules extracted. The indices reduction process is shown in Fig. 9.6.

Fig. 9.6.   **Indices reduction process**

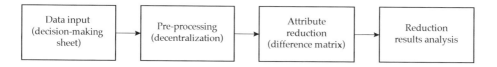

Data discretization processing on the basis of Boolean logic and RS

RS theory uses a symbolized analysis method for data processing, and its attribute value usually adopts discrete numbers like 0, 1, 2. But, in practical issues, the value is successive, thus there is a need to discretize the consecutive indices' values. This chapter employs the processing method with a mix of Boolean logic and RS. This method can solve all possible breakpoint sets according to a given information system, and the new information system obtained by adopting any breakpoint set will not change the indiscernibility of the original information system. In the practical solving process, the "greedy algorithm" method is adopted and breakpoints are gradually added to the breakpoint set based on the importance of breakpoint value. The greedy algorithm method is as follows:

Set $S = (U, C \cup \{d\})$ is a decision table, the domain of discourse $U = \{x_1, x_2, ..., x_n\}$ is limited object set, $C = \{a_1, a_2, ..., a_m\}$ is the collection of conditional attributes, d is the decision attribute, V_a is the value range of attributes, $V_d = \{1, 2, ..., r(d)\}$, $r(d)$ is the number of decision classes.

(1) Building a new decision table S' based on the original decision table S.

New decision table $S' = (U', C' \cup \{d\})$, among which, $U = \{x_i, x_j\} \in |U \times U| \; d(x_i) \neq d(x_j)$, $C' = \{ P_r^a : a \in C, r$ is attribute a's rth breakpoints $[c_r^a, c_{r+1}^a]\}$. As regards, $\forall P_r^a \in C'$, if $[(c_r^a, c_{r+1}^a) \subseteq [\min(a(x_i), (a(x_j)), \max(a(x_i, a(x_j))]$ then then $P_r^a(x_i, x_j) = 1$, otherwise $P_r^a(x_i, x_j) = 0$.

(2) Building the initial breakpoint set $CUT = \emptyset$.

(3) Choosing the breakpoint with the greatest amount of "1" in all columns and add it to CUT, then eliminating the column of this breakpoint and the row where its value is "1" at this breakpoint.

(4) If the elements in decision table S' are not empty, then turn to (3), otherwise stop. At this point, CUT is the breakpoint set solved.

This decentralized processing method is to use one or several breakpoints to differentiate the otherwise indiscernible relationship of the two examples. Under the precondition of ensuring the indiscernible relationship of the information system is unchanged, this calculation method tries to separate all the distinguishable relationships among all examples with the least number of breakpoints.

In addition, among the 34 indices (attribute set p) presented in this study, some indices need to be evaluated by experts objectively. Considering the differences of experts in knowledge, understanding and preference, this study suggests that experts give vague assessment (very high; high; common; low; very low) or (very good; good; common; bad; very bad) to these kind of indices, and then adopt the Demposter-Shafer method based on the MASS function to process the assessment results and convert multi-qualitative judgments into quantitative assessment values.

Attribute reduction based on deletion method

Deletion method is the most common in reduction algorithm. Deletion method eliminates attributes from the data sheet one by one and then makes a comparison between the new information sheet after the deletion and the original information sheet in terms of the indiscernible relationship of a decision class. If the indiscernible relationship remains unchanged, then another attribute can be further deleted in the newly-formed information sheet before further comparison can be made. But if the indiscernible relationship changes, then the information sheet should be recovered to the previous one before another attribute can be deleted.

Conclusions

Enterprise continuous technological innovative capability is, in essence, the dynamic knowledge integration capacity embedded in an enterprise's value network, and the assessment of it must be open, dynamic and systematic. This chapter has analyzed the mutual embedding and coupling of economic capital and non-economic capital (human capital, organization capital, relation capital, et cetera) during the formation of continuous technological innovative capability, and has built relevant indices systems from the element, space and time dimensions. Considering the differences in importance of indices and the redundancy of some indices, this chapter has presented the indices reduction method based on the rough set theory. The differences in sample data, decentralized processing and reduction methods will influence the building of indices systems. In the follow-up studies, the author made a sampling of listed companies in the Shanghai Stock Exchange and the Shenzhen Stock Exchange, adopted the method combining questionnaire and field investigations, and selected a more scientific decentralization and reduction method to build a scientific evaluation index system for enterprise continuous technological innovativ capability which conforms to the Chinese listed companies.

10
Chapter

The Structure of Governance for Chinese Enterprises

Xu Erming and Zhang Han

Introduction

"Nothing is more important and urgent in today's business environment than the management on innovation and reform. To compete in the new environment, companies must roll out new products, services and procedures and regard innovation as a way to survive." (Storey, 2000)

Porter (1990), a famous American scholar in strategy management, believed that the development of a country's competitiveness contains three stages: factor-driven stage, investment-driven stage and innovation-driven stage. Since the 1950s, developed countries have been committed to the industrialization and modernization, taken science and innovation as the dominant strategy, invested hugely in R&D, and completed the transition from investment-driven stage to innovation-driven stage, thus taking the lead in developing among the innovation-driven countries in the world. At the moment, industrialized countries rely greatly on innovation while the emerging industrialized countries and developing countries are in the process of accelerated utilization of information technology (King, Gurbaxani, McFarlan, Raman, and Yap, 1992).

In the past two decades, China's economy has been growing at an average annual growth rate of 9.4% and as of 2004, its per capita GDP exceeded USD1, 000, indicating that China has begun to enter into the take-off stage of industrialization. That being said, China's development is still driven by investment factors and is faced with daunting and urgent resource scarcity and ecological pressures. In the next 15 years, strengthening independent innovation and building an innovation-driven country will be one of China's basic national policies and strategies. At the core of the construction of a national innovation system, enterprises must become the key players of innovation, build up their original innovative capability, harness innovation ability, as well as bring and absorb re-innovative capability. In this case, "how enterprises can better innovate" has become one of the major research topics for scholars.

In 1912, Austria economist Joseph Schumpeter put forward the concept of "innovation" in his masterwork *Economic Development Theory* (Germany version)[1]. He noted that innovation means the new arrangement of production factors, i.e. "to establish a new production function," with the aim of obtaining potential profits. Entrepreneurs, as innovators, seek to promote economic development by creatively breaking down market balance. Peter Drucker (1985) pointed out in his book *Innovation and Entrepreneurship* that innovation is a means used by entrepreneurs to show their entrepreneurship and render resources a new capacity to create wealth. In another words, through innovation, enterprises are well-positioned to attain new resources to acquire core competitiveness and form a sustainable competitive advantage.

In mature industry-based economies, profits brought about by innovation activities in the global market made by business organizations in the pursuit of competitive edges are on the increase (Barlett and Ghoshal, 1989). Enterprises are more reliant on innovation to acquire profits. As the major source for competitive advantages, innovation represents the strategic transformation (Cooper and Schendel, 1976), becomes part of enterprise strategy (Ettie, Bridges, and O'Keefe, 1984), and accordingly turns into the responsibility of top managers.[2] As a result, the process of innovation strategy selection has become a process for companies to elevate their innovative capability and efficiency. Only by adopting innovation strategies can enterprises obtain unique competitiveness and finally realize sustainable growth.

The role of governance structure in business strategy affords no ignorance (Xu Erming and Zhang Han, 2007). According to the theory of entrustment-agent, shareholders are major stakeholders with the highest power in a corporate governance structure, the decisions made by the board of directors (BOD) and mangers are for the benefit of shareholders, and all mechanisms are designed to safeguard the interests of shareholders. Shareholders are providers of important sources for enterprises (capital, land, technology and reputation, among others) and one of the original driving forces for enterprises to survive and realize strategies (Wang Zhihui, 2002). In particular, a positive corporate governance structure could add to the stability of the exterior environment of enterprises, and effectively restrain the opportunist activities of managers while consolidating the abilities of enterprises to grow and thrive through establishing long-term partnerships. In addition, the channel to obtain innovation resources, cost, competition capacity, capital and human resources support, and the acquirement of innovation achievements in the decision-making of innovation strategies are all subject to the corporate governance structure.

Beyond that, the impact of governance structure on the selection of innovation strategy is not just limited to resources. The survival of enterprises in a social environment will surely be subject to the impact from institutional environment. While seeking economic benefits, enterprises must take into account the legality of their business. Hoskisson and his colleague (2000) believed that in the transitional economy, the institutional theory focusing on the relationships among organizations must be given priority, followed by the RBV theory of the specific ability of enterprises. Peng (2003) also argued that during a period of economic transition, the original system of enterprises has broken down while the new market economy system has yet to be established. Under such circumstances, different companies may select different development paths to obtain resources through either networking or market competition. Different governance structures lead to a divergence in costs and profits. When it comes to the legality of innovation activities, corporate owners and BOD members may exert different institutional pressures on the company due to the impact of the paths they have chosen, while the governance structure may affect the decision-making of innovation strategies. In this connection, exploration into the impact of corporate governance structure on business innovation is a research topic with great significance.

Literature Review and Theoretical Model

Literature review

Factors affecting business innovation

The first thing to be considered in selecting innovation strategies is the relationship between technologies, other enterprise resources, and the business' competitive capacity. Many studies have provided analysis into issues from this perspective (Penrose, 1959; Wernerfelt, 1984; Barney, 1986; Dierickx and Cool, 1989; Peteraf, 1993; Parahalad and Hamel, 1990; Teece, Pisano, and Shuen, 1997). They argued that companies with advantageous resources are blessed with more competitive advantages than those without, making it difficult for the latter to compete against the former. However, resources and capacity are not necessarily the only reasons that define the competitive advantages of companies — environment is also an important factor.

The innovation knowledge of enterprises is always stored in the minds of individuals in a default or implicit manner, which could not or is difficult to be coded. Rather, knowledge is a dynamic, cumulative, and evolving process which changes along with individual creativity and differs as organizations exchange views. At the same time, innovation exists in an uncertain environment and is subject to limited information and capacity; and an enterprise's innovation decision is based on "bounded rationality" (Simon, 1956, 1965).[3] These features of innovation knowledge are the difference between business innovative capability and the comprehension in introducing new opportunities. As such, these factors become the prerequisite upon which research on innovation activities must focus.

Essentially, three factors are at play in innovation strategy selection. First, capital and talent factors within companies, i.e., factors that have direct impact; second, systemic factors including social capital, organizational factors and social culture; third, exterior factors, that is, factors in the businesses' macro environments, including political and institutional factors.

Interior factors

In previous years, scholars carried out research on interior factors that affect business innovation strategy, analyzing from the perspective of talent and capital support, and focusing on technological innovation.

(1) Capital factors. Investment is a realistic foundation to achieve technological innovation. The operation and expansion of any material production comes with the integration of capital and labor force. Investment means that the formation of capital with the materialization of scientific and technological achievements is, to some degree, the formation of capital. In this case, investment activities have become the junction of science and economy, i.e., the foundation for technological innovation.

(2) Talent factors. Hoffman (1993) argued that the key factors which make innovation successful are personnel and technology. Technological innovators are a major force of technological innovation activities; without them, technological innovation is impossible. To begin with, the features of researchers are rather important (Miller, Burke, and Glick, 1998; Daellenbach, McCarthy, and Schoenecker, 1999). Any technological innovation is a systematic process which requires the cooperation of the entire innovation team. Innovators who can make breakthroughs in theory and basic approaches, and who have keen market sense and unique creativity, matter a lot to an innovation's success. Secondly, entrepreneurship is also significant. Joseph Schumpeter pointed out that "the function of entrepreneurs is to reform or revolutionize production mode through a new invention, or generally, to use a potential technology that has yet to be tested to produce a new commodity or an old product via a new method, or through exploring into new sources of raw materials supply or new markets of products, or through reorganizing industry...such a function is not about inventing new products or conditions that enterprises can develop and utilize, but lies in making things happen." Hambrick (1983a) believed that decision makers are important sources for strategic decisions and the key for advocating innovation.

In addition, knowledge dissemination (Marinova, 2004; Gibbons and Johnston, 1974; Faulkner, 1994), geographical location (Hassink and Wood, 1998), and technological management–aided innovation (Damanpour, 1987) are also tangible influencing factors that must be taken into account in business innovation.

Systematic factors

Innovation is not just the work of R&D department, and an enterprise's relevant support system is necessary for a successful innovation; it is unviable if an enterprise only takes into consideration the resource input of technological innovation without considering the corresponding social and cultural development. After in-depth exploration, many scholars believe that the R&D capacity and internal system of enterprises play an important role; many scholars have carried out research on factors that affect business technological innovation including corporate organization, core competences, competence matching, human resources (HR) development, et cetera (Zahra, 1991; Stopford and Baden-Fulle, 1994; Fujita, 1997; Balkin, Markman, and Comez-Mejia, 2000).

(1) Social capital. As technological competition intensifies, enterprises have taken note of the importance of social networking for enterprise innovation. Only by continuously improving relationships and building confidence with governments, research institutions, and other social networks can technological innovation of enterprises be accelerated. This reflects the highly-connected and complicated relationship between science, technology and knowledge on the one hand, and social and cultural development on the other.

(2) Organizational factors. First, organizational structure. According to the argument of Neil Kay in 1998, technological innovation is featured by time-lag, uncertainties, costliness, and by other unspecified features. The organizational structure of enterprises is exerting impacts on the entire technological innovation by relying on these four features. According to the influential degree from strong to weak, the major influencing factors of organization structure include: information flow, experience accumulation degree, flexibility, openness, organizational efficiency and logistics (Li Fan and Nie Ming, 1999). Second, information system. Information is an important resource for technological innovation, and a strong information system is a necessity for a company to succeed in innovation. Competition among enterprises in the ever-intensive market competition is defined by competition for technological innovation, which is largely dependent on the acquisition and utilization of advanced technologies and information about market demand. Whether and when important information is obtained has a direct bearing on the result of the formation and success of technological innovation. Utterback (1975) argued that it comes as a comprehensive utilization of information obtained by technical means which meets the market demand, and the impediment for information dissemination is the primary obstacle for technological innovation. Beyond that, Cohen and Levinthal (1990) also believed that the capacity of enterprises in absorbing information has a great impact on innovation.

(3) Corporate culture. Major players in technological innovation are enterprises'

members or active players. Corporate culture impacts technological innovation in the contemporary world when science and technology keep advancing with each passing day, and it exerts influence mainly via affecting the major players in technical innovation.

Exterior factors

Over the years, scholars have shifted their research focus to the macro environment from internal environment for business technological innovation.

(1) Policy factors. Policies about technology and innovation have played a due part in different technological, institutional and policy settings. Business innovation develops along with changes in innovation policies as a result of an imperfect market and is also subject to static policy factors, such as policy that seeks to achieve the optimal investment scale by balancing competition investment and corporate horizontal investment. The function of policies is summarized as: to bring requirements for technological reform into the economic system by changing the demand mode or adopting technological revolution. Governments participate by encouraging effective technological reform and promoting growth and welfare (Bartzokas, 2001).

(2) Institutional factors. Human society carries out activities within certain institutions. "Institution is a rule widely recognized in the economic society. Rule is the action taken by people or a certain way of behavior." (North, 1990) The institutional theory strengthens some core issues which are important in strategic evolution but are ignored in other theoretical perspectives, including the adaptability, role of convention, irrationality and social decisiveness of enterprises' strategic behavior. According to institutional theory, companies operate within a social framework which is composed of a social norm, values and a host of widely-recognized assumptions that define which kind of economic acts are suitable and acceptable under this framework. During this process, the legality, resources, and survival ability of enterprises will be enhanced. In this connection, the conformity of a company with a social expectation largely determines the success and survival of the company.

We found that existing research on factors affecting the selection of enterprise innovation strategy are launched from the perspective of how to stimulate business innovation and add to innovation efficiency. However, the selection of innovation strategy is not only subject to the limitations of technology, information, and income, as emphasized in the neoclassical model, but also subject to the constraints of various social structures that reflect the attributes of people, such as norms, customs and conventions (Powell and DiMaggio, 1991). Institutional analysis showed that what motivates organizational behaviour is not just economic maximization, but also the pursuit of social fairness and obligations (Meyer and Scott, 1983). As such, both theoretical and business communities have placed a great deal of emphasis on the relationship between an enterprise system and innovation, from the perspective of institutional.

Role of governance structure

Corporate governance is closely linked with corporate strategy. It is an essential institutional arrangement whose implementation could be realized through certain specific institutional measures, including an incentive system, distribution system, personnel system and financial system, all belonging to an organizational management system. In this sense, an organizational management system is a specific tool for corporate governance and exists in the institutional structure with corporate governance at its core. Corporate governance also defines the framework and channel of operation and management, and the characteristics of governance modes largely affect those of the corporate management. Finally, corporate governance could not create economic benefits for enterprises on its own, but affects these benefits indirectly through influencing corporate strategy; because of this, corporate governance becomes meaningful (Hua Jinyang, 2003). Therefore, corporate governance affects the efficiency of the formulation and execution of strategy, and also reflects corporate value (Magretta, 1998).

Corporate governance affects corporate strategy through two aspects (Fig. 10.1): first, to affect strategic management through its impact on the BOD structure; second, the operational efficiency of corporate governance could exert an impact on strategic management. Under a certain governance structure, the operational efficiency of corporate governance affects corporate performance. The operational efficiency is mainly manifested in corporate governance and within the function of BOD, in particular.

Fig. 10.1.   **Analysis on the impact mechanism of corporate governance on corporate performance**

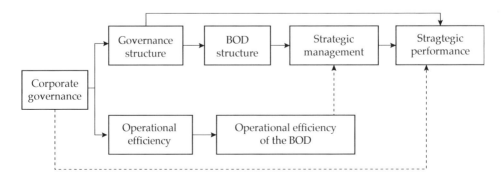

The impact of institutional theory on strategy selection was first studied by Child (1972), and then developed by Child (1997) and Oliver (1997) later. However, studies in this field are mainly either based on a static institutional environment, or within a certain industry to explore the influence of enterprises on the formation of the institutional framework of the industry (Hoffman, 1999). Neither of these two types of studies has addressed the impact of institutional transition faced by transition economies on the selection of business strategy. Peng (2000, 2003) put forward a two-stage model to focus on this issue. No matter how different the transition economies in the speed and source of transition are, they have one thing in common: the economic system of those countries is changing while the remnants of the traditional economic system still exists, and a new system has yet to be fully put into operation. According to Peng (2003), when a country is in the process of transition: (a) there are now enough rules to govern all transformations (North, 1990); (b) it falls short of strong and reliable implementation of the existing rules (Stiglitz, 1999); (c) organizations are inertial, have many limitations, and lack adaptability (Newman, 2000; Oliver, 1992).[4] There are two kinds of strategies in transition economies. The first is a strategy based on networks, which focuses on intangible assets formed among managers, and between enterprises and various players (Powell, 1990). The second is based on market, which pays attention to competitive resources and capacity, including quality, accounting, and marketing, which are independent from the network, relationships, and links of companies (Barney, 1991). See Table 10.1.

Table 10.1. Enterprise growth: three strategy selection models

Strategy selection	Organizational mode	Institutional premise
Internal and general expansion	Grade-based enterprises	Competent managers
Merger and acquisition	Market	Functional strategic element market (such as capital market)
Relations among organizations (such as network)	Blend (neither grade-based enterprises nor market)	Mutual trust and understanding data

Source: Mike W. Peng and Peggy Sue Heath, "The growth of the firm in planned economies in transition: institutions, organizations, and strategic choice," *Academy of Management Review* 21, no. 2 (1996): 492–528.

According to institutional theory, the strategic selection of companies is subject to exterior pressure (Meyer and Scott, 1983; Pfeffer and Salancik, 1978). In order to survive, they must reflect exterior requirements and expectations (Meyer and Rowan, 1977; Pfeffer and Salancik, 1978). Tolbert and Zucker (1983) believed that organizations in the same field are subject to different institutional pressures. Due to differences in the power of regulators, the part of an enterprise influenced and the degree of pressure on enterprises differs among different institutions. Meanwhile, institutional pressure varies along with changes in recognition, beliefs, and the degree of control on moral standards in different places and times.

Many specific strategies (such as contracts, joint selections, joint ventures and alliances) include the building of an organizational structure which, in its own right, is the carrier of collective behaviors and adaptive behaviors operating at different analysis layers (Astley and Van de Ven, 1983; Astley, 1985). A strategy suitable for an institutional structure may be forbidden in another structure (Kim and Kim, 1985; Hambrick, 1983a). The responses of enterprises vary between obedient and resistant, passive and positive, subconscious and manipulative, impotent and influential, habitual and opportunistic (Meyer and Rowan, 1977, Pfeffer and Salancik, 1978). Institutions often put limits on the direction and content of business transformation (Zucker, 1987). Oliver (1991) put forward five strategies to cope with institutional stress: acquiescence, compromise, avoidance, defiance and manipulation.

Organizational transformation may be triggered by market forces and non-market forces; the forces may be consistent or conflicting (D'Aunno et al., 2000; Oliver, 1992). In this case, the key to select a strategy in a transitional economy is to examine how enterprises should perform when the market economy system has yet to be fully established. Governance structure could greatly affect the strategy management process of a company (Eisenmann, 2002). According to the research of Oliver (1991) on factors influencing strategy selection, we conclude that the institutional impacts of governance structure on innovation strategy decision-making are as follows:

1. Causes: exploring why organizations are subject to institutional rules or expectations.

2. Structure: finding out who is exerting institutional pressure on organizations.

3. Content: providing research on what the companies' institutional pressures are.

4. Control: studying how the institutional pressure is exerted.

5. Environment background: discussing the background where institutional pressure is exerted.

The resource effect of corporate governance is: different owners have different capabilities and forces to influence business activities due to their different appendant resources ("appendant resources" hereof refer to resources possessed by owners, which are not directly invested into companies through equity, but could be used by owners to serve corporate activities) (Xia Dong, 2006). In this connection, the ownership structure determines the necessary supply of innovation resources (Hua Jinyang, 2003).

First, different ownership structure means different modes of raising capital and different amounts of available capital.

Second, owners' intervention on business activities has affected the HR supply of managers or other innovators.

Third, owners will affect the innovation structure of enterprises.

Theoretical model and assumptions

To sum up, Chinese enterprises may suffer from the impacts in government structure in two aspects: lawful pressure from the institution, and the ability to acquire and maintain resources. The difference in corporate ownership structure in these two aspects may propel companies to take different independent innovation actions, which would result in different innovation efficiencies and effects (Fig. 10.2).

Fig. 10.2. Governance structures with different functions

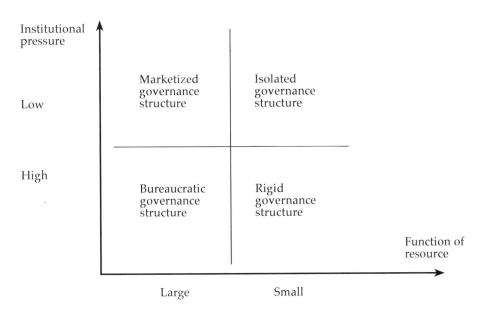

(1) Marketized governance structure: the most beneficial for business innovation.

(2) Bureaucratic governance structure: due to high institutional pressure, companies are sometimes forced to take some innovative actions that confirm to the lawful requirements from shareholders, but may not produce sound results. However, companies may acquire important resources necessary for innovation through the governance structure. Therefore, such a structure still makes sense for business innovation.

(3) Isolated governance structure: although the governance structure of a company is seldom impacted by corporate institutions, this kind of governance structure could not bring necessary innovation resources for enterprises; thus, it offers no direct driving force for innovation.

(4) Rigid governance structure: such a structure would not provide innovation resources for enterprises; instead, it may force them to take actions which may go against innovation under institutional pressures. It is the most dangerous structure for the independent innovation strategy of companies.

State-owned equity is the most representative of institutional pressure. Equity structure, as the most fundamental institutional arrangement of enterprises, is the influencing factor of institutions for rules. As China has undergone economic transition, state-owned enterprises (SOEs) have transformed into enterprises with state-owned equity, and the original system of enterprises has broken down. However, the new market economy system has yet to be completely established and the state-owned equity from the traditional planned economy system has formed its own distinctive institutional features. The ownership structure has great strategic meaning for R&D investment of enterprises (Hill and Snell, 1989; Williamson, 1964), because different investors have diverse targets for investment (David, Kochhar and Levitas, 1998; Kochhar and David, 1996). This leads to starkly different investment amounts that accordingly affect investors' willingness to challenge managers and encourage innovation with their power. Investment goals of state-owned shareholders influenced by corporate institutions are bound to exert impacts on the strategic behaviours of enterprises (Yuan, Sun, and Yi, 2006; Ma, Yao, and Xi, 2006).

The results of institutional influences on state-owned equity are reflected in two aspects: first, insufficient ability to absorb new knowledge; second, the tendency of "expansion impulse" formed in the traditional planned economy still remains. Therefore, we will focus on the influence of institutional factors on the changing amount of state-owned equity on independent enterprise innovation, and by adopting resource theory.

First, it is much easier for state-owned shareholders to acquire resources with lower costs or rare resources from the government, compared with other shareholders. Gautam Ahuja (2000) believed that the connection between enterprise and its stakeholders was the major source for innovation resources and information. People recognized that state-owned enterprises are still dominant, and resources that state-owned shareholders could offer are incomparable for many other companies. The relationship between state-owned enterprises and the government makes it easier for SOEs to obtain resources and information necessary for innovation. Meanwhile, these companies are protected by the country, and suffer little market pressure (Meyer and Lu, 2005; Nee, 1992). When the market is underperforming, this monopoly could better protect the innovation results of enterprises (Lunn, 1985). Independently performing innovation is more conducive to maintaining a "soft" relationship within companies, thus forming core competences and competitive advantages.

Second, Cassiman and Veugelers (2002) believed that exterior information flow is linked with cooperative R&D, and those companies taking available exterior information flow as important innovation input are more likely to conduct cooperative R&D. SOEs often obtain their main resources and information from the government, and from the perspective of institutional theory, such a dependence formed in a planned economy system restrains other companies from obtaining information from other channels through cooperation. In addition, cooperative R&D may add to the risk of diffusing corporate advantages.

Third, according to Cohen and Levinthal (1990), the absorption ability is always the key to obtaining exterior resources. If a company fails to develop its ability to absorb at the very beginning, its recognition of technical opportunities will not change, as they may not detect that the signals are different from those they experienced before. As a result, when new opportunities come along, they may not pay attention to, nor will they invest in the ability to absorb such information. This phenomenon is termed as "lockout." State-owned shareholders came from the traditional planned economy, and the institutional restrictions make them lack the corresponding experience necessary to discover an innovation opportunity and derive information from the market, The lockout phenomenon is severe, making it difficult for enterprises to acquire exterior innovation possibilities in a timely manner.

As such, our hypotheses are as follows:

H1: the proportion of state-owned equity is in positive correlation with internal R&D of enterprises;

H2: the proportion of state-owned equity is in negative correlation with cooperative R&D of enterprises;

H3: the proportion of state-owned equity is in negative correlation with technology purchasing of enterprises.

Research Method

Sampling

The Chinese government has categorized high-tech industries into chemicals, pharmaceuticals, machine building, communications, and electronics. We selected 541 of the publicly-traded companies from these 5 industries as our samples to verify our hypotheses.

According to the China Securities Regulatory Commission, listed companies are required to disclose certain "strategies" in their annual report, including investment in M&A, joint venture establishment, and technology and new products. For this study, we analyzed the annual reports between 2000 and 2005 of all sample companies. Then, we extracted every strategy (event) of each innovation as an observed value. Given the different research needs and practical limitations, our sample size does not include all listed companies. Our research focuses on the impact of the ownership structure on enterprise innovation, and therefore requires the selected enterprises to have certain number of innovation projects. However, there is virtually no technical innovation in such industries as finance, tourism and commerce. In this case, when selecting samples, we first removed special treatment (ST) and particular transfer (PT) enterprises; second, our research focused on industry and manufacturing sectors and excluded finance (including banking and securities), hospitality, commerce and other service sectors; third, given the stability and consistency of sample data, we selected all major investment in innovation projects of listed companies from 2000 to 2005; finally, in consideration of company changes (for instance, 000003 SJT of Shenzhen Gintian Industry (Group) Co., Ltd. was delisted after 2002 and no data and information could be found after the year 2002, so it was excluded) and data loss, the final sample size of innovation companies was 541, 97.6% of all companies of the selected industries (541/554), includes altogether 8,553 pieces of data, conforming to the statistical requirements of a large sample size.

Our research includes sample companies listed in Shenzhen and Shanghai because it is difficult to collect data of general companies in China, and listed companies are important sources for collecting data of Chinese companies (Li et al., 2003).[5]

First, as a company group, listed companies account for a small proportion in either developed or developing countries, but they hold an outstanding status in their respective national economic systems. The number of listed companies in China rose from 10 at the beginning of the establishment of China's securities exchange in 1990 to more than 1280 in 2002 (the SME board was opened in 2004 and at the end of that same year, the number of listed companies surged to 1459). They play a vital role as the pioneers in reform and in the establishment of a modern enterprise system during the economic transition process in China.

Second, listed companies have much clearer property relations, and a stricter organizational mode and management system.

Third, they are subject to the supervision of the public and rules of the stock market, and need to disclose information about their production and operation to the public and relevant stakeholders. This brings convenience to empirical studies and is convenient for launching time series analyses and longitudinal studies.

Fourth, most of China's listed companies emerged from SOEs and possess the features of many companies in economic transition, so they are characteristic of representative companies for studying state-owned shareholders.

Fifth, information of listed companies is open, which is easy for data collection.

Sixth, listed companies cover a wide range of industries and virtually all sectors in the national economy. In addition, they are often large companies and extend their reach all across China, which makes them typical representatives (see Table 10.2).

In this connection, listed companies are ideal objects to study ownership structure and the relationship between the selection of enterprise innovation strategy and achievements.

Table 10.2. Regional distribution of listed companies in 2002 (only listed in A-share market)

Region	Number of enterprises	Region	Number of enterprises	Region	Number of enterprises
Jiangxi	23	Gansu	18	Anhui	37
Liaoning	55	Guangdong	65	Beijing	76
Inner Mongolia	19	Guangxi	22	Fujian	44
Ningxia	12	Guizhou	16	Qinghai	9
Hainan	23	Hunan	40	Hebei	31
Shandong	69	Jilin	35	Henan	31
Shanxi	22	Jiangsu	77	Heilongjiang	33
Shaanxi	28	Yunnan	19	Hubei	59

(Cont'd)

Region	Number of enterprises	Region	Number of enterprises	Region	Number of enterprises
Shanghai	138	Zhejiang	63	Tibet	8
Shenzhen	71	Chongqing	25	Xinjiang	27
Sichuan	64	Tianjin	21	—	—
Total	1, 280				

Source: Yao Jun, "Empirical Study on the Diversification, Structure and Performance of Enterprise Group in Transitional Period," PhD thesis of South China University of Science and Technology (2004).

Sample description

Table 10.3 reflects the industrial distribution of sample enterprises.

Table 10.3. Industrial distribution of sample enterprises

Industry	Shanghai Stock Exchange	Shenzhen Stock Exchange	Total
Machinery Industry	121	75	196
Pharmaceutical Industry	54	24	78
Communications Industry	52	32	84
Electronics Industry	28	17	45
Petrochemical Industry	81	57	138
Total	336	205	541

Table 10.4 indicates the enterprise distribution based on the proportion of state-owned equity of the sample enterprises. We can see from Table 10.4 that among the production-oriented listed companies, most of sample enterprises have state-owned shares at their core positions. This has indicated that this study regarding the influences of state-owned shareholders on enterprise innovation strategies is of great importance to Chinese companies. In addition, enterprises without state-owned stocks are emerging. Thus, to the study of innovation strategies of non-state-owned enterprises is also beneficial for the development of Chinese companies.

Table 10.4. Ownership nature of sample enterprises

Proportion of state-owned stocks	0	Less than 30%	30%–50%	More than 50%
Number of enterprises	153	96	93	199
Proportion in the sample	28.3%	17.7%	17.2%	36.8%

Table 10.5 is reflects the classification of sample enterprises according to the concentration of ownership. Most production-based listed companies in China are controlled by large stock holders. Of these, the top 10 stock holders possess 50–70% of the shares (292 enterprises, or 54% of the total sample). We can learn from the shareholding proportion of the largest shareholders that, although the enterprises with their largest shareholder possessing less than 30% shares constitutes a larger proportion (41.6%) in the sample, the enterprises with their largest shareholder possessing 30–50% shares still occupies another larger proportion. This indicates that the Chinese companies are dominated by the largest stake holder.

Table 10.5. Ownership nature of sample enterprises

Shareholding proportion of top 10 shareholders	Less than 30%	30%–50%	50%–70%	More than 70%	Proportion of stock right the largest shareholder	Less than 30%	30%–50%	50%–70%	More than 70%
Number of enterprises	14	88	292	147	Number of enterprises	225	192	103	21
Proportion in the sample	2.6%	16.3%	54.0%	27.2%	Proportion in the sample	41.6%	35.5%	19.0%	3.9%

We can find from Table 10.6 that the diversification degree of Chinese production-based listed firms is relatively even, and the number of highly diversified enterprises is comparatively larger. Most of the companies are young—five to fifteen years old—because most of them were established after China's Reform and Opening-up program was initiated in 1978. The same is true for the distribution of listed age: most became listed five to fifteen years ago; the majority was listed less than ten years ago. The two Chinese stock exchanges, Shanghai Stock Exchange and Shenzhen Stock Exchange, were established in 1990.

Table 10.6. Distribution of diversification degree and age of sample
enterprises

Item		Number of enterprises	Proportion in the sample
Diversification degree	0	80	14.8%
	<0.5	100	18.5%
	0.5–1	151	27.9%
	>1	210	38.8%
Age of enterprises (year)	<5	0	0
	5–10	228	42.1%
	10–15	239	44.2%
	>15	74	13.7%
Listed age (year)	<5	92	17.0%
	5–10	258	47.7%
	10–15	183	33.8%
	>15	8	1.5%

Research method

This research employs discrete-time-event-history techniques to analyze data
and estimate regression model (Allison, 1984; Yamaguchi, 1991). The data in
this study is an integration of different amounts of time, regards each project
as a unit, and every enterprise is observed to get multiple numerical values.
This research mainly uses SPSS software (11.5 version) in statistical analysis
and verification, and multiple regression analysis methods are used in verifying
the logistic regression model (i.e., Enter). Because this study is to verify the
pre-assumption and has acknowledged the important degree of variables, it is
proper to use Enter (Agresti and Finlay, 2002).

Measurement of variables

Dependent variable: innovation approaches. We classified them into three types:
self-dependent innovation (SI) represents the innovation activities of internal
investment; cooperation innovation (CI) means to collectively invest in a unique
innovation activity with others who may be investors, business partners and/

or research institutes; technology purchasing (P) indicates to acquire innovation results directly from others. Those variables are evident in the annual report of companies. We use the three dummy variables to quantify the selection of innovation methods.

Independent variable: concentration ratio of state-owned shares. The research is designed to verify the impact of ownership structure, especially the concentration ratio of state-owned shares on innovation approaches. In this sense, independent variables are the state-owned shares' concentration ratios (SC)—the percentage of state-owned shares in total capitalization.

Control variable: the first control variable to be considered is the company scale. We use the natural logarithm of the main business income in the year of innovation to measure it. Then, we draw into the age of the company and the year of being listed. Next, we take into account the impact of business diversification on the innovation performance of enterprises. Finally, we take into consideration the term of investment. Generally, the longer the investment term of the project, the higher the cost of the project and the larger the impact on business performance. In addition, we regard the researched industries as a control variable. We use four dummy variables to describe the samples from the five industries.

Argumentations

The matrix of samples is appraised to ensure that data can meet basic statistical assumptions. The research sample is large enough to verify the assumptions, and the related coefficients between dependent variables show that there is no multi-collinearity between them. Table 10.7 indicates the basic description statistics and the correlation coefficients of all variables used in the study.

We conclude with six models to report the test results of the assumptions of the impact of stock-owned shares on the innovation method selection (see Table 10.8). Model 2 verifies H1. The model demonstrates that state-owned shares and internal innovation have an evident positive correlation, so H1 is verified. Model 4 verifies H2. The assumption attains evident correlation, so H2 is verified. We further use Model 6 to check H3. Model 6 indicates that there is no clear relationship between state-owned shares and technology purchasing. In this case, H3 has no support.

Findings of the empirical analysis showed that state-owned shares prefer self-dependent innovation of internal investment and do not take the initiative to partner with others, while technology purchasing showed no special preferences.

Table 10.7. Description statistics and correlation between variables

	Mean value	Standard deviation	1	2	3	4	5	6	7	8	9	10	11	12	13
Pharmaceutical industry	0.190	0.394	1												
Petrochemical industry	0.220	0.412	-0.257**	1											
Electronics industry	0.060	0.246	-0.128	-0.139**	1										
Machinery industry	0.370	0.482	-0.373**	-0.403**	-0.201**	1									
Investment term	1.512	1.021	0.041**	-0.044**	0.000	0.050**	1								
Natural logarithm of the scale of main business	11.377	1.220	-0.043**	0.016	0.005	0.019	0.045**	1							
Company age	11.052	3.685	-0.020	-0.012	-0.005	-0.024*	-0.123**	0.078**	1						
Listed age	8.214	3.128	-0.091**	0.006	0.021	0.033**	-0.158**	0.196**	0.658**	1					
Business diversification	0.593	0.454	0.014	0.019	-0.041**	-0.025*	-0.008	-0.063**	-0.021	-0.100**	1				
State-owned shares	0.300	0.269	-0.011	0.092**	-0.039**	0.049**	0.057**	0.107**	-0.252**	-0.084**	-0.075**	1			
Internal R&D	0.810	0.393	-0.114**	0.072**	0.094**	-0.028*	0.228**	0.067**	-0.018	-0.068**	-0.076**	0.075**	1		
Cooperative R&D	0.162	0.369	0.041**	-0.161**	-0.071**	0.093**	-0.216**	-0.053**	0.050**	0.100**	0.045**	-0.059**	-0.846**	1	
Technology purchasing	0.072	0.259	0.035**	-0.074**	-0.026*	0.053**	-0.113**	-0.022	0.032**	0.045**	-0.004	-0.024*	-0.026*	-0.123**	1

Notes: $^{*}p<0.1$, $^{**}p<0.05$, $^{***}p<0.01$.

Table 10.8. Results of logistic regression analysis

Variables	Model 1	Model 2	Model3	Model4	Model5	Model6
	Self-dependent R&D		Cooperative R&D		Technology purchasing	
Constant terms	1.300***	−1.320***	1.649***	1.626***	−0.808	−0.905
Pharmaceutical industry	−0.924***	−0.944***	0.153	0.159	0.328*	0.314*
Petrochemical industry	0.207*	0.164	−1.602***	−1.574***	−0.501**	−0.503**
Electronics industry	0.947***	0.970***	−0.917***	−0.941***	−0.266	−0.265
Machinery industry	−0.561***	−0.566***	0.364***	0.362***	0.568***	0.566***
Investment term	1.753***	1.747***	−2.228***	−2.222***	−0.931***	−0.924***
Natural logarithm of the scale of main business	0.137***	0.114***	−0.131***	−0.108***	−0.118***	−0.120***
Company age	0.018	0.028**	0.004	−0.002	0.015	0.020
Listed age	−0.055***	−0.052***	0.054***	0.049***	0.032	0.031
Business diversification	−0.456***	−0.424***	0.375***	0.368***	0.033	0.034
Proportion of state–owned shares	—	0.542***	—	−0.460***	—	0.268
Wald/2	866.602***	862.435***	980.905***	974.887***	208.338***	204.332***
Log–Likelihood	4,829.763	4,780.464	4,253.799	4,211.424	2,684.975	2,659.810

Notes: $*p<0.1$, $**p<0.05$, $***p<0.01$.

Conclusions

This research is aimed at verifying the impact of state-owned shares on companies' selection of innovation methods. We found through our analysis that state-owned shares actually have important impacts on innovation methods. Enterprises dominated by state-owned shares are supportive of self-dependent innovation of internal investment. This finding is of great importance for the theoretical development of state-owned shares and companies' innovation strategies. State-owned shares play the role of a shareholder who obtains residual claims that could deeply affect the innovation strategies of the enterprises. Our explanation of the complicated relationships between state-owned shares and enterprise innovation could adopt the institutional theory (DiMaggio and Powell, 1991). Against the backdrop of a traditional planned economy system, China held a monopolized role in resource allocation through its national plan. Previous studies on the relations between China and SOEs deem the country went beyond the institutional restriction of autonomous administration in strategic decision-making (Lu, 1996). During economic restructuring, the country or government began to change its role from a single unit to a multi-functional entity. For instance, in large enterprises with a dominance of state-owned shares, the country or government serves as a large shareholder or major investor, so it has obtained the direct control right of companies' strategy management. Given the existence of an institutional vacuum, and under imperfect market conditions, the government has maintained some controlling rights over resource allocation. Compared to non-SOEs, SOEs could enjoy more key resources provided by the government. China's economic reform is trying to transforming traditional SOEs into shareholding companies, but state shareholders have inherited the power to plan and exert control on companies.

Research has shown that during the process of China's economic transition from a top planner to an investor, the country was subject to many restrictions from the traditional planned economy and the country's multi-functional roles. The traditional administrative control may disappear along with the reform of SOEs, but multiple functions of the country still play an important role in companies' strategy management. In this sense, SOEs are only partially market-oriented. As shown in our study, SOEs are more supportive of self-dependent innovation of internal investment, although cooperative innovation is presently the main trend.

Our research makes two major contributions to the theoretic development in this regard. To begin with, there exists no similar study to help understand the complex relations between corporate governance and innovation in the transitional economy. Such complex relations are the result of the multi-functional features of the government. Previous research in this respect pointed out the importance of institutional investors for innovation strategy, but a comparative study of Lee (2005) noted that institutional investors differ in different societies. Our studies pointed out such a fact: if a country invests in companies, it would be regarded as an investor, but different from the standard institutional investors due to its multi-functional roles of policy making, resource allocation, external control, and going beyond the rules of economic activities.

Secondly, state-owned shares of listed companies may engender severe agent issues. Previous studies on SOEs mainly focused on the agent relationship between the country as a main investor and manager. The advantageous position of a country in the ownership structure of publicly traded companies will trigger other agent issues: the government may act as a large shareholder to grab the wealth of other investors (Fan, Wong and Zhang, 2005). With regard to how such an agent relationship could affect business innovation, we need to carry out specific research in the future.

Beyond that, our studies have important practical significance. Our studies indicated that the concentration ratio of state-owned shares in listed companies should be reduced, and the companies could attract more independent investors — institutional investors in particular. This could add to the input of companies in cooperative innovation and accordingly elevate the competitive advantage of companies. The government should take steps to improve market conditions as a way to support all forms of innovation, regardless of whether the companies are dominated by state-owned shares, since this could bring new views and original ideas that companies greatly need.

It should be noted that, despite of the supportive role of state-owned shares in the internal innovation of enterprises, state-owned shareholders have brought institutional pressure on the innovation activities of enterprises instead of efficiency. As the world economy develops and innovation activities become increasingly complicated, many innovations cannot be realized by individual companies; rather, companies must partner with others to achieve innovations. But state-owned shares have become a major obstacle; in order to promote cooperation and communication between companies and promote more effective innovations, Chinese companies must work to reduce the proportion of state-owned shares in the long run.

Our studies have offered opportunities for future research. It is definitely necessary to launch long-term studies on business to explore the relationship between state-owned shares and business innovation strategies. The features of innovation have determined that its influences on business capacity and performance are long-term and subtle. A long-term observation of enterprises and long-term comparisons of the relationship between different ownership structures and innovation strategies could help us gain a thorough understanding about the relationship between ownership structure, innovation, and business performance. In particular, if state-owned shares have undergone changes during the studies, information available would be more useful and meaningful.

11
Chapter

The Organizational Model of Independent Innovation
Wang Fengbin

Introduction

As enterprises' demand for new business growth points enhance and innovation makes a more evident contribution to economic development and business profit, innovation activity has been widely regarded as the task of the entire enterprise rather than just the work of the research and development (R&D) team. The innovation process, from the formation of the product concept to the profits from the market, requires the efforts of many. Because they are scattered throughout different levels of enterprises or domains of products or functions, coordinating and integrating these participants throughout the entire innovation process has become vital for the success of the independent innovation of enterprises. It is evident that the organization in the innovation process has become an indispensable part for the building of the independent innovation system of enterprises.

If enterprises want to obtain business upgrading and profits such as "founder's gain" through innovation activities, they must guarantee that new products they have developed through R&D are successful, not only technologically, but also commercially and economically. The product innovation process can be divided into two stages—to propose a new product concept and to put the concept into practice. Although these two stages are different, respectively reflecting the exploration of new knowledge and exploitation of existing knowledge, they must be integrated to strike a balance; only in this way can effective innovation be made a reality (Tushman and O'Reilly, 1997; He and Wong, 2004; Rirbinshaw and Gibson, 2004).

"Ambidexterity" in this chapter means an organization possesses ability to both explore and utilize acquired knowledge. This chapter is based on literature reviews about the paradox and its solutions raised in the studies of the "ambidexterity" of enterprises, in particular, Haier Group. It seeks to analyze how enterprises form an organization mode of ambidexterity at both the team and individual level in terms of knowledge composition of R&D team players as well as internal and external communications. This study seeks to explain the paradox relationship of knowledge exploration and utilization from the perspective of the unity of opposites.

The structure of this chapter is as follows: after the introduction, the following parts summarize relevant literatures with respect to the paradox relationship of knowledge exploration and utilization, take stock of the existing understanding of scholars about the paradox in innovation activities, conclude and analyze several strategies for solving the paradox, and then generalize the issues this research seeks. In general, existing literatures carry out normative studies at the abstract organization structure and try to provide an organization mode or strongly tied and closed network structure in knowledge exploration, while in knowledge utilization, the options are a mechanical structure of organization or weakly tied and bridged network structure. However, the implicit issue of how to link the 'structural gap' between the two organizational modes in these studies has yet to be answered explicitly. While drawing inspirations from paradox management thinking, this study seeks to address these "how questions." This chapter explains the value of the case study approach and specific implementation process, before describing in detail the case study of new product R&D of Haier air conditioner and then carrying out a comparative study.

Literature Review I: Paradox in Innovation and its Solutions

Innovation process is an integration of two stages—innovation production and innovation application. Management master March (1991) pointed out, from the perspective of organizational study, that generating innovations could be tracked back to the exploration of an organization into new knowledge, which includes various activities of conceiving an innovation, such as search, variation, adventure, experiment, improvisation, flexibility, discovery, and creation. I innovation application refers to the exploitation of existing knowledge by an organization which could be manifested by such terms as refinement, selection, production, efficiency, choosing, implementation and execution. They are different activities of different natures which have the effects of increasing diversity (McGrath, 2001) and selective retention (Dopfer, 2004) respectively, bringing out two different results of stimulating creativity and elevating efficiency.

The generation and application of a concept are two starkly different processes, but they are integral elements for a successful innovation. Technological invention in a laboratory only underscores a new innovation result, but fails to note whether the result could be transferred to a readily marketable commodity and bring a final economic success. To do pioneering work within an enterprise is a complete process, from generating a new concept to obtaining profits from the new pioneering work. The management team must balance the tension formed between the requirements of creativity and efficiency, and the different subjects' behavior logic in knowledge exploration and utilization. This tension is called "dilemma" or "paradox" by Western academics (Poole and Van de Ven, 1989; Handy, 1994; Mitroff, 1995; Bouchikhi, 1998) and is termed as "contradiction" by Chinese scholars (Li Zhanxiang et al., 2000).

Given the paradox or contradiction in the innovation process, the concept of "either A or B" has become more widely accepted in both the business circles and management academics communities supported by the "contingency theory." For instance, when it comes to the explanation of the relationship between organizational structure and the performance of researchers or research organizations, scholars feel that efficiency and results are meaningful measurement scales and organic organization structure is often associated with the target realization, while mechanical organization is related to the high efficiency of resource utilization (Kedia et al., 1992; Hoyt and Geroff, 1999). Daft (2003) noted clearly, from the perspective of organization design, that an organic structure, such as decentralization of powers and autonomy of employees, is conducive to stimulating the innovation concept, but is unsuitable for the large-scale application of the new concept. Other scholars specializing in "innovation tension" (Gassmann, 2006; Perez-Fgeije and Enkel, 2007) argued that there is a strong tension between the resource efficiency of R&D activities driven and pushed by customer demand and organizational discipline and attained through the process management on the one hand, and the creativity obtained through personnel management propelled by technological possibility and supported by organizational slack on the other. Under this condition, enterprises select "either this mode, or that mode" according to the contingency factors (such as industry dynamics). As stated in "The Creative Tension during the Innovation Process," (the design of) innovation control system varies according to the contingent industry dynamics. For enterprises which are faced with rapidly changing environment, their control system should support creativity, while for those operating in a slowly changing landscape, their innovation control system should enhance resource efficiency (Perez-Fgeije and Enkel, 2007).

Admittedly, it may be called the wisest and most viable strategy to address the paradox or contradiction of knowledge exploration and utilization if different contingency methods are adopted in innovation management based on real conditions. However, the implication of this strategy is that there is an antagonistic relation between knowledge exploration and utilization, and the only way to solve this contradiction is to introduce the contingency factors in order to adopt different organizational modes during different occasions. Of course, contingency happens among different enterprises and among different business departments within an enterprise. However, if different organization modes are used in different business departments within an enterprise, it may be difficult to reconcile the conflicts between them. Given this, Drucker (1987) advocated that an innovative business should be organized independently. He pointed out that "in an enterprise where innovation is necessary, product development is a key activity that can offer profits. It shall not be subject to any other activities, but be greatly valued as an independent innovation unit...in order to pursue innovation, a separate structure shall be established outside of the business structure of a going concern." Christensen (1997) also believed that innovation always exists in an "autonomous" organization. For example, an established business could form an independent and external company through stripping or absorbing the risk investment. This step can speed up innovation, and particularly at the early stage of innovation, it can stimulate the creativity of the new business developers. However, due to the great difficulty in integrating with the existing business, the independent enterprise system, only linked through investment, becomes less important (Iansiti and Levien, 2006). As for the traditional R&D system (which is starkly different from the independent enterprise system), product R&D is restricted within the R&D department of an enterprise, ignoring the coordination with other departments. Iansiti and Levien believed that such a R&D organizational mode goes against the efforts to promote the major destructive revolution, thus, is an extreme mode that is not recommended for utilization.

These studies demonstrate that knowledge exploration and utilization in the organization of innovation activity are a paradox or contradiction. The ignorance of this paradox will only throw enterprises into a status of low production capacity. There are three viable strategies to handle the paradox (Fig. 11.1).

Fig. 11.1.   Solutions to the paradox

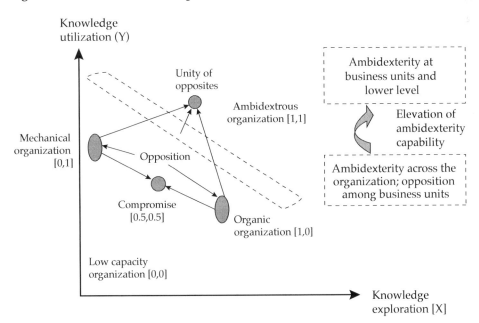

Opposition strategy

To put it simply, the opposition strategy is to address the paradox through the method of "either A or B."[1] For instance, innovation activity is organized through an independent company structure which separates the innovation activities with the regular activities of an enterprise. It seems that this strategy has realized the integration between knowledge exploration and utilization, judging from the level of the entire organization. However, when it comes to business units, some (newly-established independent companies) are committed to knowledge exploration while others (existing units of the parent company) focus more on knowledge utilization. In fact, the two fail to reach an integration and are in an opposite status of "either A or B." This can be manifested by the fact that it is more likely for newly-established units to adopt an organic organization mode and choose knowledge exploration over utilization; on the contrary, existing units within the parent company employ mechanical organization mode and choose knowledge utilization over exploration. In this connection, the subjects of activities (business units) lower than the organization level had to select one from "either A or B." The two cannot be obtained at the same time. As is shown in Fig. 11.1, the units employing organic organization mode are in [1,0] while those embracing mechanical organization mode are in [0,1].

Contingency strategy

Chinese scholar Yuan Yongzhi (2001) regards the specific environment in which where enterprises are" as the contingency factor for which the dual organizational structure in the innovation design must be taken into account. He believed that (old) enterprises at the stage of incremental innovation shall adopt the mindset of maintaining some qualitative changes in overall quantitative changes, while those at the stage of radical innovation are always new enterprises who shall embrace the thinking mode of keeping quantitative expansion in qualitative changes. With regard to "old enterprises," he uses two non-intersecting circles to represent the dual pattern of organization mode of incremental innovation and organization mode of radical innovation. As for "new enterprises," he makes the judgment that they shall immediately enter into the stable growing stage, start incremental innovation and then re-start a new incremental innovation after a radical innovation in the start-up stage. Then he shows, through two overlapping circles, that the organization mode of incremental innovation is included in the organization mode of radical innovation. After variables of different innovation stages are introduced, the comparative graphical representation reflects the difference of new and old enterprises when dealing with the relationships of conflicts or integration of two innovations. Old enterprises need to have some qualitative changes in overall quantitative changes, but the organization modes of two innovations are exclusive. New enterprises which need quantitative expansion in qualitative process inclusively incorporate the organization mode of incremental innovation into the original organizational mode of radical innovation. But with regard to what is behind this difference, the author provides no explanation. Nonetheless, the article implicitly noted that in terms of inclusiveness, new enterprises are stronger than old ones, which are highly rigid.

Compromise strategy

Making the best of both worlds means to make knowledge exploration and utilization integrated, according to a certain proportion within a specific subject. The result of compromise is, if optimistic, to make the absent nature enhanced by sacrificing another nature in the organization; but if pessimistic, it means that both knowledge exploration and utilization are reduced to some degree. Unit or organization in [0.5, 0.5] and its surrounding areas in Fig. 11.1 realize the "increase" in one dimension through the "decrease" of another dimension. That being said, due to the non-linear function, a certain proportion of "decrease" in one dimension does not necessarily mean the same proportion of "increase" in another dimension. Many rigid and inflexible big companies, in the face of innovation pressure, take measures to reduce their mechanical features of the structure, replace complex regulations and rules with simple "mission statements," and promote the organizations to embark on organic structure through the disintegration of administrative bureaucracy. That being said, the result is not always enhanced innovative capability as expected, but chaos in the organizations. An effective innovation calls for an "organized disorder" (Daft, 2003) rather than a so-called "semi-orderly" condition by linear or summation logic.

Strategy of unity of opposites

Philosophically, this strategy is to address the relationship of two contradictory aspects through "equilibrium between yin and yang" in Chinese Taoist philosophy. Chinese ancient philosophy believed that all creatures in the world are composed of yin and yang. "Yang" represents strong, dynamic, rising, stimulated, open and positive factor while "yin" stands for delicate, static, decreasing, quiet, convergent and conservative factor. These opposite factors (binary) are contradictory and interdependent, and co-exist in harmony and peace. Based on the specific explanation of traditional philosophy of yin and yang, Ju Qiang (2003, 2007) noted that the operation of the enterprise system is similar to that of the human body and is binary in the operational mode. The two are opposite yet integrated, restrictive but indispensable. A good coordination between the dual factors means a smooth development of the company; otherwise, the enterprise will develop slowly, stop or even die out. In this case, he argued that the normal operation of a society or organization depends on the relative balance and co-existence of the two factors, namely, yin and yang.

We can therefore deduce: different from the middle course of seeking compromise, enterprises which pursue the strategy of "unity of opposites" can find the most suitable relative balance in two opposite factors (yin and yang). As shown in Fig. 11.1, we may take knowledge exploration as one side of the conflict (yang) and knowledge utilization as the other side (yin). They are two complementary factors and can maintain the relationship of both opposition and unification, like yin and yang.

The ambidextrous organization that this article analyzed refers to a dual or binary organization which can unite knowledge exploration (yang) and knowledge utilization (yin). Duncan (1976) came up with the idea of ambidexterity for the first time in "The Dual Structure of Innovation Design." After two decades of having this work largely ignored, Tushmann and O'Reilly (1996) jointly published the article named "Ambidextrous Organization: Management on Evolution and Revolutionary Transformation" in *California Management Review* and published the book *Innovation Leads to Success* in 1997, which aroused the attention of western business circles and academic communities on the existing contradictions in the organization model of business innovation. It has recently become one of the hot spot research topics.

However, the existing literature focuses more on the potential pros and cons of such organization than the management difficulty and how the "ambidextrous capability" is formed. Most of the research discusses the abstract organization structure. A South Korean scholar and two American scholars jointly issued a review article of theoretical deduction in *The Academy of Management Review*, which, from the perspective of strategic human resources (HR) management, focused on the relationships (divided into start-up type and cooperative type corresponding to exploration learning and utilization learning) between the employees with core knowledge and their internal and external partners (referring to employees in a traditional sense, and co-operators in partnership enterprises respectively). These relationships were examined in three dimensions: structure, sentiment, and recognition, and reveals how specific HR management helps the natural formation of the expected relationship (Kang et al., 2006). However, the authors only described the content of HR practice, targeting at individual start-up and cooperative relationship and maintaining the standpoint of "either A or B." For instance, in order to promote the formation and development of start-up relationships, the authors suggest managers taking flexible working structure, results-based incentives and cross-professional skill development as the ideal HR practice; on the contrary, in order to facilitate the formation and development of a partnership, the authors propose an inter-dependent working structure, incentives conducive to the cultivation of community culture, and broad capability development that can promote the formation of common framework knowledge as more viable HR practices. The demonstration of these practices, together with the association analysis between these and the three dimensions of the relationship, showcase how managers promote the creation of a bigger value through managing the partners inside and outside of enterprises. However, the picture they show us is still a pattern of "either A or B," choosing from the two opposite HR management practices, two opposite relationships, and the two incompatible knowledge exploration and utilization.

The inspiration this study offers us is that core employees could form a either strong or weak relationship with partners inside and outside of enterprises. This view has coincided with the concept that some scholars have proposed when they analyzed from the corporate level to the business level or below. Together, they guide us to focus more on the "ambidextrous capability" of employees rather than the "ambidextrous (binary) structure" of the organization. It is in the analysis of the lower level subjects that we extend the definition of "ambidextrous organization" to "dual structure" and "binary structure" to "ambidextrous capability," and explore the strategies to achieve a balance between knowledge exploration and utilization during the innovation process in the lower level.

Literature Review II: From Dual Structure to Ambidextrous Capability

Multi-faceted challenges amid a contemporary competition environment urge enterprises to create two organizational conditions, i.e., organic and mechanistic conditions, to obtain creativity and maintain high efficiency at the same time. In order to realize the two targets which are not contrary, many companies adopt dual structure. Dual structure means that two organizational structures and management processes, which are conducive to the stimulation and application of innovation, are incorporated within a company. Such an organization can operate in an organic way when stimulating new thoughts while it also can operate in a mechanistic manner when implementing and executing new thoughts. For now, many large organizations have put in place creative department to assume innovation works. R&D, engineering design, systemic analysis, and other independently established departments put forward innovation ideas for other departments to adopt. These departments, which initiate innovations, are often organized in an organic manner to promote the smooth generation of new thoughts and technologies. Meanwhile, those departments which apply innovation always use a mechanistic organization structure, more suitable for highly efficient production.

Studies conducted by Nonaka (1994, 1995) on large Japanese companies such as Canon and Toyota about punctuated technical innovation showed that these companies have established a team (referred to as an "internal incubator") composed of younger staff who are responsible for R&D in technical innovation. The team in the "internal incubator" is often loosely formed and pays attention to the formation and progress when technical development enters into a new cycle. In order to ensure it is free from the impediment of the original organization, the "internal incubator" is always physically separate from it. (Shi Jiangtao, and Cao Bing, 2006).

Overall, both domestic and overseas scholars have advocated that dual organization is an organization form that enables large entities to carry out gradual and radical revolution at the same time. Under such a form, the original organization of a company is committed to gradual innovation while the team engaging in radical innovation is a highly independent internal organization whose capacity, structure and culture are highly inconsistent with those of the original organization (Duncan, 1976; Tushman and O'Reilly, 1996; Yuan Yongzhi, 2001; Zhang Hongshi, Chenjin, 2005; Zhang Xi, 2006; Li Yuling, 2006). So will the two highly inconsistent departments or organizations be coordinated to avoid isolation or confrontation? In the following paragraphs, we will summarize the strategies offered by scholars to deal with conflicts arising from the dual structure based on literature about the paradox-resolving methodology and innovation paradox (Poole and Van de Ven, 1989; Gharajedaghi, 1999; Adler et al., 1999).

Spatial separation and the role of top management

With regard to the issue of spatial separation existing in dual structure, Tushman and his partners came up with the concept that an ambidextrous organization must be equipped with a strong integration capability (Tushman and O'Reilly III, 1996, 1997, 2004), strategic integration capability (Benner and Tushman, 2003), or the capability to manage strategic conflicts (Smith and Tushman, 2005). They argued that, in face of innovation, enterprises shall separate a new exploratory department from the traditional knowledge utilization department and allow the new department to put in place different procedures, structures, and cultures, but it shall be subject to the management of the top executives. Only by being managed by the closely linked top executives can the separately established department make the organization committed to utilize not only existing potential, but also new opportunities, thus, serving for both short-term efficiency and long-term innovation.

It should be noted that traditional organization logic integrates the separated dual structure through centralized power (i.e., top management as the centralized agency) to achieve the integration of the dual structure. This reflects the top-down coordination and management process. Those scholars who echo ambidextrous structure but discard centralized power are committed to seeking the power to unify the dual structure at a lower level of the organization. Such a trend is actually a reflection of the attention on the coordinated role of distributed agency. The following discussions about switching structure and ambidextrous structure demonstrate the tendency of anti-centralization.

Temporal separation and switching structure

With regard to ambidextrous organization,[2] some scholars (McDonough and Leifer, 1983; Daft, 2003) proposed that through switching structure, those engaging in conventional works could serve as innovators at some time. New United Motor Manufacturing, Inc. (NUMMI), an arm of Toyota in Fremont, California, established a cross-functional, independent group named the "navigator team," which is responsible for devising production procedures for new cars and trucks. After a new type of vehicle is put into production, team members will return back to the workshops and engage in daily production and work. Another example is a building material manufacturer based in Oregon that establishes more than 150 temporary teams every year, which are composed of staff from different departments. Staff at these organizations participate in innovation teams then return to working in a mechanistic fashion at their 'regular' positions; thisis called "switching structure." That being said, it is not an easy task for staff to switch smoothly between the "mechanistic–organic–mechanistic" forms. Therefore, scholars summarize the potential problem in this structure as "temporal separation." The authors of "Temporary Separation for the Sake of Success" put forward the strategy of re-integration after temporary separation to encourage organizations to use cross-sectoral coordination after initial exploration (Siggelkow and Levinthal, 2003). But they only answer why a temporary separation is needed and when to integrate through computer simulation, without touching upon how the structure switching is realized.

Ambidextrous capability

Given the real difficulty in the switching structure, or structural inertia, Gibson and Rirbinshaw, scholars of London Business School (2004) came up with the idea of a day-to-day balance between innovation and regular activities. They believed that enterprises with ambidextrous capability are not the institutions which use organic and intensive structure in one instance while using mechanical and extensive structure in another. Rather, the two contrary structures co-exist from time to time. They started analysis at the business unit level and advocated the formation of ambidextrous capability (the capability integrated by both alignment and adaptability through positive interaction that is conducive to the elevation of performance of business unit) cultivated in the ambidextrous situation. Such an organization ambidexterity generates through duality, so some domestic scholars also named it "situation duality" (Zheng Ping, 2007).

Under the dual structure advocated by Duncan, Tushmann and other American scholars, enterprises only have the ambidextrous feature or capability at the level of the entire organization. Some business units may only have the alignment beneficial for elevating efficiency while others only have adaptability conductive to innovation. In this sense, even when an enterprise is ambidextrous at the organizational level, they are opposite and far from being integrated at the business unit level. Studies were carried out from the situation and developed the analysis to the level of business units. They advocated that business units can form situational ambidextrous capability through the interactive combination between situational conditions for performance management (composed of discipline and extension) and situational conditions for social aspects (composed of support and trust) as a way to increase the performance of business units from the cultivation of situational conditions.

Undoubtedly, building an ambidextrous situation may affect every staff member in a business. As such, we can form data at the macro level by using individual staff members, reflecting the organizational features of the business, through the aggregation of low-level data satisfying the requirement of homogeneity.[3] As for how heterogeneous but complementary elements interact with each other, it still calls for further explorations.

Research Method

How will the paradox or contrary aspects in the innovation process be coordinated and integrated? Under which situation will the lower layer units and staff be committed to the exploration and utilization of new and existing knowledge? This study carried out qualitative research based on the specification of research questions, and tried to present an abstract theoretical explanation on how subjects participate in the formation of an ambidextrous organization and what is the role of top management in the process, through in-depth analysis on typical cases selected according to "theoretical sampling" (Eisenhardt and Graebner, 2007).

Different from the method of deductive inference used by most theoretical studies, our studies establish relevant theoretical viewpoints refined through analysis of relevant literature and based on both first-hand and secondary data, and the summary of typical experiences. When selecting enterprises for case studies, we took into consideration the organization's features, and the case information was obtained by combining the methods of file review, interview and secondary data.

Research background: profile of Haier Group and the innovation of the air conditioner

Haier Group, headquartered in Qingdao City, Shandong Province, China, is one of the three largest electrical appliance manufacturers in the world. Since the globalization strategy was carried out at the end of 1998, Haier has embarked on a major organizational revolution themed by "globalized competition, networked structure and re-engineered procedures." After seven years of development, its global turnover recorded RMB103.4 billion (USD12.8 billion) in 2005. Back in 1999, Haier produced 8600 types of household appliances in 42 categories, and by 2005, it had extended to 13,200 types of products of 86 categories. At the end of 2005, when celebrating the 21st anniversary of its founding, Haier Group put forward the strategy of "globalized brand." It now strives to elevate the competitiveness of its products and business operation by transforming its patents into differentiated products to meet the demands of customers in different countries and regions, under the production pattern of "large scale, more varieties and new products."

As it works to become an executor of independent brand strategy with differentiated advantages from low-cost suppliers within the global arena, Haier Group is committed to the development of the system for independent R&D and innovation. And like a pyramid, at the top is the central research institute with its mission of providing core technical support for Haier Group to build a world-renowned brand; at the middle is the product R&D department affiliated to various product departments; at the bottom are the centers for the R&D of production techniques scattered among different manufacturing units. The Air Conditioner Development Department, founded in 1991, is located at the middle of the pyramid and takes charge of R&D of household air conditioners (including wall-mounted split, cabinet and all-in-one air conditioners), dedicated air conditioners, dehumidifiers, fan heaters, air cleaner and other products. By focusing on independent innovation and differentiated products under the theme of "healthy, energy-saving and environmental protection," researchers of the Development Department have gained the trust of customers from 137 countries. In March 2006, Haier air conditioner won the Germany IF Industrial Design Award which represents the highest industrial design level in the world. It was the first time in the award's 52-year history that a Chinese home appliance manufacturer won. The patented technology "Double fresh air," developed independently by Haier, is the first of its kind in the world, which reflects that Haier has created a healthy household air conditioner standard with world-class technology.

Presently, the annual production capacity of Haier air conditioner is ten million sets. According to the information released by China Industrial Information Issuing Center under the National Statistic Bureau in March 2006, Haier had 21% of the sales volume and sales value in 2005 — a double winner. In 2005, Haier air conditioner recorded a 100% increase in foreign exchange earnings in the North American market compared with that of the previous year, becoming its top air conditioner. Based on its new product development speed and technological competitiveness, Haier air conditioner has become a domestic and international mainstream brand. It has planned to raise its air conditioner sales value to RMB30 billion, becoming the largest supplier in the world. The organization of the development process of Haier air conditioner is used as a major objective for research in our study.

Process of data collection

We started our data collection work back in 2000. At that time, Haier Group organized several seminars on the reform of the business process with a market chain as the link in a bid to attend the Seventh National Award Selection on Innovation Results of Modern Enterprise Management sponsored by China Association of Enterprises. Attracted by its reform achievements, the author of this chapter guided an MBA student to complete the master's thesis "Process-based Organization Transformation: Haier's Improvement in Process to Enhance Competitiveness" based on the research and study of its reform practices. Two years later, the author guided another MBA student to launch an investigation and analysis on the relationship between the development strategy and organizational transformation of Haier Group. Through these preliminary investigations, the author has grasped the relationship between the R&D department and the products and business department before and after the re-engineering of the market chain and different modes of cross-functional integration during the product development process. As Haier's reform is continues, the development team of Haier air conditioner has evolved into "SBU operation"[4] which reflects the idea of "complete result-oriented management." Based on this, the research team of this study launched a survey at the beginning of 2006 into the files of "SBU operational profit and loss statement of model manager" under the air conditioner development program set up in 2004 and 2005. The team gained access to network data which covered two years of project managers (model managers) and the personnel who participated in product development work, shared profits and formed working relationships. In the meantime, in order to obtain relational data outside of workflow, a major team player in the survey team launched an interview with managers of the product development team and with some model managers. There was also a two-day supplementary interview with the help of some members of the case study group at the end of June 2007. In order to explain the first-hand data, we read relevant reports in *Haier People* (one-volume edition) from 2004 to 2006, an internal document of Haier Group. With regard to policy or institutional regulations, we verified by examining the materials from the R&D promotion department and other product developments of Haier Group, and referred to the media reports on Haier Group in recent years. The data gathered from various channels are mutually complementary, making the data more reliable.

Haier's Market Chain Re-Engineering and New Product Development Team under "SBU"

Re-engineering of Haier market chain: the deconstruction of business process

Haier's manufacturing bases in China were among the first to embark on global development. In order to adapt to globalized development and the strategic transformation of global market competition, Haier used the market chain as the link while introducing international large suppliers (sub-suppliers) and large merchants (clients) via a networked structure, and constructing a strategic alliance with them. The re-engineering steps it took included three aspects:

First, separate the businesses of marketing, purchasing and accounting originally affiliated to each business division (department) and integrate them into the independently-operated commercial distribution (marketing system) promotion department, logistics (supply system) promotion department, and capital flow promotion department as a way to realize unified marketing, purchasing and settlement within the Group;

Second, centralize the original functional management resources. For instance, separate human resources, technological quality management, information management, equipment management, legal affairs and security, among other functions, from each business division (department), and form independently-operated service providers (units with an internal profit center);

Third, link the specialized processes through market chain, and design standards of remuneration (R), reimbursement (R) and trip (T) to realize the objectives of the signing parties.

The re-engineering of the market chain of the Haier Group has been improved gradually through implementation, summary and improvements alongside the re-engineering process, but looking back into its eight-year reform process, we find that the organizational structure, composed of the main process (named core process at the early stage) and the supportive process, is the fundamental direction of the organizational revolution of the strategic innovation. Under this structure, the target of the main process is to create the market and be responsible for customer orders throughout the process, realizing profits in the process of market value creation. The supportive process is required to transform into a profit center and is encouraged to provide public services (services for the third parties), but there remains a clear prerequisite, i.e., the objective of the main process is the ultimate goal of the supportive process (the "T model budget" that reflects the idea of countdown implemented in 2005 by Haier Group, the time (T) of the supportive process must be identified by attaching it to the target of the main process). If we want to review whether the service offered by the supportive process is valuable, we need to consider whether it can add value to the main process. Its value lies in the promotion and development of the work of the main process, which is closely linked with it.

All-inclusive SBU System: formation of independent innovation subject

Prior to the reconstruction of the market chain, the economic relationship between the R&D department and product division of Haier Group was only reflected in output. The traditional system, with "output" as the standard, made the R&D

department separate from the market. After the reconstruction at the end of 1998, Haier R&D company began to settle with the product division through intra-corporate price; that is, the newly-added profit brought about by newly developed products were used as a base number which was multiplied by the proportion predetermined through negotiation between the two sides, and the result represents the reward the R&D company shall obtain. As the "All-inclusive SBU System" is implemented in the R&D system of various products after 2002, the practice changed: the R&D team received commission in proportion to the surplus of the break-even point, based on the profits their new products gained in the market. Accordingly, the market result is integrated into the performance measurement of the R&D system.

As shown in Fig. 11.2 and Fig. 11.3, within Haier Group after the reconstruction of the market chain, there established a market relationship of paying rewards based on market results between the product division (internal client) responsible for order execution and the R&D department which provided the new product design (internal supplier). The market result of the new product was assessed after confirmation by the "gate departments" (such as the commercial distribution department, capital flow promotion department, and after-sales service department —later changed to customer service company) (Fig. 11.3). Such a process-connected economic relationship stimulates developers to shift their focus from commercialization of technologies to the overall result of product development. The attention on the profit of the entire process that has facilitated Haier Group to spread the SBU operational system to the area of product development in 2002 and put in place the SBU operation profit and loss system, under which the model manager of the development department is responsible for the entire process of the products it has developed.

At the early stage of implementing the SBU operational profit and loss system, model developers were required to be accountable for the entire process: "information collection, product plan, industrial design, structure, design, development of system and electrical control, verification, product certification, testing, trial-production, hit the market, new product follow-up, and quality improvement." It was rewarded with the net profit commission for the portion surpassing the break-even point (no commission for the part within the break-even point). The commission amount was reduced by 10% on a monthly basis; the term of commission was within 18 months after the product was launched. Later, given the fact that different measurements existed for the market result of newly developed products, Haier air conditioner headquarters began to implement differentiated commission measures for three types of products:

Fig. 11.2.   The economic relationship between the R&D department and product division before market-chain reconstruction

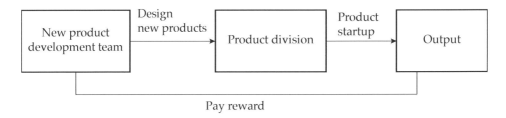

Fig. 11.3.   The Economic relationship between the R&D department and product division after market-chain reconstruction

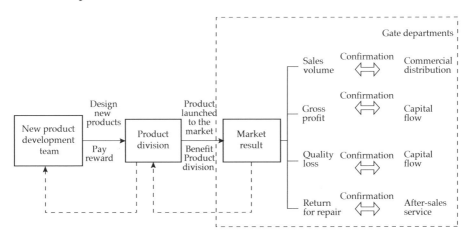

"Rare flowers," "Grassland," and "Uncommon trees," while exploring into the SBU operational system of model managers. "Rare flower" developers focus on creating profits; their commission is based on the profit surpassing the budget, with a basic requirement for sales volume. "Grassland" developers focus on quantity, with their commission calculated according to sales volume, adopting a policy of small profits and a quick turnover but ensuring a certain profit margin; commission is earned only when products reach the designated sales volume. "Uncommon trees" are those developers who focus on cultivating new types of products with high technical content (similar to uncommon trees like evergreen trees); the group encourages these developers to build up a technological reserve and become "excellent raisers."

Such a differentiated incentives and disincentives system as a result of diverse values created by products means that developers must strike a balance between meeting existing demand and creating a future market while considering power or resource allocation. Under market pressure, every SBU unit of Haier Group (individuals, teams or entities) has no leaders or subordinates, only market goals and relations. Haier Group CEO, Zhang Ruimin, offered a new explanation for SBU operation: decision-makers (leaders) provide market space and a resource platform (status quo, advanced information, a problem-solving channel and supportive process) while SBU makes innovations to address customers' complaints and become inspired by their satisfaction. Spiritual drive is the loyalty of users, while material stimulation is profit-based commission.

Under the guidance of strategic targets identified by decision-makers, SBU is tasked with the mission to "vie for resources, create orders and add value." For instance, the same components and parts of a refrigerator can be used to build two types of refrigerators. If one refrigerator meets the demand of users and is able to create profits for the enterprise, naturally, the model manager would earn a reward; the other, failing to address users' needs, would not earn its manager any rewards.

Based on this "complete result-oriented management," Haier has implemented an incentive and disincentive system under which the remuneration of both the product manager responsible for obtaining orders and the model manager in charge of creating orders is almost completely linked with the market result. A slight difference from the SBU incentives of the main process is that the incentives and disincentives for supporting or promoting Haier's process is based on the payments made by its internal clients; that is, to claim "remuneration" (R) based on internal transfer price. For instance, as the marketing platform, the commercial promotion department charges its product department a platform fee, which is used as service compensation, and the product department and the commercial promotion department simulate a market settlement based on "supplier price." The product and business division is the affiliation of the product department, which is in charge of order execution. If the order in execution receives customer complaints regarding product quality, the product and business division will be punished for quality loss. Likewise, if the product and business division discovers that the quality issue was not caused during the manufacturing process, but was caused during the development process, the division can claim "reimbursement" (R) against the model manager. If disputes arise and the responsibility cannot be identified, then the quality standard set up by the quality assurance company (as shown in the information platform of Haier, established via the SAP system) will become the basis for judgment. If no claim or claim remuneration occurred, it means that there is a flaw in the "gate," or the "trip" (T). Therefore, the "RRT" standard has become the top mechanism for various processes to communicate and coordinate after the reconstruction of Haier's market chain. Companies can act upon the real-time work data obtained from the information system to calculate the service fees for the upstream process or supportive process acting as suppliers (informatization, "three codes" and the integration between personnel and orders are basic steps taken by Haier to implement the "complete result-oriented management").

More importantly, the fact that Haier cultivates individuals, teams, or entities capable of independent innovation and responsible for operational results via the SBU system is not just limited to those in the main process and supportive process. In order to enable managers at various levels to operate products and create values in an innovative manner while focusing on the market, Haier designs the SBU remuneration of managers as follows:

(Management post) Monthly SBU salary = Basic salary + Commission – Loss

In this remuneration scheme, management posts are divided into S, B and U grades and each grade is sub-divided into three grades in accordance with the market size the manager operated, so there are 9 total grades. Then, each grade is further divided into large, medium and small levels—altogether 27 levels— according to market risk. This remuneration system shows that the largest value creator will get the highest remuneration.

The income of managers is, to some extent, connected with the profit of the product (model) made, so within the product department, the model manager (U grade) in the development department is called "small" model manager, while the directors of the development department (B grade) and product development (S grade) are termed as "medium" model manager and "large" model manager respectively.[5] These SBUs, large or small, are accountable for indicators of "price, profit and quantity" ("3A") of different contents, forming an all-inclusive SBU operation.

Composition of the development team and cross-functional relationship

The model manager is responsible for the development of new products throughout the process, but this does not necessarily mean that he should do everything in every link; rather, he can arrange resources of every aspect. The development of an air conditioner involves various stages. Stimulated by the target for market results, new product developers of Haier Group realized that they must enhance their coordination and communication capacity to construct a SBU team composed of relevant staff involved in the development process of new products. Any project team without boundaries could obtain relevant capital support; it is impossible to realize the project target by relying solely on a single department.

Haier's theory for building a project team without boundaries is to let employees vie for projects through a bidding process rather than to let the leader designate works. Composition and qualifications of the new product development team of Haier air conditioner is shown in Table 11.1.

Table 11.1. Composition and qualifications of the new product development team

Composition of the development team		Qualifications
Model developers (sharing 65% of commission from total profit)	Project lead (model manager)	Graduated from junior college or above; more than two years of working experience; familiar with development process; majored in refrigeration, heating, ventilation and air conditioning (HVAC), machinery, electrical or other relevant disciplines; strong capability in coordination and communication.
	Model design manager	Graduated from junior college or above; more than two years of working experience; familiar with development process; majored in refrigeration, HVAC, machinery, electrical or other relevant disciplines. [strong capability in coordination and communication]*
	Module manager for complete machine structure design; module manager for refrigerating system design; module manager for electronic control design	Graduated from junior college or above; more than two years of working experience; familiar with development process; majored in refrigeration, HVAC, machinery, electrical or other relevant disciplines.

(Cont'd)

Composition of the development team		Qualifications
Assistant developers (sharing 35% of commission from total profit)	Platform support manager	Graduated from junior college or above; more than two years of working experience; familiar with development process; majored in refrigeration, HVAC, machinery, electrical or other relevant disciplines.; strong capability in coordination and communication.
	Appearance design manager, certification manager and presswork manager	Graduated from junior college or above; more than one year of working experience; familiar with development process;
	Testing manager and test clerk	Graduated from technical secondary school or above; familiar with production process; can operate forklift and do welding work.

Source: internal material of Haier Group (2005).
Note: [ ]* in the Table is added by the author based on interviews.

Based on a summary of years of practical expertise and with reference to international experience, the new product development process of Haier is often composed of a team of two parts: core players are called "model developers" including the model manager as the project lead[6], one or two model design managers, several module managers; and "assistant developers" providing platform support, appearance design, certification, printing materials, experiment, and trial-production, among other assistant works. The distribution proportions of total commission by the two parts of stuff are 65% and 35% respectively.[7]

In the development of Haier Group household air conditioners, there are 13 project teams, each having 10–15 team players. Specifically, module managers include persons engaging in the design of the complete machine structure, design of the refrigeration system, and electronic control, among others. Together they bid for the development project with the designated chief project lead, based on their professional expertise, knowledge of the development process, and history of previous cooperation (according to the Haier regulations, the development of one creative idea shall be bid by at least two project teams). After successful bidding, they will work under the "model manager" to carry out individual or cooperative R&D works according to the timetable specified by "T model" budget.[8] To be specific, in the "T-90" (budgets of three quarters), the model manager, module managers and other team players participate in the project design plan, product design, product test and various development works. In the process of "T-180" (half-a-year budget), module managers and the model manager take part in the product innovation development, feasibility analysis, and product project approval , and formulate the project design plan of the team according to the planning requirements of the department for "product scheme for industrial leadership or market leader," to participate in project bidding. In addition, in the process of "T-360" [9] (three-year rolling budget) led by the department director with the planning director, market director and financial director as key players, after the general development target and market orientation are defined, the development director will become involved at the analysis stage on the status quo of the product mix; the model manger will come in at the analysis on competitiveness stage, and module managers and the quality director will join in at the stage of analysis on development orientation (including users' demand analysis, quality function disposition, analysis, and the identification of the product pattern of three quarters).

Apparently, Haier's developers gain profits from serving the market through innovation of three quarters, but we can find through the "T mode" budgets of three quarters, half a year and three years that model developers (core players of the project team), including the model design manager, model manager and module managers extensively participate in the phase of product development (including strategic vision and product planning). Such participation determines that the Haier product planner can share the profit commission of the project team as a module design manager, while the product department focuses on preparing for technical innovation for three years or more for future development, which is conducive to the integration between marketing driving and technical push.

The cross-participation in various development processes of different budget periods helps professionals with technical knowledge establish a development idea oriented toward market demand and aimed at the top competitive edge. Manager Zhang of the domestic air conditioner department, a winner of "Excellent Model Manager 2004" from Haier Group, said that the inspiration of product innovation always comes from users and markets. Director Ji of Haier corporate culture center, noted that because Haier calls the lead of the development team a "model manager," it shows that the development work has changed from developing products to developing markets. He feels that to design products with high market competitiveness while solving the difficulties faced by users is their main task. Their consumer research, for instance, noted that some problems included getting fresh air during windy, sandstorm weather without opening the window, and averting the problem of worsened indoor air quality triggered by poor ventilation as a result of having windows and doors closed for an extended period of time. In response, Haier developers created the 2007 "Fresh Air" air conditioner featuring "double fresh air," "air ionization purification," and "anion technology," thus, meeting consumers' demands for grade A indoor air quality (including air oxicity, cleanliness and freshness).

Haier CEO Zhang Ruimin noted when he delivered a speech at the "21st Anniversary of the Founding of Haier & Seminar on Global Brand Strategy of Haier Group" at the end of 2005, that there are only four to five famous international household appliance brands and the win-win capacity is reliant on whether you can create demands better and faster than your rivals; but all of this depends on people's creativity. The core at the stage of brand globalization of Haier is to effectively manage people: "enable everyone to do, be able to do, and be able to do a good job."

In order to ensure that the development team can use their initiative and creativity in making orders the primary concern, the product department of Haier grants the model manager "four designated rights" (quantity, price, time, and people), and urges the model manager to be responsible for "four news" (the utilization of new materials, new technology, new techniques and new equipment in the newly developed products). The product department links the profits for the model manager and the team under his management and the market result of developed products, and implements the SBU system, responsible for any quality loss as a result of a product design flaw. Correspondingly, the development team is in charge of the entire process of product development and promotion. In another words, when establishing the project, they collect market demand from various channels, including door-to-door surveys at users' homes during weekends, conducted by the model manager and his module managers, and feedback information provided by the product manager, client service manager and call center of the Group, to clearly understand users' demand. Based on this information to develop a product, the team must also ensure the product is designed on time. At the market development and promotion stage, the model manager should maintain close contact,[10] formal or informal, with product managers dispatched to local companies, and urge them to place orders for the type of product he has developed. In addition, in order to address users' problems in technology and provide competitive solutions, the model manager can demand the deployment of personnel from the other departments or even from outside of the Group according to the specific needs of the product and each module. In some cases, the development team can invite domestic and overseas experts to carry out subject research and pay a research fee. (After the research is completed, the relevant intellectual property right will belong to Haier). The cross-functional composition of the project team and the cross-communication beyond the product department or even beyond the Group have not only enlarged the information source, but also helped obtain diversified knowledge.

Knowledge and capacity structure of the development team

As durable consumption goods, the air conditioner's basic function is providing refrigerating in summer, and heating in winter. It demonstrates diversified and differentiated function in the speed of refrigeration (or heating), energy efficiency, silence, health, and appearance. The refrigerating system of an air conditioner is composed of a compressor and condenser, which are installed outdoors, as well as an expansion valve (or other throttling elements) and evaporimeter, which are installed indoors. In the process of the refrigeration cycle, the compressor plays an important role, so it is called the "heart" of the air conditioner. But only by integrating with other elements (condenser, evaporimeter, etc.), can the air conditioning system play the role of refrigerating (or heating). Therefore, knowledge of the relationships among various elements and functions will facilitate designers to provide products or solutions for customers through the combination of major parts and components as well as supporting elements. Dutton and Thomas (1985) said "get through learning by studying," i.e., by changing different conditions and parameters, professional knowledge is gradually obtained after repeated trials and simulations, and through failures and successes. What the knowledgeable professionals have learned is the principle and theory behind the operation of a particular technology; it is referred to as know-why knowledge. All the model developers and platform supports of Haier's project teams must have graduated from junior college or above with more than two years of working experience and be equipped with technological knowledge in relevant areas such as refrigerating, HVAC, machinery and electronics, thus ensuring the availability of the know-why knowledge necessary for the development of Haier's new products.

Customers buy and use products or technical systems out of a specific purpose. During use, they will recognize what problems the product (or technical system) has and what still needs to be addressed. This process of recognizing what the product could (not) do the customers is called know-what knowledge. Some of this knowledge is explicit and can be codified and transferred to the client information system of an enterprise through complaints or positive feedbacks from customers, where it will be stored and shared by relevant personnel. However, some of the learned knowledge is implicit and tacit and cannot be codified, which requires the staff with the closest customer contact to use their experience, skills, professional knowledge and inspiration to understand. Therefore, listening to the opinions of many product users will help manufacturers obtain information about the configuration of the product system expected by the market (Garud, 1996). In the creative development during the process of "T-180," the planning director coordinate with the market director, model manager, module managers, and expert team to analyze the data of the defective rate of $N-2$ to N years of the Haier Group, provide insights into the trend of the defective rate of $N-2$ to N years, and clarify the principles of determining the defective rate of $N+1$ to $N+3$ years, before identifying the target of defective rate of $N+1$ to $N+3$ years and the competitive selling point (point of demand fulfillment), so that the creative ideas of products can truly satisfy future market demand. And after the innovative concepts of a product system are formed, the user testing on a functional module will be launched by the planning director, according to the pre-determined standard for a user group at the testing stage. If problems are discovered during the testing process, it must be rectified and re-tested according to the feedback of users. If the product function conforms to the demand or preference of the users, the expected amount of users' orders will be estimated according to the test data. When this amount surpasses the sales target of the new product, the design program will be formulated. In addition, Haier provides project initiation training at the beginning of the "T-90" process in order to realize the match between products and market demands. This training is given to the bid-winning project team, in an effort to ensure that developers clearly understand the design plan of the "T-180 process."

To realize the expected product function in the product design, the developers require background knowledge or previous experience in order to turn a principle or theory into practice. Such knowledge is the result of the accumulation of practical experience. Judging from the entire process of product development, the developers can obtain knowledge through practice in at least two links. The first link is development: the development of a certain model or a series of products at the initial stage will provide experience for the development of similar products at later stages. Household appliance-related development, more often than not, involves higher tooling charges (as high as RMB500,000 to RMB1 million). After finishing the design of the product structure, performance, appearance and system, the development team needs to produce a functional prototype to identify potential problems in the initial mould and find solutions to address these problems before officially starting production. Once a mould is produced, the process' operation instruction will be accepted among the project team, and the model manager;

the product will then be produced according to the process operation sample. The project team will then carry out testing and simulate user experiments. All of these stages require direct practical experience to ensure that the design of the new products' performances is perfectly developed with low costs. Second is the production link. A small amount of new products are required for market warm-up, as are large amounts of order production. The result of practical learning at the production stage is that the production technique is improved along with the accumulation of product quantity. However, the practical learning result at the development stage is the accumulation of experience of developing various models, which differs from that at the production stage. The combination of these different learning results will booster experiences in the previous for similar work at the later stages.

Beyond that, product development is a process to help users address the problems, whether identified or not, through innovative technologies or the combination of them. Understanding users' demand is actually a recognition of users or the types of users group; knowing who possesses the technology and capacity needed to address a certain problem is considered "know-who" knowledge.[11] In particular, it is vital to know which intermediary agent or agency possesses the required capacity. When Haier was expanding their household appliance market and the product departments were busy with surveys of the product's "three information" (information about market, rivalry, and self), Haier found that their commercial distribution engineering department could play the role of an information intermediary in integrating the internal information of the group. While considering the market trend that current purchasers had begun to transform from households to real estate developers, the commercial distribution promotion department established an engineering department responsible for the market channel construction of engineering products. In the summer of 2006, a manager of the commercial air conditioners got 18 pieces of engineering information, and a Beijing industrial engineering manager obtained 22 pieces of project information, one of which was the same. After sharing the information, the commercial air conditioner department identified the highlights of its product development work. In this case, the commercial air conditioner department made a breakthrough in their regular practice, discovering that the existing commercial channel only provided household clients' information. The department no longer relies on its subordinate product managers to obtain information pertaining to its engineering products, so that it can expand the market ahead of other product departments. After this cooperation between the product department and commercial distribution engineering department was promoted, other product development departments also began to use the commercial distribution engineering department as a market information transmitter.

When each module manager is committed to developing the product's technological function and utilizing "know-how" knowledge (or cooperating with experiment and testing personnel to use the knowledge), the module manager should acquire the framework knowledge about how the design components can be integrated into a product that possesses the new function or feature. Framework knowledge means a systematic understanding about how a set of parts and components are combined or assembled (know-how), how each part generates needed product function through interaction (know-why), and what kind of product is needed by customers (know-what). That being said, it is nearly impossible for any model manager to comprehensively and thoroughly understand all technicalities, even though he may have a technical background. So when it comes to the balancing selection of various knowledge, Haier's model manager must give up the right to assign works to his members, as a person-in-charge of the SBU team and the principle of complete result-oriented management, but strictly distributes the profit to team players according to their respective contribution to the project. The model manager prefers to consider their leading role from the perspective of interpersonal relationships, to stabilize the key members of the development team. As a result, each module manager is not only committed to the development and design of the module of his own team, but also able to jointly address the interface problems between modules and satisfy users' demands. At the same time, with regard to obtaining exterior resources, the development team works to integrate creativity and resources in the global arena, and create advantageous resources through independent innovation to obtain more quality exterior resources. For instance, the compressor is regarded as the heart of an air conditioner, and its quality and function have a direct bearing on the quality and service life of the air conditioner. It has been the hope of Haier's development team to seek cooperation with a well-known Japanese company and obtain the support of a competitive sub-supplier. However, the Japanese company was initially very "stingy"; the compressor it provided was far from the budget (demand) either in quality or in quantity. But in 2005, after Haier rolled out the world's first direct-current frequency conversion "Double Fresh Air" air conditioner, which was ultra-thin, quiet, and super-efficient, and whose sales volume surpassed tens of millions, the Japanese company suddenly became "generous" in annual cooperation talks and took the initiative to offer very advantageous resources to Haier.

While attracting sub-suppliers and retailers with a global competitive edge based on the self-improvement spirit and its own competitiveness, Haier's product developers also pay attention to the cooperation with the research institute, eight global design centers, and other product development departments within the

Haier Group. As the key members of the development team, the model manager, model design manager and platform support manager are required to have outstanding coordination and communication skills. What merits our attention is that before the reconstruction of the market chain, employees of Haier were not actively cooperating with each other. As the operation of SBU scheme strengthened, the management bodies with independent interests come to realize that resources should not become available only after signing a contract, or even via a complaint.; Instead, Haier's departments realized they should cooperatively address their bottlenecks and improve their competitive edges to obtain resources, and before demanding that others provide various resources for them, they should consider what they could offer as an exchange. Fair exchange has thus become an unwritten principle to obtain both internal and external partnerships. A model manager said, "I believe people with independent interests can cooperate after getting to know one another. Wheeler-dealing and departmental selfishness are not inevitable outcomes of the rebuilding of the market chain. Of course, fair exchange will not come by waiting, but must be competed for. The prerequisite for competing is that you have a competitive edge and ensure people that establishing a partnership with you could get a win-win result."

This reflects that resource value-based fair exchange is becoming the standard by which different departments within Haier Group treat their relationships. And the formation of such a standard is partly the experience of gradual learning in situational activities (the opportunist behavior of harming others for one's own benefit is unacceptable among colleagues), and is also partially related to the SBU values Haier executives advocated (the "values integration office" established in the general office of the product department is committed to cultivate employees' thoughts and conducts after the reconstruction of market chain). Additionally, Haier has established multi-tiered technical platforms for encouraging various SBU departments to share technical knowledge or resources. For instance, Haier established a product information management platform inside the product development department (known as "product development module," or PDM), and established a basic (benchmark) base, technical module base and standardized parts base, among others, at the Group level so that institutional conditions were made possible for knowledge sharing among various models, series and product departments. And the platform support manager with strong coordination and communication skills as an assistant developer, provided a convenient channel for the development team to obtain the necessary knowledge through the platform. With the support of information technology, the platform made it possible for the developers to obtain knowledge accumulated by former developers by relying on the search function of the platform, regardless of who owned such knowledge. As a result, knowledge sharing and exploitation in a comprehensive way across the entire Group (visible) has been made a reality. In this sense, "knowing who knows what" has been simplified to "know what can be obtained through where."12 As the relational network of developers has been transformed from "who" to "where," the skills to handle this relationship has changed from interpersonal skills to the capacity to enter the technical platform.

Table 11.2 summarizes why the development team of Haier, with the model manager at the core, is able to promote the exploration and exploitation of knowledge in the context of cross-functional cooperation, from the perspective of knowledge classification, and the sources and roles of the diverse knowledge of the development team. It can be seen from Table 11.2 that heterogeneous knowledge composition enabled the SBU team inspired by the market result target to directly or indirectly connect different knowledge suppliers while forming a strong network within the development team. As such, they could compensate for the limited knowledge and resources of the cohesive network (team) and enable the air conditioner department to develop new products under the theme of "healthy,

Table 11.2. Knowledge type and the heterogeneous knowledge composition of new product developers

Knowledge Type	Knowledge owner	Knowledge source	Learning mechanism	Role
Know-what knowledge	Planning director/ Model planning manager	Contact with customers; feedback from customers	Learning by using	Knowledge exploration regarding product function
Know-why knowledge	Model manager/ Module manager	Product design principle and theory	Learning by researching	Knowledge exploration about product structure and module technology
Know-who knowledge				
1. Know "who knows who"	Model manager	Direct and indirect social networking experience	Learning by networked connection	Knowledge exploration based on extensive connection
2. Know "who knows what" or "where has what"	Module manager, and platform support manager	Technological modularity and standardization of parts and components	Learning by substitution	Knowledge exploitation based on modular technology and standard components platform

(Cont'd)

Knowledge Type	Knowledge owner	Knowledge source	Learning mechanism	Role
Know-how knowledge				
1. Development skills 2. Production skills	Module manager, experiment manager and test production personnel	Direct working experience	Learning by practicing (learning by doing)	Knowledge exploitation based on learning curve
3.Interpersonal skills	Model manager, model planning manager and platform support manager	Direct social networking experience and the capacity to enter into a technology network	Learning by situational activities	Knowledge exploitation based on the enhancement or changes in interpersonal relationships, or the connection of technical nodes

energy-saving and environmentally friendly."

Discussions

Scientific Research Management carried an article recently, which noted, based on the studies on the case of Haier's business process innovation since 1999, that the process rebuilding and market chain of Haier are recognized both domestically and internationally. But judging from the basic operation behind the process, R&D is only recognized as a support process while marketing (reflected as commercial distribution) is the major process of enterprise. This demonstrates that its technical innovation and market innovation do not have the coordination features, only functional integration; if integrating market information into the R&D function, the result will be used to support the launch of new products. At the same time, the implementation of Haier's internal market mechanism creates a market exchange relationship between R&D and marketing . Taking the R&D process as a support process will not benefit the formation of the cultural concept in the company that technical innovation is the innovative value of company. In the long haul, it may make the business strategy focus more on marketing than on the expansion and development of core technical strength of the company (Zhu Ling et al., 2006).

It was found after this case study on Haier that, generally speaking, Haier's R&D only serves as a support process and it is incorrect to judge its R&D

capacity according to this. The product development department of Haier air conditioners has applied for altogether 716 patents, of which 60 are patents of innovation. This shows that the system supporting its innovation is vigorous. As with the experience of the air conditioner department transforming their patent technology into a marketable and profitable commodity (2007 "Fresh Air" air conditioner) and making it an industry, the washing machine department has developed a double engine machine with no need for washing powder. The electric heating department has developed the "wall call" technology, used for water heaters, which has become the first national standard originating from a corporate patent in China. All of these indicate the unique capacity of Haier in technical innovation.

Our thorough case study on Haier air conditioner development indicates that Haier Group regards their product directors, development directors, and model managers (collectively referred to as "top three managers") as large, medium, and small model managers, different only in the scale of responsibility. The model manager, responsible wholly for the development process, has representative of the "main process" of the group, and the commercial distribution department has become a "support process," providing a commercial platform for the main process (despite the fact that at the early stage of the market chain reconstruction, Haier regarded the commercial distribution and logistics promotion departments, responsible for unified marketing and purchasing, as the "core process" paralleling with the product department).

Of course, after the reconstruction of Haier Group's market chain, purchasing, marketing and other functions were unified to the group level, the business division of the manufacturing stage in the product department was transformed from collecting processing fees to original equipment manufacturer (OEM). Therefore, the design, production and sales of a product are separated in Haier Group. The difference from the developers of functional organization who only focus on design rather than marketing and who are only responsible for leaders rather than market results, is that the model manager and his team under the SBU operation system in Haier are responsible for all problems in the product life cycle and for whether the product they developed can create corporate value. The model manager of a product must take into account every factor of the entire development process; otherwise, any problem in any link will directly have an impact on the market result of the product.

When promoting the establishment of SBU system by mainly relying on top three managers in 2005, Zhang Ruimin, CEO of Haier Group, noted that the Group must provide better resources and construct better platforms for various products in order to be part of global competition; operators of various products must compete for resources to be market leaders. It is irresponsible to rely solely on resources integration, because resources are limited and after others have done the integration work, you will not be able to do the same. Therefore, vying for resources is a must. It is during the process of trying to compete for and integrate various resources relevant to the development process that the product model manager forms a cross-functional product development team and relies on internal and external networks. With the support of the modular design of household products, of the technical and other resource platform, and based on the communication and coordination skills (interpersonal skills) of major developers in the team, the product innovative capability has been enhanced in the process of utilizing the existing knowledge of the Group.

At Haier, a model manager has competitiveness, vitality, and independent innovation, and must also be willing and able to draw support and wisdom from outside his own team. The development team, with the model manager at the core, is a team without constraints. In the process connection, major members of the development team often participate in the strategic thought (T-360 process) and product planning (T-180 process) at the early stage of the production process before the project is officially set up. After winning the bid of the project, the development team will, based on users' demand for product, think about how to meet their demand using the T-90 process. They will also be responsible for the realization of product marketing and promotion, and for turning customers' complaints or suggestions into new demands. The construction and operation of the team enable Haier's development department to engage in independent technical innovation at an average speed of 1.7 inventions and patents in each working day.

When interviewed by senior editor Cohen of *Harvard Business Review* in early 2007, Zhang Ruimin, CEO of Haier Group, said that technical backwardness is only an apparent phenomenon—the difference lies in the management level. During the process when the purchasing and marketing functions are centralized to the Group level, and the product department is divided into the market department, development department, manufacturing department, or business department, it was important to avoid the separation of the process and also to make the new product development process efficient (including reducing development cycle and development cost) and creative (innovation in product function or structure). Haier, under the SBU system which enables every staff to become an independent innovation player, is exploring an innovative organization mode, about which Zhang Ruimin stated, "I hope Haier's management does not rely on personnel, and I hope to build a system and a platform for sustainable development."

After the process reconstruction (Haier CEO, Zhang Ruimin, made clear recently that the functions separated from the business division or business department from 1998 shall be termed as "unified process" or "unified function"), top three managers (large, medium and small model managers) in the headquarters of product development are persons responsible for the main process. Since they are put in charge of market results, they must possess the right to dominate and coordinate certain activities, including parts and components purchasing, product marketing, and human resources after these functions are separated from the business department. In this sense, the relationship between the product department and other departments in the new organization reflects a higher requirement for a close connection. In the product development process, the model manager must contact with external departments to obtain better innovation results. And the development team, with the model manager at the core, must work together with the SBU team, which enables the developers to span across time and spatial boundaries and engage in innovation activities from the initial creative concept to finally transforming the ideas into products with market values. The model manager who prefers to establish a stable, long-term connection in new product development with core members (model planning management, module manager and other assistant developers), enables the team players with different knowledge to work together for a long time, focusing on the in-depth development and accumulation of their specialized knowledge, and also enjoying seamless communication through cross-disciplinary cooperation (although apparently they demonstrate the inability to designate personnel or allocate profits according to contribution). This is definitely beneficial for the new product development based on tacit knowledge. As for the codified, explicit knowledge stored and communicated, Haier promotes the comprehensive sharing or allocating of competitive resources via the establishment of multi-tiered technical platforms and the emphasis on the market chain connection between departments of different functions abiding by the principles for fair exchange.

Conclusions

In the functional organization where the overall result is achieved only by close connection of various processes (functions), top executives are the persons responsible for coordinating various activities in the innovation process. At Haier, model managers, big or small, serve as the organizers of the innovation process. Siggelkow and Rivkin (2006) who are adept in carrying out complex adaptive system research using the multiple-agent simulation method, believed that the decentralization of powers in the exploration process (to transfer power to lower organizational level), may bring either good or bad results to the organization. Their simulation results showed that only when the interdependence among departments is reduced to a loose connection level, can the units at the lower level focus on knowledge exploration activities, so it is not necessary to strike a balance between knowledge exploration and exploitation (Gupta et al., 2006). However, based on the research of Haier's closely connected organization system established after the reconstruction of market chain, this chapter provides insights into the fact that the SBU team, inspired by the market results and with the model manager at the core, is able to realize the "ambidextrous organization" at a lower level during the process of new product development with the balance between knowledge exploration and utilization.

Tushman and his partners who have published a series of studies on the ambidextrous or biplanar organization use the concept of ambidextrous organization to refer to the organization form which reconciles exploration and utilization the two inconsistent and paradoxical demands through establishing an inconsistent structure within the organization. They believed that in an organization which is able to use its own knowledge while exploring new knowledge, a more complex organization structure is needed, which, in turn, requires high strategic integration capacity of the top management team (Benner and Tushman, 2003). Other researchers who echo or follow this idea also endorsed that if an organization is ambidextrous, it must be able to explore new knowledge by a loosely connected network or an organic structure, while fully utilizing the existing knowledge through a tightly connected network or a mechanical structure. And the top management team, which is composed of the mutually complementary top executives (in both insight and capacity) and is responsible for the coordination or integration of the two capabilities, plays a role as a bridge between the loose network of knowledge exploration and the dense network of knowledge utilization, thus, enabling the generation and application of innovation to unify. Obviously, the ambidexterity they mentioned can only exist in the entire organizational level and at business units or an individual level; knowledge exploration or utilization is in opposition to each other, so there is no substantial difference from the concept of dual structure put forward early by Duncan (1976). The major contribution of those scholars was adding a centralized power of top management team to make possible the strategic integration of the dual units different in process, structure and culture. That being said, from the angle of business unit or lower organizational level, the confrontation between knowledge exploration and utilization is difficult to be overcome.

Conversely, there is a switching structure design which reflects this balance. It enables staff at the lower level to engage in tasks of a different nature, or a knowledge-exploratory task, while assuming a knowledge utilization task. If regarding staff as the unit of analysis, it is an ambidextrous organization. But judging from a certain point in time, this staff only assumes one task, either knowledge exploration or knowledge utilization, so they still show inconsistent or confrontational status.

Based on the Haier Group case studies after the rebuilding of the market chain, we found that having adopted the system of SBU, new product teams with different knowledge bases, from different functional fields have demonstrated their initiative in carrying out innovation activities with the balance between knowledge exploration and utilization. Haier's experience shows that ambidexterity is not only a concept at the organization level, but also a concept at individuals levels (a team and even a model manager). The entire SBU system advocated by Haier enables versatile individuals or teams with visionary insight and diverse knowledge to bring out their best divergent thinking (Haier's CEO Zhang Ruimin believed that this thinking is different from the usual restrained thinking of top executives), thus, helping Haier obtain the ambidexterity competitiveness featured by differentiation and low cost global competition. Case studies showed that a team with diverse knowledge bases[13] is, on the one hand, forming a cohesive status during the stable operation of team members, and on the other, with the aid of the professionals in the team (who have outside contacts), and several key players who have a wide knowledge base, the SBU team can obtain the support from external personnel or institutions in the fair exchange via exchanging their own competitive resources, using the advantageous resources of other parties.

Our studies proved that it is biased to regard knowledge exploration and utilization as an opposite dual structure. Ambidexterity is not the two poles of a continuum, but the two aspects of one thing (Adler et al., 1999; Gupta et al., 2006). From the angle that a dual feature can be compatible and can achieve a balance at various organizational levels, this study believes that a complex adaptive system featured by such a multi-tiered self-similarity (such as the large, medium and small model managers of Haier) is an ambidextrous organization, rather than a dual or biplanar structure.

Our focus has been to demonstrate why and how ambidexterity can appear in a lower organization. As for why, under other situations, it is a phenomenon only at the entire organizational level, we need to carry out further studies on the difference of various situational conditions including the external environment for the survival of the organization and the internal environment affected by managers. The explanations regarding these differences will help us clarify that the ambidexterity at the entire organizational level is caused by the fact that the organization or managers intend to meet the requirement of an objective environment at the lowest cost or at the minimum level, and is also the result of the failed cultivation of the subjective environment that is beneficial for the ambidextrous capability of staff. But this means that when top managers do not serve as the integrator of the dual structure in an ambidextrous organization, they may play an indirect role. Research on the shifting role of top managers and explanations of the differences of situational conditions are topics for further study in the research area of ambidexterity organization.

Postscript

This book is a valuable result of joint cooperation and research. The division of labor of various parts is as follows: Introduction and Overview by Yi Zhihong; The International Environment and Industrial Policies for Independent Innovation by Gu Kejian and Li Gang; The Financing Environment for Independent Innovation by Guo Yan, Yi Zhihong and Li Yan; Technological Innovation through Resource Integration by Song Hua, Yi Zhihong and Yu Kangkang; The Operation of the Innovation-Oriented Service Supply Chain by Song Hua; Performance of Enterprise Informatization under Institutional Innovation by Song Hua, Wang Lan and Wang Xiaoliu; The Marketing Channel Innovation of Chinese Enterprises by Huang Jiangming, Dai Yingqiong and Zhang Junhai; Proprietary Brand Innovation and the Performance of Chinese Enterprises by Liu Fengjun, Wang Liuying, Jiang Tao and Ou Dan; Developing Evaluation Indicators for the Continuous Technological Innovative Capability by Wang Huacheng and Guo Chunming; The Structure of Governance for Chinese Enterprises by Xu Erming and Zhang Han; The Organizational Model of Independent Innovation by Wang Fengbin. Yi Zhihong is responsible for the final compilation and editing before finalizing this book. Song Hua and Huang Jiangming have assumed huge organization and coordination responsibilities during the project's research process and Han Yan has undertaken many tasks for the smooth completion of this research project.

We have been strongly supported by the scholars of the Planning Division and the Science and Technology Division of Renmin University of China in our project research. When we set up and submitted this research project, Prof. Du Houwen and Deng Ronglin offered us valuable suggestions. During the whole process, many professors and PhD candidates of the Business School of Renmin University of China participated in our discussions and provided valuable opinions for revision. We hereby express our heartfelt thanks to them.

The publication of this book (Chinese version) was supported by China Renmin University Press. Without the commitment of editors at the press, it would be impossible for our book to be published. We hereby express our sincere appreciations to them.

Yi Zhihong

Notes

Chapter 2

1.   Just as Findlay (1981, 427) made it clear: "Mill is the first economist to take into consideration the influences of technical progress on terms of trade." H.G. Johnson (1963, 97) also pointed out: "The classical analysis of the influences of technical changes on trade started from Mill but also ended with his analysis of the implications of the improvement of production mode for the changes of terms of trade." Caves (1960, 152) and Batra (1973, 128) also mentioned Edgeworth and Bastable when talking about research in this regard, but none of them are before Mill.

2.   In fact, the analyses before Mill all presumed that technical progress happens in either export industries or general industries, but failed to assume the circumstance that it especially happens or mainly happens in import-competing industries.

3.   The first two kinds of technical progress refer to the "export-oriented technical progress" and the "import-oriented technical progress" used by Hicks.

4.   Even though they have discussed many aspects of technical progress, they always assumed that technical progress will increase but not decrease the import of the technologically-advanced countries. They also asserted that these technical progresses will spread to other countries.

5.   Ironically, what was used by Hicks is not the theoretical model of his own; instead, he used the constant cost model of Ricardo. He also presumed that each country only makes export-oriented products and non-tradable goods. And his successors, including Johnson, Corden, Findley and Gruber, also employed the H-O model framework.

6.   If capital-labor ratio is a fixed value, the ratio of the marginal production of capital to the marginal production of labor will remain unchanged, which means the technology substitution rate of capital and labor will remain unchanged and technical progress will be Hicks Neutral Technical Progress.

7.   Under the initial relative factor price, the optimal capital-labor ratio remains unchanged after technical progress happens.

8. The neutral technical progress refers to the technical progress without factor intensity changes under the conditions of identical factor price ratio; intensive-factor-using technical progress refers to the factor density of the commodities of technical progress will change under the conditions of identical factor price ratio, and it prefers to use intensive factors; intensive-factor-saving technical progress refers to the factor density of commodities of technical progress will change under the conditions of identical factor price ratio, and it prefers non-intensive factors.

9. In fact, according to the standpoint of Bloomfield (1984), when analyzing the influences of growth driven by technical progress on terms of trade, Schroepfer (1833) and Torrens (1835) had used the concept of "immiserizing growth."

10. A common consensus of the international economics community is that Jones is the first scholar to systematically expand the H-O framework and analyze the influences of technical progress on terms of trade under the conditions of technological differences.

11. Technological densities refers to the quantity of technical knowledge and the actual capacity to use technologies in production.

12. The difference between large countries and small countries in the sense of international economics refers to the price-maker and price-taker respectively in the international market.

13. Bhagwati has always been dividing industries of a nation into the import industry and the export industry when analyzing the influences of economic expansion on terms of trade.

14. The merit to adopt the "relative-relative" assumption is that we can refrain from being separated from the economic reality. After all, it is impossible to have only one industry undergoing technical progress while other industries are totally static. They are only different in aspects like speed.

15. Product innovation was first put forward by Schumpeter (1912). Following that, many scholars and organizations of all kinds made their own definitions. We tend to accept the viewpoint of Baumol (2004) in line with the research theme of this study.

16. One example of product innovation-based technical progress increasing the product non-budgetary constraint is the intensified technical trade barriers rising in recent years. As a matter of fact, the emergence of technical trade barriers is more or less related to the product innovation-based technical progress in export destinations. Of course, some even directly involves technological innovation, but we will not analyze it in this section.

17. As the model only studies the data of Chinese industries, we choose the deterministic effect model.

18. The textile and garment industry and the mechanical and electronic industry are the large export industries. In this connection, we especially include these two industries in line with the four major classifications at the United Nations Conference on Trade Development.

19. As the Paasche Index is easily overestimated while the Laspeyres Index is easily underestimated, we have chosen the Fisher Index. The latter is the geometric average of the former two and is also known as the optimal index.

20. See note 17 above.

21. The varying-coefficient model is used to analyze the influences of the process of innovation-based technical progress of each listed company on the cost-price ratio, while the varying-intercept model is used to make a global analysis of the influences of the process of innovation-based technical progress of these 12 foreign trade companies on the cost-price ratio.

22. The finding of the existing exchange rate research is that developing countries basically use a fixed exchange rate system, and even though the country called its system a floating exchange rate system, it is still a fixed exchange rate system; developed countries basically use a floating exchange rate system.

23. In the formula, TOT refers to the terms of trade index, P_E, P_I, P_E^*, P_I^* pte, pti respectively represent the export price index before the fluctuation of the exchange rate, the import price index before the fluctuation of the exchange rate, the export price index after the fluctuation of the exchange rate, the import price index after the fluctuation of the exchange rate, the export price transmission elasticity of the exchange rate and the import price transmission elasticity of the exchange rate.

24. The 16 trade partners include Chinese Hong Kong (0.230 322), Japan (0.196 801), the U.S. (0.152 570), Germany (0.084 016), Taiwan (0.047 109), France (0.046 697), Italy (0.042 933), the UK (0.036 844), Canada (0.027 657), South Korea (0.026 481), the Netherlands (0.022 859), Belgium (0.019 968), Singapore (0.019 258), Australia (0.017 827), Switzerland (0.015 870), and Spain (0.012 788). Data inside parenthesis are the weights of relevant countries when calculating the effective exchange rate of RMB in real terms.

25. These four countries and regions respectively represent different types of import source and the trade weights show an echelon distribution. The data of unit labor cost of these four countries and regions are the only complete series in the international labor statistical database.

Chapter 3

1. Based on the needs of this study, this section mainly elaborates and analyzes the lifecycle theory of independent innovation from two aspects—namely, the financing demand and risk features.

2. Including Haidian Park, Fengtai Park, Changping Park, Electronic Zone Technology Park, Yizhuang Science and Technology Park, Desheng Park, Jianxiang Park, Shijingshan Park, Tongzhou Park, Yonghe Park, and Daxiang Biomedical Base. The leading function of Haidian Park is the high and new technology's R&D, radiation, incubation, and business and trade center, and the major function of other parks is the development base of high and new technology industries.

3. Seed Fund adopts the investment method of "incubation plus venture capital" which means the venture capital development center joins the recognized scientific and technological enterprise incubator in the park to make investments in the high-tech enterprises inside the incubator, and at the early stage of innovation in the park, under the same conditions.

4. Follow-up investment fund means that the venture capital development center provides supporting equity investment based on a certain proportion of the real investment amount of venture capital firms, making joint investment under the same conditions.

5. Refers to Beijing Century Real Technology Co., Ltd.

6. Refer to Beijing Time Technologies Co. and Sinosoft Co., Ltd., with the capital increment of RMB50 million and RMB60 million and the price earning ratio of 12 and 20 folds respectively.

7. Only the well-operated enterprises being invested could exit through going public, while many venture capital transactions that exit through acquisition do not report the value at the time when they exist, and most transactions reported by them are successful transactions. Therefore, selection bias may occur if we just measure the returns of venture capital transactions of the reported listed companies that exit through acquisition.

8. Ratchet clause usually sets the enterprise performance targets in advance, such as sales amount and net assets. If an enterprise could not meet the predetermined targets in the future, both sides would adjust the shares of entrepreneurs to protect the interests of investors. On the contrary, some investment agreements also provide that entrepreneurs may also reap more proportions of shares if the future performance of an enterprise is much better than the expected achievement.

9. Some local governments regulate that venture capital institutions can also enjoy the tax breaks for high-tech enterprises if they invest over 50% of their capital in high-tech enterprises.

10. For instance, Israeli YOZMA Program started in 1992 and planned to be implemented for seven years, but, in 1997, the Israeli government completed the mission of guiding private capital, and thus accomplished the privatization of venture capital funds two years ahead of the schedule.

11. Refer to the Law of the People's Republic of China on Commercial Banks, Section 4, Article 39 (revised edition, 2003).

12. Considering the difficulties faced by many enterprises in getting loans, this should be a realistic circumstance.

13. Gazelle is a species of antelope that is good at jumping and running. Thus, venture capital community usually calls high-growth small and medium-sized enterprises as "Gazelle Enterprises." And in one region, the more gazelle enterprises there are, the more dynamic the innovation is, and the faster the region develops.

14. Of course, up to now, equity investment beyond credit services of the Silicon Valley Bank could not be realized due to the restrictions of China Banking Regulatory Commission.

15. State Economic and Trade Commission of the General Office of the State Council, *Notice on Policy Suggestions on Encouraging and Promoting the Development of SMEs* (August 24, 2000).

16. People's Bank of China, *Suggestions on Further Improving the Financing Services for SMEs* (June 1998), *Notice on Expanding the Interest Rate Fluctuation Degree for Loans to SMEs* (October 1998), and *Guidance for Strengthening and Improving the Financing Services for SMEs* (November 1998).

17. The viewpoint of Marshall about the contradiction between the economy of scale and monopoly. Marshall argued that free competition will result in the expansion of production scale, form the economy of scale, increase the market share of product and inevitably lead to market monopoly. But monopoly is bound to put a stop to competition and slow business activities when they develop to a certain degree, which will lead to irrational allocation of resources. The problem becomes: how to realize an effective and rational balance between market competition and the economy of scale, and generate the maximum production efficiency. "Marshall Conflict" is applicable to the industries witnessing increasing returns (decreasing-cost industry), such as the telecommunications industry and the banking industry.

18.   Direct financing includes issuing corporate bonds and stocks in the open market. For the part of small-scale technological innovation enterprises, issuing corporate bonds is not an economical choice due to the problems in information transmission of the economy of scale, credit rating and bond price (Hakansson, 1999). Therefore, this chapter will mainly discuss issues about the stock issuance of innovative enterprises.

19.   Refer to BVCA, NVCA and Venture Economics for relevant statistical data.

20.   Many studies on the comparison between the main board market and the GEM have been published in China.

Chapter 7

1.   Questions in this part of investigation mainly referred to the questionnaires of Wilkinson (1981), Soumava Bandyopadhyay and Robert A. Robicheaux (1998), and Zhuang Guijun (2004) with some appropriate modifications.

2.   The fourth part of the questionnaires adopts the Likert Scale, and we will not analyze the Cronbach's α value.

3.   Sales policies of Jiangxi Gree Sales Company.

4.   Ibid.

5.   Ibid.

6.   "The 6-Year Free Warranty Period of Gree Air Conditioners," http://news.sina.com.cn/s/2005-01-28/06084970051s.shtml.

7.   "Gome versus Gree—Two Voices Unveil the Conflicts of the Sales Channel Reform in Household Appliance Industry," http://www.sznews.com/n1/ca854886.htm.

8.   Ibid.

Chapter 9

1.   Mao Ning, and Wang Chen, *Knowledge Assets: Soft Finance* (Beijing: China Economic Publishing House, 2005), 25.

Chapter 10

1.   Schumpeter's *Economic Development Theory* was published in German in 1912. After being published in English, this book won widespread attention from academics.

2.   Richard C. Hoffman, and William Harvey Hegarty, "Top management influence on innovations: effects of executive characteristics and social culture," *Journal of Management* 19, no.3 (1993):549–574.

3. Fulvio Castellacci, Stine Grodal, Sandro Mendonca, and Mona Wibe, "Advances and Challenges in Innovation Studies," *Journal of Economic Issues* Vol. XXXIX, no.1 (March 2005).

4. Peng M. W., "Institutional transitions and strategic choices," *Academy of Management Review* 28 (2003):275–296.

5. Yao Jun, "Empirical Study on the Diversification, Structure and Performance of Enterprise Group in Transitional Period," PhD thesis of South China University of Science and Technology (2004).

Chapter 11

1. Gupta and others believed that through conversion in the time course (refers to the analysis in the following parts of the chapter), organizations prefer to be engaged in the knowledge utilization of incremental improvement in a certain period (often a relative longer period), while embarking on exploratory tasks that may bring about innovation breakthroughs in another period (often a shorter period that has broken the original balance). Such a "punctuated balance" is viewed as a way to take exploration and utilization as two extreme statuses that are a repulsive continuum, instead of two dimensions that are compatible at same time point (Gupta et al., 2006). In this sense, from the perspective of synchronicity, it could be seen as a confrontational strategy.

2. Some scholars define ambidexterity as a synchronic pursuit of knowledge exploration and utilization. According to the definition, switching structure is not a variance of ambidexterity, but a totally different mechanism called "punctuated equilibrium" which means that an organization could pursue either knowledge exploration or utilization at a certain point of time (Gupta et al., 2006).

3. From this perspective, Giddens and Rirbinshaw carried out special verification on the validity of data aggregate in the studies on the causes, effects, and intermediary role of ambidextrous organization to ensure that the score given by the staff who provide data of business units in the sample enterprises meets the statistical requirement of homogeneity within the group, and heterogeny among different groups. The reason is that in the studies across the levels of staff and enterprises, only groups (business units) meeting the homogeneity requirements could form causal models between reason variable (situation), intervening variable (ambidextrous capability) and outcome variable (performance of business unit), and carry out estimation of regression or related coefficient via mean value of variables.

4. Different from "SBUs" used by American General Electric company in the 1970s, which refers to the business units in business division and the level above, "SBU" used by Haier Group refers to individuals, teams or entities who are equipped with independent innovative capability and responsible for business performance (profits).

5. With regard to the relationship between large and small model managers, Haier's CEO, Zhang Ruimin, explained: "the director of product development is no longer responsible for the business division; what he is in charge of is to create order. He has the model manager under his management, so in this sense, he is a large model manager. A small model manager is responsible for one or two product models. A large manager has a dozen product models under his management. He acts as a commander – to be responsible for the market across the country, i.e., to decide which one could be launched, and where a certain product model should be rolled out. And a model manager might say 'my model is very well, and has produced 100,000 products.' But for the department, this number is not enough and the whole department requires 1 million. Thus only when each model manager makes 100,000 products, the total number of the department could reach 1 million. If this number has not been achieved, the director (large model manager) should shoulder the responsibility for failing to organize the small model managers."

6. In the product department of Haier, no model managers are life tenure. The practice is that when a developer identifies the complaints of users at the earliest time, and develops solutions which could be developed after argumentation, he will be eligible for the title of model manager.

7. According to Haier's SBU operation system, full members of the development team, including the model manager, get commission directly from the profit created by the product they have developed (SBU operational profit and loss statement shows the data of income of the development team, income distribution of model developers and assistant developments on top of the original data input and profit calculation). As for the starting and terminating point in time of the commission, it shall be identified according to the principle of Product Lifecycle Management (PLM). The proportion of commission will be identified by the team members and model manager at the time of project establishment with an agreed coefficient, based on the technical grade of the members (accredited according to the standard of the Human Resources Center), the work load in the project, and the responsibilities and contributions. Those managers providing supporting facilities in the process do not belong to the full

members of the development team. They can make an agreed time frame for their work with the model manager by signing a contract and their remuneration will be calculated according to the value their service generated.

8. Haier's technical innovation is divided into three levels according to the designated budget time: "three quarter innovation" refers to the innovation made in the previous quarter, current quarter and the next quarter of each peak season, with the aim of serving the current market; "innovation for three years and above" is to make technical preparations for future development; "fundamental innovation" is an innovation for long-term development, including advancing research on basic theory, and on products, techniques and core technologies. The Research Institution and Technology Center will be responsible for the fundamental innovation, while the first two levels of innovation will be taken charge of by product department and its affiliated development department. Under the coordination and guidance of the R&D promotion department of the Group, the product department will develop process specifications for "T-360" (three-year rolling budget), "T-180" (half-a-year budget) and "T-90" (three-quarter budget) according to the features of the product it develops, to ensure the development work is in line with the standard of "market acceptance check of bill of material (BOM)." Haier believes that the enterprise institution and other institutes are not designed to obtain a Nobel Prize, but to serve as the technology reserve for the Group to obtain long-term profit. Therefore, they work under the principle of "complete result-oriented management." The outcome measurement includes short-term and medium-term time dimensions and that of more than 10 years, among others.

9. The external expert team outside the department takes part at the stage of the competitiveness analysis and development direction analysis of "T-360" process in order to ensure a wider insight for the front end work of the development process.

10. Such a relationship is usually two-way. For instance, in the monthly regular meeting of product managers, the product managers stationed in 42 trading companies nationwide will be sent back to the headquarters of the product department. They will first give the model manager feedback of users' opinions and problems they encountered in the sales so that developers immediately understand the market conditions and upgrade the product to meet users' demand in the future. These product managers working nationwide are affiliated to the market department or market

support department under the product department in administrative relationship. They do not have formal relations with the model manager, but their connection in interests forms the double cross relations between them.

11. More generally, "know-who knowledge" refers to knowledge the participant (a certain node) gets from the various indirect paths in the network. Given that the node in the knowledge network could be a person or the technology carrying knowledge (such as the "materials," including reference basis, technical modular, standardized parts and components which can be used jointly, and other written technical data). In addition, the knowledge carried by the node can be divided into "know-who" and "know-material" knowledge (including the materials apart from know-who knowledge), thus when the participant (a certain node) gets knowledge not from the nearby node, the node connected by the indirect road and the combination of different knowledge types in the connection have formed certain types: know who knows who, know who knows material, and know where one can contact who. The know-who knowledge, in a narrow sense, refers to the first type.

12. "Where" herein means the technical carrier which serves as knowledge path. For instance, the model manager who developed the "double engine" washing machine came up with the idea of developing the machine when looking into the introduction of a double-engine cleaner of a company in an airplane magazine. As a result, he developed the innovative product that can avoid the "twinning" issue in the working process of the cylinder washing machine. The magazine became (although unconsciously) an intermediary node for the model manager to obtain the technology of the cleaner company. The establishment of technology platform of Haier is designed to help the development team know where they can get what technology.

13. Tayor and Greve (2006) referred to the in-depth diversity, which is different from surface diversity of the demographic characteristics, as cognitive diversity.

References

Chapter 1

English materials:

Almeida, Paul, and Anupama Phene. 2004. "Subsidiaries and knowledge creation: the influence of the MNC and host country on innovation." *Strategic Management Journal* 25.

Corsoetal. 2000. "Knowledge management in product innovation: an interpretative review." *International of Management Review* 3.

Frambacha, Ruud T., and Niels Schillewaert. 2002. "Organizational innovation adoption: a multi-level framework of determinants and opportunities for future research." *Journal of Business Research* 55: 163–176.

Freeman, C. 1982. *The economics of industrial innovation*. Cambridge: MIT Press.

Goesetal, J. B. 1997. "Interorganizational links and innovation: the case of hospital services." *Academy of Management Journal* 40.

Govindarajan et al. 2004. "Strategic innovation and the science of learning." *MIT Sloan Management Review*, winter.

Grev et al. 2001. "Innovations as Catalysts for Organizational Change: Shifts in Organizational Cognition and Search." *Administrative Science Quarterly* 45.

Hall, Richard, and Pierpaolo Andriani. 2002. "Managing knowledge for innovation." *Long Range Planning* 35.

Hall, Richard, and Pierpaolo Andriani. 2003. "Managing knowledge associated with innovation." *Journal of Business Research*, (56).

Hicks, John R. 1963. *The theory of wages*. London: Macmillan.

Johnson et al. 2002. "Knowledge, innovation and share value." *International of Management Review* 4.

Jones, Oswald. 2005. "Manufacturing regeneration through corporate entrepreneurship." *International Journal of Operations and Production Management* 25 (5): 491–511.

Kamien, M., and N. L. Schwartz. 1975. "Market structure and innovation." *The Journal of Economic Literature* 23.

Majchrzak. 2004. "Knowledge reuse for innovation." *Management Science* 50.

Mansfield, E. 1968. "Industrial research and technological innovation." *Econometrics* 29.

Mensch. 1979. *Stalemate in technology: Innovations overcome the depression*. Cambridge: Ballinger.

Powell, Walter W., Kenneth W Koput, and Laurel Smith-Doerr. 1996. "Interorganizational collaboration and the locus of innovation." *Administrative Science Quarterly*, (3).

Simpson, Barbara, and Michael Powell. 1999. "Designing research organizations for science innovation." *Long Range Planning* 32 (4).

Skilton et al. 2002. "Technological knowledge maturity, innovation and productivity." *International Journal of Operations and Production Management* 22.

Stuart, Toby E. 2000. "Interorganizational alliances and the performance of firms: a study of growth and innovation rates in a high-technology industry." *Strategic Management Journal*, (21): 791–811.

Tatikonda et al. 2001. "Integrating operations and marketing perspectives of product innovation: the influence of organizational process factors and capabilities on development performance." *Management Science* 47.

Tidd, J. 2001. "Innovation management in context: environment, organization and performance." *International Journal of Management Reviews* 3.

Tsai. 2001. "Knowledge transfer in intra-organizational networks: effects of network position and absorptive capacity on business unit innovation and performance." *Academy of Management Journal* 44.

Translated materials:

Chandler, Alfred D. 1987. *Kan de jian de shou: meiguo qiye de guanli geming* 看得見的手︰美國企業的管理革命 (The Visible Hand: The Managerial Revolution in American Business). Beijing: The Commercial Press.

North, Douglass C. 1991. *Jingji shi zhong de jiegou bianqian* 經濟史中的結構變遷(Structure and Change in Economic History). Shanghai: Joint Publishing Shanghai Branch.

Schumpeter. 1990. *Jingji fazhan lilun* 經濟發展理論 (The Theory of Economic Development). Beijing: The Commercial Press.

Chinese materials:

Cai Bing 蔡兵. 2006. "Zizhu chuangxin nengli buzu yu zizhu chuangxin de wenhua, zhidu zhangai" 自主創新能力不足與自主創新的文化、制度障礙 ("Insufficient Independent Innovation Capability and Independent Innovation Cultural and Institutional Barriers"). *Xueshu yanjiu* 學術研究 (Academic Research), 2.

Chang Xiuze 常修澤. 1994. *Xiandai qiye chuangxin lun* 現代企業創新論 (On

Modern Business Innovation). Tianjin: Tianjin People's Publishing House.

Chen Jinhua 陳錦華. 2006. "Zizhu chuangxin yu qiye fazhan" 自主創新與企業發展 ("Independent Innovation and Business Development"). *Qiye guanli* 企業管理 (Business Management), 3.

Cui Jinhua 崔金花. 2006. "Tigao zizhu chuangxin nengli de ruogan sikao" 提高自主創新的若干思考 ("Thoughts on Improving Independent Innovation Capability"). *Lilun tansuo* 理論探索 (Theoretical Exploration), 2.

Wei Jingxiang 危敬祥. 2006. "Chuangxin xing siwei he zizhu chuangxin" 創新性思維和自主創新 ("Innovative Thingking and Independent Innovation"). *Jiangxi shehui kexue* 江西社會科學 (Jiangxi Social Sciences), 2.

Zhao Gengshen 趙更申. 2006. "Butong zhanlue daoxiang dui zizhu chuangxin yu hezuo chuangxin de yingxiang yanjiu" 不同戰略導向對自主創新與合作創新的影響研究 ("Research on the Influence of Different Strategic Orientation on Independent Innovation and Cooperative Innovation"). *Dangdai jingji kexue* 當代經濟科學 (Modern Economic Science), 3.

Chapter 2

English materials:

Asimakopulos. 1957. "A note on productivity changes and the terms of trade." *Oxford Economic Papers, New Series* 9 (2):225–233.

Benham, Frederic. 1940. "The terms of trade." *Economica* 7: 37–60.

Bhagwati, Jagdish. "Immiserizing growth: A geometrical note." *The Theory of Commercial Policy Essays in International Economic Theory*I. Cambridge: The MIT Press.

Dixit, Avinash K., and Joseph E. Stiglitz. 1977. "Monopolistic competition and optimum product diversity." *The American Economic Review* 67 (3):297–308.

Gould. J. R. 1981. "On the interpretation of inferior goods and factors." *Economica, New Series* 48 (192):397–405.

Grossman, G. M., and E. Helpman. 1991. "Endogenous product cycles." *The Economic Journal* 101 (408):1214–1229.

Hicks, J. R. 1953. "An inaugural lecture." *Oxford Economic Papers, New Series* 5 (2):117–135.

Kemp, Murray C. 1955. "Technological change, the terms of trade and welfare." *The Economic Journal* 65 (259):457–473.

Krugman, Paul. 1979. "A model of innovation, technology transfer, and the world distribution of income." *The Journal of Political Economy* 87 (2):253–266.

Rivera-Batiz, Luis A., and Paul Romer. "International trade with endogenous technological change." NBER Working Paper Series, no.3594. http://www.nber.org/papers/w3594.

References

Robertson, D. H. 1938. "Changes in international demand and the terms of trade." *The Quarterly Journal of Economics* 52 (3):539–540.

Translated materials:

Aghion, Philippe, and Peter Howitt. 2004. *Neisheng zengzhang lilun* 內生增長理論 (Endogenous Growth Theory). Beijing: Peking University Press.

Baumol, William. 2004. *Ziyou chuangxin jiqi: ziben zhuyi de zengzhang qiji* 自由創新機器：資本主義的增長奇跡 (The Free-Market Innovation Machine: Analyzing the Growth Miracle of Capitalism). Beijing: China Citic Press.

Grossman, and Helpman. 2003. *Quanqiu jingji zhong de chuangxin yu zengzhang* 全球經濟中的創新與增長 (Innovation and Growth in the Global Economy). Beijing: China Renmin University Press.

Krugman, Paul, and Maurice Obstfeld. 2002. *Guoji jingji xue* 國際經濟學 (International Economics). 5th ed. Beijing: China Renmin University Press.

Chinese materials:

Gu Kejian 谷克鑒. 2000. "1990–1998 nian guomin yu waizi bumen chukou bodong chayi de shizheng fenxi—HBS tuiduan zai zhongguo de yanzheng yu tuozhan" 1990–1998年國民與外資部門出口波動差異的實證分析—HBS推斷在中國的驗證與拓展 ("Empirical Analysis of Difference in Export Fluctuation between National and Foreign Sector between 1990 and 1998—Test and Extension of HBS Deduction"). *Guanli shijie* 管理世界 (Management World) 2.

Gu Kejian 谷克鑒. 2000. "Zhongguo duiwai maoyi fazhan zhong de jingzheng zhengce xuanze" 中國對外貿易發展中的競爭政策選擇 ("Competition Policy Choice in the Development of Chinese Foreign Trade"). *Zhongguo shehui kexue* 中國社會科學 (Social Sciences in China) 3.

Gu Kejian 谷克鑒. 2001. "Guoji jingjixue dui yinli moxing de kaifa yu yingyong" 國際經濟學對引力模型的開發與應用 ("Development and Application of the Gravity Model in International Economics"). *Shijie jingjie* 世界經濟 (The Journal of World Economy) 2.

Gu Kejian 谷克鑒. 2005. "Zhongguo liyong waizi shijian de gongneng pingjia yu zhanlue xuanze" 中國利用外資實踐的功能評價與戰略選擇 ("Function Assessment and Strategic Choice of China's Practice in the Utilization of Foreign Capital—Empirical Description based on Economics and Management"). *Caimao jingji* 財貿經濟 (Finance and Trade Economics) 3.

Gu Kejian 谷克鑒. 2006. *Zhongguo de jingji zhuanxing yu maoyi liudong* 中

國的經濟轉型與貿易流動 (China's Economic Transition and Trade Flow: Theoretical Examination and Quantitative Study based on Institutional and Technical Factors). Beijing: China Renmin University Press.

Lu Feng 盧鋒. 2006. "Wo guo gongzi yu laodong li chengben biandong ji guoji bijiao—renminbi shiji huilü changqi zoushi yanjiu zhi er" 我國工資與勞動力成本變動及國際比較—人民幣實際匯率長期走勢研究之二 ("Changes of Wage and Labor Cost in China and International Comparison—Research on the Long-term Trend of Renminbi Real Exchange Rate II"). Beijing daxue zhongguo guoji yanjiu zhongxin taolun gao xilie 北京大學中國國際研究中心討論稿系列 (Discussion Paper Series of China Center for Economic Research, Peking University), no. C2006008, (13 April): 1978–2004.

Lu Feng 盧鋒. 2006. "Woguo laodong shengchanlü zengzhang yu guoji bijiao—renminbi shiji huilü changqi zoushi yanjiu zhiyi" 我國勞動生產率增長與國際比較—人民幣實際匯率長期走勢研究之一 ("China's Labor Productivity Growth and International Comparison—Research on the Long-term Trend of Renminbi Real Exchange RateI"). Beijing daxue zhongguo guoji yanjiu zhongxin taolun gao xilie 北京大學中國國際研究中心討論稿系列 (Discussion Paper Series of China Center for Economic Research, Peking University), no. C2006004, (3 April): 1978–2004.

Lu Xiangqian 盧向前, and Dai Guoqiang 戴國強. 2005. "Renminbi shiji huilü bodong dui woguo jinchukou de yingxiang: 1994–2003" 人民幣實際匯率波動對我國進出口的影響：1994–2003 ("Influence of Renminbi Real Exchange Rate on China's Import and Export: 1994–2003"). *Jingji yanjiu* 經濟研究 (Economic Research Journal) 5.

Lu Feng 盧鋒, and Han Xiaoya 韓曉亞. 2006. "Changqi jingji chengzhang yu shiji huilü yanbian—ba la sa-sai miu er sen xiaoying jiashuo jiqi jingyan zhengju" 長期經濟成長與實際匯率演變—巴拉薩-塞繆爾森效應假說及其經驗證據 ("Long-term Economic Growth and Real Exchange Rate Evolution—Balassa-Samuelson Effect Hypothesis and the Empirical Evidence"). Beijing daxue zhongguo guoji yanjiu zhongxin taolun gao xilie 北京大學中國國際研究中心討論稿系列 (Discussion Paper Series of China Center for Economic Research, Peking University), no. C2006002, (20 February).

Wang Ping 王平, and Qian Xuefeng 錢學鋒. 2007. "Cong maoyi tiaojian gaishan kan jishu jinbu de chanye zhengce daoxiang" 從貿易條件改善看技術進步的產業政策導向 ("Looking at the Industrial Policy Orientation from Terms of Trade Improvement"). *Zhongguo gongye jingji* 中國工業經濟 (China Industrial Economics) 3.

Zhao Yumin 趙玉敏, Guo Peixing 郭培興, and Wang Ting 王婷. 2002. "1993–2000 zhonguo maoyi tiaojian bianhua gaikuang" 1993–2000中國貿易條件變化概

References

况 ("Overview of China's Terms of Trade Changes: 1993–2000"). *Jingji yanjiu cankao* 經濟研究參考 (Review of Economic Research) 69.

Zhao Yumin 趙玉敏, Guo Peixing 郭培興, and Wang Ting 王婷. 2002. "Gaishan maoyi tiaojian de jianyi" 改善貿易條件的建議 ("Suggestions on Improving Terms of Trade"). *Jingji yanjiu cankao* 經濟研究參考 (Review of Economic Research) 69.

Chapter 3

English materials:

Ang, James S., Rebel A. Cole, and James W. Lin. 2000. "Agency costs and ownership structure." *Journal of Finance*, (LV):1.

Berger, A. N., and G. F. Udell. 2002. "Small business credit availability and relationship lending: the importance of banking organization structure." *Economical Journal*, (112):477.

Bester, H. 1985. "Screening vs rationing in credit markets with imperfect information." *American Economic Review*, (75):4.

Chan, Yuk Shee, Daniel Siegel, and Anjan V. Thakor. 1990. "Learning, corporate control and performance requirements in venture capital contracts." *International Economic Review*, (31):2.

Cochrane, John H. 2001. "The risk and return of venture capital." NBER working paper, no.8066, 002.

De Bondt, R., P. Slaets, and B. Cassiman. 1992. "The degree of spillovers and the number of rivals for maximum effective R&D." *International Journal of Industrial Organization*, (10):35–54.

Gilson, Ronald J., and Bernard Black. 1999. "Does venture capital require an active stock market." *Journal of Applied Corporate Finance*: 36–48.

Gompers, Paul A., and Josh Lerner. 2002. *The venture capital Cycle*. Rev. ed. Cambridge: MIT Press.

Gompers, Paul A., and Josh Lerner. 1998. "The determinants of corporate venture capital successes: organizational structure, incentives, and complementarities." NBER working paper, no. 6725.

Jensen, Michael C., and William H. Meckling. 1976. "Theory of the firm: Managerial behavior, agency costs, and ownership structure." *Journal of Financial Economics*, (3).

Katsoulacos, Y., and D. Ulph. 1998. "Endogenous innovation spillovers and technology policy." *Journal of Industrial Economics*, (46):333–357.

Mayers, Stewart C. 1974. "Interactions of corporate financing and investment decisions: implications for capital budgeting." *Journal of Finance*, (29).

Nesheim, John L. 2000. *High tech start up*. New York: The Free Press.

Petersen, Mitchell A., Raghuram G. Rajan. 1995. "The effect of credit market competition on lending relationships." *Quarterly Journal of Economics*, (110).

Sah, Raaj K., and Joseph E. Stiglitz. 1986. "The architecture of economic systems: hierarchies and polyarchies." *American Economic Review*, (76).

Schumpeter, Joseph. 1934. *The theory of economic development*. Cambridge: Harvard University Press.

Stein, J. C. 2002. "Information production and capital allocation: decentralized versus hierarchical firms." *The Journal of Finance*, (LvII):5.

Stiglitz, J., and A. Weiss. 1981. Credit rationing with imperfect information. *American Economic Review*, 3.

Chinese materials:

Chen Shulong 陳樹隆. 2001. "Lun xian jieduan tuichu chuangye ban mianlin de fengxian" 論現階段推出創業板面臨的風險 ("On Risks for Introducing in GEM at the Present Stage"). *Caimao jingji* 財貿經濟 (Finance and Trade Economics), 7.

Chen Xiaohong 陳曉紅et al. 2003. *Zhong xiao qiye rongzi chuangxin yu xinyong danbao* 中小企業融資創新與信用擔保 (SME Financing Innovation and Credit Guarantee). Beijing: China Renmin University Press.

Cheng Siwei 成思危. 2003. *Fengxian touzi lun cong* 風險投資論叢 (Venture Capital Treatise). Beijing: Democracy and Construction Press.

China Venture Capital Research Institute. 2004. "Zhongguo gaoxinqu fazhan baogao" 中國高新區發展報告 ("Report on the Development of High-Tech Areas in China"). *2003 Zhongguo fengxian touzi nianjian* 2003 中國風險投資年鑒 (2003 China Investment Yearbook), ed. Cheng Siwei 成思危. Beijing: Democracy and Construction Press.

Fu Jiaji 傅家驥. 1998. *Jishu chuangxin xue* 技術創新學 (Technological Innovation). Beijing: Tsinghua University Press.

Guo Bin 郭斌, and Liu Manlu 劉曼路. 2002. "Minjian jinrong yu zhong xiao qiye fazhan: dui wenzhou de shizheng fenxi" 民間金融與中小企業發展：对温州的实证分析 ("Private Finance and SME Development: Empirical Analysis of Wenzhou"). *Jingji yanjiu* 經濟研究Economic Research Journal, (6).

He Guanghui 何光輝, and Yang Xianyue 楊咸月. 1999. Fengxian touzi, er ban shichang ji duo cengci ziben shichang fazhan 風險投資、二板市場及多層次資本市場發展 (Venture Capital, the Second Board Market and Multi-level Capital Market Development). *Caijing yanjiu* 財經研究 (Journal of Finance and Economics), (5).

References

Hua Rong 花榮. 2000. "Guanyu woguo chuangye ban shichang de sikao" 關於我國創業板市場的思考 ("Some Thoughts on GEM of China"). *Nanfang jingrong* 南方金融 (South China Finance), (11).

Huang Junhui 黃俊輝, and Wang Huanchen 王浣塵. 2002. "Fengxian touzi yu chuangye ban shichang" 風險投資與創業板市場 ("Venture Capital and GEM"). *Shangye jingji yu guanli* 商業經濟與管理 (Business Economics and Administration), (3).

Li Yang 李揚, and Yang Siqun 楊思群. 2001. *Zhong xiao qiye rongzi yu yinhang* 中小企業融資與銀行 (SME Financing and Bank). Shanghai: Shanghai University of Finance and Economics Press.

Li Zhiyun 李志赟. 2002. "Yinhang jiegou yu zhong xiao qiye rongzi" 銀行結構與中小企業融資 ("Banking Structure and SME Financing"). *Jingji yanjiu* 經濟研究 (Economic Research Journal), (6).

Lin Yifu 林毅夫, and Li Yongjun 李永軍. 2001. "Zhong xiao jinrong jigou fazhan yu zhongxiao qiye rongzi" 中小金融機構發展與中小企業融資 ("Development of Chinese Small Financial Institutions and SME Financing"). *Jingji yanjiu* 經濟研究 (Economic Research Journal), (1).

Lin Yifu 林毅夫, and Sun Xifang 孫希芳. 2005. "Xinxi, feizhenggui jinrong yu zhong xiao qiye rongzi" 信息、非正規金融與中小企業融資 ("Information, Informal Finance and SME Financing"). *Jingji yanjiu* 經濟研究 (Economic Research Journal), (7).

Liu Manhong 劉曼紅, and Hu Bo 胡波. 2001. "Zhongguo fengxian touzi diaocha baogao" 中國風險投資調查報告 ("Investigation Report on China's Venture Capital"). Zhongguo ruan keti xiangmu yanjiu baogao 國家軟課題項目研究報告 (Research Report on National Soft Subject Projects).

Liu Manhong 劉曼紅. 2003. *Zhongguo zhong xiao qiye rongzi wenti yanjiu* 中國中小企業融資問題研究 (Studies on SME Financing in China). Beijing: China Renmin University Press.

Luo Danyang 羅丹陽, and Yin Xingshan 殷興山. 2006. "Minying zhong xiao qiye fei zhenggui rongzi yanjiu" 民營中小企業非正規融資研究 ("Research on the Informal Financing of Private SMEs"). *Jinrong yanjiu* 金融研究 (Journal of Financial Research), (4).

Sheng Lijun 盛立軍. 1999. *Fengxian touzi: caozuo, jizhi yu celue* 風險投資：操作、機制與策略 (Venture Capital: Operation, Mechanism and Strategy). Shanghai: Shanghai Far East Publishers.

Shi Jinchuan 史晉川 et al. 2002. *Zhidu bianqian yu jingji fazhan* 制度變遷與經濟發展 (System Changes and Economic Development: Research on Wenzhou Mode). Hangzhou: Zhejiang University Press.

Shi Jinchuan 史晉川, and Ye Min 葉敏. 2001. "Zhidu niuqu huanjing zhong de jinrong anpai: wenzhou anli" 制度扭曲環境中的金融安排：溫州案例 ("Financial Arrangement in System Distorted Environment: Wenzhou Case"). *Jingji lilun yu jingji guanli* 經濟理論與經濟管理 (Economic Theory and Business Management), (1).

Shi Jinchuan 史晉川, Sun Fuguo 孫福國, and Yan Gujun 嚴谷軍. 1997. Shichang Shenhua zhong minjian jinrong de xinqi 市場深化中民間金融的興起 (Emergence of Private Finance in Market Deepening). *Jingji yanjiu* 經濟研究 (Economic Research), (12).

Wang Guogang 王國剛. 2000. "Sheli 'chuangye ban' zhong yinggai zhuyi jiejue de jige renshi wenti" 設立 "創業板" 中應該注意解決的幾個認識問題 ("Several Cognitive Problems That Need to Be Addressed in the 'GEM'"). *Guoji jinrong yanjiu* 國際金融研究 (International Finance Research), (10).

Wang Jingtian 王競天, Li Zhengyou 李正友, and Feng Xuefei 馮雪飛et al. 2001. *Zhong xiao qiye chuangxin yu rongzi* 中小企業創業與融資 (SME Innovation and Financing). Shanghai: Shanghai University of Finance and Economics Press.

Xing Tiancai 刑天才. 2003. "Woguo duo cengci ziben shichang tixi de goujian yu fazhan silu" 我國多層次資本市場體系的構建與發展思路 ("The Building and Development Train of Thought of the Capital Market System in China"). *Caijing wenti yanjiu* 財經問題研究 (Research on Financial and Economic Issues), (3).

Yang Xiaokai 楊小凱. 2004. *Yang Xiaokai tan jingji* 楊小凱談經濟 (Discussion about the Economy by Yang Xiaokai). Beijing: China Social Sciences Press.

Ye Chunhe 葉春和. 2000. "Chuangye ban shichang: wanshan woguo zhengquan shichang tixi jianshe de biran xuanze" 創業板市場：完善我國證券市場體系建設的必然選擇 ("GEM: Inevitable Choice for Improving the Development of Chinese Securities Market System"). *Guoji jinrong yanjiu* 國際金融研究 (International Finance Research), (11).

Zhang Jie 張杰. 1998. *Zhongguo jinrong zhidu de jiegou yu bianqian* 中國金融制度的結構與變遷 (Structure and Change of Chinese Financial Institution). Taiyuan: Shanxi Economic Publishing House.

Zhang Jie 張杰. 2000. "Minying jingji de jinrong kunjing yu rongzi cixu" 民營經濟的金融困境與融資次序 ("Financial Difficulties and Financing Order of the Private Economy"). *Jingji yanjiu* 經濟研究 (Economic Research Journal), (4).

Zhang Jie 張捷. 2002. "Zhong xiao qiye de guanxi xing jiedai yu yinhang zuzhi jiegou" 中小企業的關係型借貸與銀行組織結構 ("The Relationship Lending of SMEs and the Organization Structure of Banks"). *Jingji yanjiu* 經濟研究 (Economic Research Journal), (6).

Chapter 4

English materials:

Capon, N., and R. Glazer. 1987. "Marketing and technology: a strategic coalignment." *Journal of Marketing* 51(3):1–14.

Chatterji, D. 1996. "Accessing external sources of technology." *Research Technology Management* 39(2):48–56.

Cohen, W. M., and R. C. Levin. 1990. "Absorptive capability: a new perspective on learning and innovation." *Administration Science Quarterly* 35(1):128–152.

Council of supply chain management professional. 2006. "Supply chain management/logistics management definitions." http://www.cscmp.org/Website/AboutCSCMP/Definitions/Definitions.asp.

De la Mothe, J., and G. Paquet. 1998. *Local and regional system of innovation.* Norwell: Kluwer Academic Publishers.

Dodgson, M, D. M. Gann, and A. J. Salter. 2002. "The intensification of innovation." *International Journal of Innovation Management* 6 (1): 53–83.

Duysters, G., and J. Hagedoorn. 2000. "Core competencies and company performance in the worldwide computer industry." *Journal of High Technology Management Research* 11(1):75–91.

Edquist, C. 1997. *Systems of innovation.* London: Printer.

Eisenhardt, K. E., and J. A. Martin. 2000. "Dynamic capabilities: what are they." *Strategic Management Journal* 21:1105–1121.

Esienhardt, K. M. 1989. "Building theories from case study research." *Academy of Management Review* 14(4):532–550.

Fleming, L., and O. Sorenson. 2003. "Navigating the technology landscape of innovation." *MIT Sloan Management Review* 44(2):15–23.

Freeman, C. 1974. *The economics of industrial innovation.* Penguin: Harmondsworth, U.K.

Freeman, C. 1991. "Networks of innovators: A synthesis of research issues." *Research Policy* 20:499–514.

Gibb Dyer, W. J. R., and Alan L. Wilkens. 1991. "Better stories, not better constructs, to generate better theory: a rejoinder to eisenhardt." *Academy of Management Review* 16(3):613–619.

Granstrand, O., E. Bohlin, and C. Oskarsson. 1992. "External technology acquisition in large multi-technology corporations." *R&D Management* 22(2):111–133.

Halfat, C., S. Finkelstein, W. Mitchell, M. A. Peteraf, H. Singh, and D. J.

Teece. 2007. *Dynamic capabilities: understanding strategic change in organizations*. Blackwell: Oxford, U.K., Winter SG.

Huber, G. P. 1991. "Organizational learning: the contributing processes and a review of literatures." *Organization Science* 2(1):88–117.

Iansiti, Marco, and Jonathan West. 1997. "Technology integration: turning great research into great products." *Harvard Business Review* 75(3):69–80.

Jonash, R. 1996. "Strategic technology leveraging: making outsourcing for you." *Research Technology Management* 39(2):19–25.

Kamien, M., and N. Schwartz. 1975. "Market structure and innovation." *Journal of Economic Literature* 23(1): 1–37.

Kline, S., and N. Rosenberg. 1986. "An overview of innovation." *The positive sum strategy*.eds. Landau R. and Rosenberg N. Washington: National Academy Press.

March, J. G., and H. A. Simon. 1958. *Organizations*. Wiley: New York.

Nelson, R. 1993. *National innovation systems: A comparative analysis*. New York: Oxford University Press.

Nelson, R. R. 1982. *An evolutionary theory of economic change*. Harvard University Press: Cambridge, MA., Winter SG.

Noori, H. 1990. *Managing the dynamics of new technology*. New Jersey: Prentice Hall.

Rothwell, R., and W. Zegveld. 1985. *Reindustrialization and technology*. London: Longman.

Steinmueller, W. E. 2000. "Will new information and communication technologies improve the "codification" of knowledge?" *Industrial and Corporate Change* 9: 361–376.

Teece, David J. 2007. "Explicating dynamic capabilities: the nature and microfoundations of (sustainable) enterprise performance." *Strategic Management Journal* 28:1319–1350.

Tsai, K. H., and J. C. Wang. 2005. "External technology acquisition and firm performance: A longitudinal study." *Journal of Business Venturing* 7:91–112.

Yin, Robert K. 2003. *Case study research: design and methods*. California: Sage Publications.

Translated materials:

Janszen, Felix. 2002. *Chuangxin shidai: wangluo hua shidai de chenggong moshi* 創新時代：網絡化時代的成功模式 (The Age of Innovation: Success Mode of the Network Era). Kunming: Yunnan University Press.

References

Schumpeter, Joseph Alois. 1990. *Jingji fazhan lilun* 經濟發展理論 (The Theory of Economic Development). Beijing: The Commercial Press.

Chinese materials:

Ding Juan 丁娟. 2002. "Chuangxin lilun de fazhan yanbian" 創新理論的發展演變 ("The Development and Evolution of Innovation Theory"). *Xiandai jingji tantao* 現代經濟探討 (Modern Economic Research), (6).

Huang Lucheng 黃魯成. 2004. "Chuangxin qunluo jiqi tezheng" 創新群落及其 特徵("Technological Innovation Community and Its Characteristics"). *Kexue guanli yanjiu* 科學管理研究 (Scientific Management Research), (8).

Chapter 5

English materials:

Akkermans, Henk, and Bart Vos. 2003. "Amplification in service supply chain: an exploratory case study." *Production and Operations Management* 12(2):204–223.

Council of Supply Chain Management Professional. 2006. "Supply chain management/Logistics management definitions." http://www.cscmp. org/Website/AboutCSCMP/Definitions/Definitions.asp.

De Waart, Dirk, and Steve Kemper. 2004. "5 Steps to service supply chain excellence." *Supply Chain Management Review*.

Dyer, W. G. Jr., and Alan L. Wilkins. 1991. "Better stories, not better constructs, to generate better theory: A rejoinder to eisenhardt." *Academy of Management Review* 16(3):613–619.

Eisenhardt, K. M. 1989. "Building theories from case study research." *Academy of Management Review* 14(4):532–550.

Ellram, Lisa M., Vendy L. Tate, and Corey Billington. 2004. "Understanding and managing the service supply chain." *The Journal of Supply Chain Management*.

Emerson, Carol J., and Curtis M. Grimm. 1998. "The relative importance of logistics and marketing customer service: A strategic perspective." *Journal of Business Logistics* 19(1):17–32.

Nahapiet, Janine, and Sumantra Ghoshal. 1998. "Social capital,intellectual capital,and the organizational advantage." *Academy of Management Journal* 23(2):242–266.

Oh, Hongseok, Giuseppe Labianca, and Myung-Ho Chung. 2006. "A multilevel model of group social capital." *Academy of Management review* 31(3):569–582.

Poole, Kevin. 2003. "Seizing the potential of the service supply chain." *Supply Chain Management Review*.

Saccani, N. et al. 2007. "Configuring the after-sales service supply chain: A multiple case study." *International Journal of Production Economics*.

Sako. 1991. "The role of trust in Japanese buyer-supplier relationships." *Ricerche Economiche*:2–3.

Sengupta, Kaushik, Daniel R. Heiser, and Lori S. Cook. 2006. "Manufacturing and service supply chain performance: A comparative analysis." *The Journal of Supply Chain Management*.

Yin, Robert K. 2003. *Case study research: Design and methods*. California: Sage Publications.

Youngdahl, William E., and Arvinder P. S. Loomba. 2000. "Service-driven global supply chains." *International Journal of Service Industry Management* 11(4):329–347.

Chinese materials:

Yang Mingming 陽明明. 2006. "Xianggang de gangkou fuwu gongyinglian" 香港的港口服務供應鏈 ("Hong Kong's Port Service Supply Chain"). *Caigou yu gongyinglian* 採購與供應鏈 Procurement and Supply Chain, (8).

Yang Zhe 楊哲. 2005. "Jiyu yidong daili de fuwu gongyinglian goujian fangzhen moxing" 基於移動代理的服務供應鏈構建仿真模型 ("Simulation Model of Service Supply Chain Building Based on Mobile Agent"). *Jisuanji gongcheng yu yingyong* 計算機工程與應用 (Computer Engineering and Applications), (10).

Chapter 6

English materials:

Attaran, Mohsen. 2004. "Exploring the relationship between information technology and business process reengineering." *Information and management*, (41):585–596.

Auramo, J., J. Kauremaa, and K. Tanskanen. 2005. "Benefits of IT in supply chain management: an explorative study of progressive companies." *International Journal of Physical Distribution and Logistics Management* 35(2):82–88.

Banner, David K. 1995. *Designing effective organization: traditional and transformation*. California: Views Sage Publications.

Barnet, J. 1997. *Gaining and sustaining competitive advantage*. Reading, MA: Addition-Wesley Publishing Company.

References

Barney, J. 1991. "Firm resources and sustained competitive advantage." *Journal of Management* 17(1):99–120.

Berger, A. 2006. "Five steps to an e-synchronized supply chain EB/OL." http://www.accenture.com/xd/xd.asp?it=enwebandxd=ideas%5Coutlook%5Cpov%5Cpov_fivesteps.xml.

Brynjolfsson, Erik and Lorin Hitt. 1995. "Information technology as a factor of production: the role of differences among firms." *Economics of Innovation and New Technology*, (3).

Brynjolfsson, Erik and Lorin Hitt. 1996. "Paradox lost? Firm level evidence on the returns to information systems spending." *Management Science* 42.

Brynjolfsson, Erik and Lorin Hitt. 2000. "Beyond computation: information technology, organizational transformation and business practices." *Journal of Economic Perspectives* 4.

Brynjolfsson, Erik and Lorin Hitt. 2003. "Computing productivity: firm-level evidence." *Review of Economics and Statistics*, (85:4).

Chopra, S., and P. Meindl. 2004. *Supply chain management: strategy, planning, and operation*. 2nd ed. New Jersey: Person and Prentice Hall.

Copacino, W. C., and R. W. Dik. 2001. "Why B2B e-markets are here to stay." *Accenture Outlook*, (2):22–9.

Cronin, M. J. 1997. "Using the web to push key data to decision makers." *Fortune* 9(29):254.

Cross, G. J. 2000. "How e-business is transforming supply chain management." *Journal of Business Strategy* 21(2):36–9.

Daekwan, K. S., C. Tamer, and J. C. Roger. 2006. "Information system innovations and supply chain management: channel relationships and firm performance." *Journal of the Academy of Marketing Science* 34:40–54.

Davenport, T. H., and J. E. Short. 1990. "The new industrial engineering: information technology and business process redesign." *Sloan Management Review* 31(4):11–27.

Elliot, R., E. D. Martin, and T. E. Philip. 2003. "Assessing the effects of operational processes and information systems on inventory performance." *Journal of Operations Management* 21:63–80.

Franke, Richard H. 1987. "Technological revolution and productivity decline: computer introduction in the financial industry." *Technological Forecasting and Social Change* 31.

Hammer, M. 1990. "Reengineering work: don't automate, obliterate." *Harvard Business Review* 68(4):104–112.

Kemppainen, K. and Vepsalainen A. P. J. 2003 "Trends in industrial supply chains and networks." *International Journal of Physical Distribution and Logistics Management* 33(8):701–19.

Klischewski, Ralf, and Ingrid Wetzel. 2003. "Service-flow management for health provider networks." *Logistics Information Management*, 16(3/4):259–269.

Lee H., and C. Whang. 1997. "Bullwhip effect in supply chains." *Sloan Management Review* 38(3):93–102.

Lee, Jooh, and Utpal Bose. 2002. "Operational linkage between diverse dimensions of information technology investments and multifaceted aspects of a firm's economic performance." *Journal of information Technology*, (17):119–131.

Lee, S., and S. H. Kim. 2006. "A lag effect of IT investment on firm performance." *The Information Resources Management Journal* 19(1):43–69.

Lee, Sangho, and Soung Hie Kim. 2006. "A lag effect of IT investment on firm performance." *Information Resources Management Journal* 19(1):43–69.

Levary, R. R. 2000. "Better supply chains through information technology." *Industrial Management* 42(3):24–30.

Loveman, G. W. 1994. "An assessment of the productivity impact on information technologies." *Information technology and the corporation of the 1990s:research studies*. Eds. Allen T. J., and Morton M. S. Cambridge, MA: MIT Press.

Mahmood, M., and S. K. Soon. 1991. "A comprehensive model for measuring the potential impact of information technology on organizational strategic variables." *Decision Science* (22):869–897.

Markus, M. Lynne, and Christina Soh. 1993. "Banking on information technology: converting IT spending into firm performance." *Strategic information technology management: perspectives on organizational growth and competitive Advantage*. Eds. R D Banker, R. J. Kauffman, and M. A. Mahrnood. Harrisburg, Pennsylvania: Idea Group Publishing.

Mentzer, J. T., and M. A. Moon. 2004. "Understanding demand." *Supply Chain Management Review* 8(4):38–45.

Nada, R. S., and P. Robert. 2005. "Modeling the Relationship between firm capability collaboration,and performance." *Journal of Business Logistics* 26(1):1–23.

Poter, M.E., and Millar, V. E. 1985. "How information gives you competitive advantage." *Harvard Business Review* 64(4):149–60.

Simchi, Levi D., P. Kaminsky, and Levi E. Simchi. 2003. *Designing and managing the supply chain: concepts, strategies, and case studies*. New York: McGraw-Hill.

Smith, Frank. 2004. "Plan with the big 'S'." *MSI* 22(11):42.

Soh, Christina, and Markus, M. Lynne. 1995. "How IT creates business value: a process theory synthesis." Proceedings of the sixteenth international conference on information systems, Amsterdam, Holand.

Swaminathan, J. M., and S. R. Tayur. 2003. "Models for supply chains in e-business."

Tippins, Michael J. et al. 2003. "IT competency and Firm Performance:is Organizational Learning a Missing Link?" *Strategic Management Journal*, (24):745–761.

Venkatraman, N. 1994. "IT-enabled business transformation: from automation to business scope redefinition." *Sloan Management Review* 35(2):73–87.

Weill, P. 1992. "The relationship between investment in information technology and firm performance: a study of the value manufacturing sector." *Information Systems Review* 3(4):307–333.

Wing L., and G. Perry. 2001. "Toward twenty-first-century pharmaceutical sales and operations planning." *Pharmaceutical Technology*:20–26.

Wolf, Emily J. 2002. "Healthcare's supply chain reaction." *Healthcare Executive*, (Sep/Oct):16–20.

Translated materials:

Yoichi Ito. 1994. *Riben xinxi hua gainian yu yanjiu de lishi* 日本信息化與概念研究的歷史 (Japan's Informatization Concept and Research History). Beijing: Social Sciences Academic Press.

Chinese materials:

Feng Rende 馮仁德, and Chen Yu 陳禹. 2003. "Jie xinxi jishu fuzhu chuangxin tisheng qiye jingzheng li" 借信息技術輔助創新提升企業競爭力 ("Leveraging on Information Technology to Assist Innovation to Improve Business Competitiveness"). *Shangye yanjiu* 商業研究 (Business Studies), (5).

Jin Jiangjun 金江軍. 2004. *Qiye xinxi hua yu xiandai dianzi shangwu* 企業信息化雨現代電子商務 (Enterprise Informatization and Modern E-commerce). Beijing: Electronic Industry Press.

Jing Jipeng 靖継鵬, and Sun Liming 孫立明. 2002. "Xinxi jishu dui qiye jingzheng youshi de yingxiang" 信息技術對企業競爭優勢的影響 ("Influence of Information Technology on Enterprise Competitive Edge"). *Qingbao kexue* 情報科學 (Information Science), (4).

Li Xueying 李學英. 2003. "Lun qiye xinxi hua zhong de she ke xinxi fuwu" 論企業信息化中的社科信息服務 ("On Social Science Information Service in Enterprise Informatization"). *Tushu qingbao zhishi* 圖書情報知識 (Knowledge of Library and Information Science), (1).

Qu Wei 曲煒. 2005. "Qiye xinxi hua dui jingzheng jizhi de yingxiang jizhi yanjiu" 企業信息化對競爭機制的影響機制研究 ("Studies on the Influential Mechanism of Enterprise Informatization on Competition Mechanism"). Kunming ligong daxue shuoshi xuewei lunwen 昆明理工大學碩士學位論文 Kunming University of Science and Technology, Master's Thesis.

Song Yuanfang 宋遠方, and Song Hua 宋華. 2005. *Yiyao wuliu yu yiliao gongyinglian guanli* 醫藥物流與醫療供應鏈管理 (Pharmaceutical Logistics and Health Care Supply Chain Management). Beijing: Peking University Medical Press.

Wang Xingming 汪星明. 2005. *Zhongguo yiyao wuliu dianzi shangwu* 中國醫藥物流電子商務 (Chinese Medicine Circulation E-commerce). Beijing: Peking University Medical Press.

Wu Jiapei 烏家陪. 1999. *Xinxi jingji yu zhishi jingji* 信息經濟與知識經濟 (Information Economy and Knowledge-based Economy). Beijing: The Publishing House of Economic Science.

Xia Lu 夏露. 2006. "Dianzi gongyinglian de yingyong youshi ji goujian yanjiu" 電子供應鏈的應用優勢及構建研究 ("Studies on Application Advantages and Building of Electronics Supply Chain"). *Wuliu jishu* 物流技術 (Logistics Technology), (8).

Xie Kang 謝康, and Chen Yu 陳禹. 1999. "Qiye xinxi hua de jingzheng youshi" 企業信息化的競爭優勢 ("Competitive Edge of Enterprise Informatization"). *Jingji yanjiu* 經濟研究 (Economic Research), (9).

Xie Kang 謝康, Zhang Haibo 張海波, and Huang Linjun 黃林軍. 2005. "Xinxi jishu yingxiang qiye xiaoji de jizhi yanjiu" 信息技術影響企業效績的機制研究 ("Studies on the Influential Mechanism of Information Technology on Business Performance"). *Jiazhi gongcheng* 價值工程 (Value Engineering), (3).

Yang Xueshan 楊學山. 2001. *Qiye xinxi hua jianshe he guanli* 企業信息化建設和管理 (Enterprise Informatization Development and Management). Beijing: Beijing Publishing House.

Zhang Hong 張弘. 2003. "Xinxi hua yu zhongguo liutong chuangxin" 信息化與中國流通創新 ("Informatization and China's Circulation Innovation"). *Caimao jingji* 財貿經濟 (Finance and Trade Economics), (10).

Chapter 7

English materials:

Anderson, J. C., and J. A. Narus. 1990. "A model of distribution firm and manufacturer firm working partnerships." *Journal of Marketing* 54(1).

Bandyopadhyay, Soumava, and Robert A. Robicheaux. 1998. "A cross-cultural study of influence strategies and satisfaction in sales channels." *Journal of Marketing Communications*, (4).

Frazier, Gary L., and John O. Summers. 1984. "Interfirm influence strategies and their application with distribution channels." *Journal of Marketing* 48.

Chinese materials:

Chen Tao 陳濤. 2003. "Guowai yingxiao qudao chongtu jiqi guanli yanjiu zongshu" 國外營銷渠道衝突及其管理研究綜述 ("Research on Foreign Sales channel Conflict and Management"). *Waiguo jingji yu guanli* 外國經濟與管理 (Foreign Economics and Management), (8).

Han Kun 韓昆, and Jin Junxi 靳俊喜. 2006. "Weili shidai de xiaoshou qudao chuangxin" 微利時代的銷售渠道創新 ("Sales channel Innovation in the Era of Meager Profit"). *Shangchang xiandai hua* 商場現代化 (Market Modernization), (3).

Jia Xin 賈欣. 2006. "Qiye yingxiao qudao biange jiqi duice fenxi" 企業營銷渠道變革及其對策分析 ("Analysis of Enterprise's Sales channels and Its Countermeasures"). *Jishu jingji yu guanli yanjiu* 技術經濟與管理研究 (Technoeconomics and Management Research), (4).

Liu Qionghui 劉瓊輝. 1999. "Shichang fazhan yu qiye fenxiao qudao chuangxin" 市場發展與企業分銷渠道創新 ("Market Development and Enterprise Distribution Channel Innovation"). *Jingji yu guanli yanjiu* 經濟與管理研究 (Research on Economics and Management), (2).

Ren Yan 任燕. 2004. "Woguo da jiadian qiye yingxiao qudao wenti tanxi" 我國大家電企業營銷渠道問題探析 ("Analysis of the Sales channels of Chinese Large Household Appliance Enterprises"). *Shangye yanjiu* 商業研究 (Business Studies), (23).

Wang Guilin 王桂林, and Zhuang Guijun 莊貴軍. 2004. "Zhongguo yingxiao qudao zhong qiye jian xinren de gainian moxing" 中國營銷渠道中企業間信任的概念模型 ("Conceptual Models for Inter-firm Trust in China's Sales channels"). *Dangdai jingji kexue* 當代經濟科學 (Modern Economic Science), (1).

Wang Zhengxuan 王正選 et al. 2006. "Woguo jiadian ye qudao biange ji celue fenxi" 我國家電業渠道變革及策略分析 ("Analysis on the Channel and Strategy of Chinese Household Appliance Industry"). *Shangchang xiandai hua* 商場現代化 (Market Modernization), (3).

Wu Xiaoding 吳小丁. 2004. "Daxing lingshou dian jinchang fei yu youshi diwei lanyong guizhi" 大型零售店進場費與優勢地位濫用規制 ("Entrance Fee of Large Retail Stores and Regulations against Dominant Position Abuse").

Jilin daxue xuebao (shehui kexue ban) 吉林大學學報（社會科學版）[Journal of Jilin University (Social Sciences Version)], (5).

Yu Kunzhang 于坤章, and Liang Chengxian 梁承獻. 2005. "Jiyu qudao quanli de fenxiaoshang xingwei yingxiang celue xuanze" 基於渠道權利的分銷商行為影響策略選擇 ("Distributors' Behavior based on Channel Power Influences Strategy Choice"). *Xiandai guanli kexue* 現代管理科學 (Modern Management Science), (8).

Zhang Junhai 張俊海. 2006. "Qudao quanli yu qudao manyi du guanxi jiqi yingxiang yinsu yanjiu" 渠道權利與渠道滿意度關係及其影響因素研究 ("Research on the Relations between Channel Power and Channel Satisfaction and Its Influencing Factors"). *Zhongguo renmin daxue shuoshi lunwen* 中國人民大學碩士論文 (Master's Thesis of Renmin University of China).

Zhang Li 張黎, and Dickson. 2004. "Qudao quanli lilun yu guowai pinpai fuzhuang zai woguo de yingxiao qudao" 渠道權利理論與國外品牌服裝在我國的營銷渠道 ("Power Theory and the Distribution Channel for Foreign-brand Apparel in the Chinese Market"). *Nankai guanli pinglun* 南開管理評論 (Nankai Business Review), (4).

Zhang Shijun 張世軍. 2007. "Haier yu geli de shichang yingxiao qudao moshi" 海爾與格力的市場營銷模式 ("The Sales channel Model of Haier and Gree"). *Shangchang xiandai hua* 商場現代化 (Market Modernization), (8).

Zhou Quanxiong 周權熊. 2007. "Woguo jiadian fenxiao qudao de yanbian yu fazhan" 我國家電分銷渠道的演變與發展 ("Evolution and Development of Chinese Household Appliance Distribution Channels"). *Shangye shidai* 商業時代 (Commercial Times), (14).

Zhou Yusheng 周瑜勝. 2003. "Gongyingshang yu lingshoushang guanxi de weiji yu duice" 供應商與零售商關係的危機與對策 ("Crisis and Countermeasures of the Relations between Suppliers and Retailers"). *Shangye jingji wencui* 商業經濟文萃 (Commercial Economics Review), (2).

Zhuang Guijun 莊貴軍, and Xi Youmin 席酉民. 2004. "Zhonggguo yingxiao qudao zhong siren guanxi dui qudao quanli shiyong de yingxiang" 中國營銷渠道中私人關係對渠道權利使用的影響 ("The Influence of Personal Relationship on the Utilization of Channel Power in Chinese Sales channel"). *Guanli kexue xuebao* 管理科學學報 (Journal of Management Sciences), (6).

Zhuang Guijun 莊貴軍, and Zhou Xiaolian 周筱蓮. 2004. "Yingxiao qudao chengyuan zhijian de chongtu yu jiejue fangfa" 營銷渠道成員之間的衝突域解決方法 ("Conflict and Resolution in Market Channels"). *Beijing gongshang daxue xuebao* 北京工商大學學報 (Journal of Beijing Technology and Business University), (1).

Zhuang Guijun 莊貴軍. 2000. "Quanli, chongtu yu hezuo: xifang de qudao xingwei lilun" 權利衝突域合作：西方的渠道行為理論 ("Power, Conflict and Cooperation: Channel Behavior Theory of the West"). *Beijing shangxueyuan xuebao* 北京商學院學報 (Journal of Beijing Institute of Business), (1).

Zhuang Guijun 莊貴軍. 2004. "Yingxiao qudao kongzhi: lilun yu moxing" 營銷渠道控制：理論與模型 ("Sales channel Control: Theory and Model"). *Guanli xuebao* 管理學報 (Journal of Management), (7).

Chapter 8

English materials:

Biel, A. L. 1992. "How Brand Image Drives Brand Equity." *Journal of Advertising Research* 6 (Nov/Dec).

Keller, Levin Lane. 1993. "Conceptualizing, Measuring, and Managing Customer-based Brand Equity." *Journal of Marketing* 57.

Nunnally, Jum. 1987. *Psychometric Theory*, 2nd ed. New York: McGraw-Hill.

Translated materials:

Aaker, David. 2004. *Chuangjian qiangshi pinpai* 創建強勢品牌 (Building Strong Brand). Beijing: China Labour and Social Security Publishing House.

Aaker, David. 2006. *Guanli pinpai zichan* 管理品牌資產 (Managing Brand Equity). Beijing: China Machine Press.

Kapferer, Jean-Noel. 2000. *Zhanlue pinpai guanli* 戰略品牌管理 (Strategic Brand Management). Beijing: The Commercial Press.

Keller, Kevin lane. 2003. *Zhanlue pinpai guanli* 戰略品牌管理 (Strategic Brand Management). Beijing: China Renmin University Press.

Lilien, Gary L., and Arvind Rangaswamy. 2005. *Yingxiao gongcheng yu yingyong* 營銷工程與應用 (Marketing Engineering and Application). Beijing: China Renmin University Press.

Ries, Al, and Jack Trout. 2002. *Dingwei* 定位 (Positioning). Beijing: China Financial and Economic Publishing House.

Ries, Al, and Jack Trout. 2002. *Yingxiao geming* 營銷革命 (Marketing Revolution). Beijing: China Financial and Economic Publishing House.

Chinese materials:

"2006 zhongguo pinpai jiazhi baogao" 2006中國品牌價值報告 ("Report on Chinese Brand Value in 2006"). December 2006. Beijing Famous Brand Asset Appraisal Co., Ltd.

"Shichang yingxiao jingdian 12 faze" 市場營銷經典12法則 ("12 Classical Rules of Marketing"). October 19, 2006. http://www.singapore-business.com.

Tencent Finance. "Zhongguo qiye shehui zeren diaocha youxiu qiye: baiwei pijiu." 中國企業社會責任調查優秀企業：百威啤酒. ("Outstanding Enterprise of The Survey on Chinese Enterprises' Social Responsibility: Budweiser Beer"). http://finance.qq.com/a/20061121/000507.htm.

"Yanjing pijiu: waizi zhengduo yinbao xiaofei shengji longtou" 燕京啤酒：外資爭奪引爆消費升級龍頭 ("Yanjing Beer: Fighting for Foreign Capital Results in Consumption Upgrading Leader"). October 20, 2006. http://www.hexun.com.

"Zhujiang pijiu duba guonei chunsheng banbi jiangshan" 珠江啤酒獨霸國內純生半壁江山 ("Zhujiang Beer Monopolizes Half of the Domestic Market Shares of Draught Beer"). September 29, 2006. http://www.spzs.com.

21st Century Business Herald. 2006. "Qingdao pijiu: dingwei nianqingren de qinggan baobei" 青島啤酒：定位年輕人的情感寶貝. ("Tsingtao Beer: Targeting the Young"). *Sohu Business* (October 31). http://business.sohu.com/20061031/n246116128.shtml.

Baidu baike 百度百科 (Baidu Encyclopedia). "Baiwei pijiu" 百威啤酒 ("Budweiser Beer"). http://baike.baidu.com/view/163777.htm.

Bing Shaoqian 邴紹倩. 2006. "Peiyu zizhu pinpai, tigao qiye jingzheng li" 培育自主品牌, 提高企業競爭力 ("Cultivating Proprietary Brand to Improve Business Competitiveness"). *Shangye xiandai hua* 商業現代化 (Business Modernization), (1).

Chen Baoming 陳寶明. 2006. "Dui woguo shishi zizhu pinpai de zhanlue sikao" 對我國實施自主品牌的戰略思考 ("Thinking of the Implementation of Proprietary Brand Strategy in China"). *Shangye yanjiu* 商業研究 (Business Studies), (18).

Chen Junhong 陳俊紅. 2006. "Zhongguo zizhu pinpai chengzhang zhangai ji duice fenxi" 中國自主品牌成長障礙及對策分析 ("Analysis of Growth Obstacles and Countermeasures of Chinese Enterprises"). *Shangchang xiandai hua* 商場現代化 (Market Modernization), (6).

Chen Liu 陳柳, and Liu Zhibiao 劉志彪. 2006. "Daigong shengchan, zizhu pinpai yu neisheng jili" 代工生產、自主品牌與內生激勵 ("OEM, Proprietary Brand and Endogenous Incentive"). *Caijing luncong* 財經論叢 (Collected Essays On Finance and Economics), (5).

Chen Qingye 陳晴曄. 2006. "Lun qiye de zizhu pinpai he hexin jingzheng li" 論企業的自主品牌和核心競爭力 ("On Business Proprietary Brand and Core Competitiveness"). *Shangchang xiandai hua* 商場現代化 (Market Modernization), (3).

References

Chen Wan'er 陳婉兒. 2006. "Qianyi qiye de zizhu pinpai zhanlue" 歡意企業的自主品牌戰略 ("A Brief Analysis of Proprietary Brand Strategy"). *Keji chuangye yuekan* 科技創業月刊 (Pioneering with Science and Technology Monthly), (11).

Cheng Zhenbiao 程振彪. 2004. *Tansuo zhongguo qiche zizhu kaifa he zizhu pinpai chuangjian zhilu* 探索中國汽車自主開發和自主品牌創建之路 (Exploring the Path of Chinese Automobile Independent Development and Proprietary Brand Building). Beijing: Beijing Institute of Technology Press.

China Food Industry Network. "Pinglun: bentu pijiu de pinpai zhanlue miju" 評論：本土啤酒的品牌迷局 ("Comment: Brands' Strategic Maze of Local Beers") *HC HTL*. http://info.food.hc360.com/2005/07/07134757356.shtml.

China News Service. 1999. "Guojia jishu jiandu ju chongshen pijiu shengchan qiye bixu shiyong guojia biaozhun pijiuping" 國家技術監督局重申啤酒生產企業必須是用國家標準啤酒瓶 ("State Bureau of Technical Supervision of PRC Restates Beer Manufacturers Must Use The Beer Bottle Of National Standard"). *Guanggong information network* (June 2). http://www.gdnet.com.cn/aspprg/gdnet/zczz/detail.asp?id=9583.

He Xiaode 何孝德. 2006. "Ruhe yunyong pinpai quanyi moxing chuangjian zizhu pinpai" 如何運用品牌權益模型創建自主品牌 ("How to Use Brand Equity Model to Build Proprietary Brands"). *Shangye shidai* 商業時代 (Commercial Times), (3).

Li Tongtong 李彤彤. 2007. "Zizhu pinpai chuangjian tanxi" 自主品牌創建探析 ("Analysis of Proprietary Brand Building"). *Shangchang xiandai hua* 商場現代化 (Market Modernization), (1).

Liu Fengjun 劉鳳軍. 2003. *Pinpai yunying lun* 品牌運營論 (On Brand Operation). Beijing: The Publishing House of Economic Science.

Lu Taihong 盧泰宏, and Kuang Danni 鄺丹妮. 1998. *Zhengti pinpai sheji* 整體品牌設計 (Integrated Brand Design). Guangzhou: Guangzhou People Press.

Lu Taihong 盧泰宏, and Liu Chao 劉超. 2005. *Zhongguo pijiu yingxiao zhan* 中國啤酒營銷戰 (Marketing Warfare of Beer Industry in China). Beijing: China Social Sciences Press.

Ma Mouchao 馬謀超. 2005. *Pinpai kexue hua yanjiu* 品牌科學化研究 (Scientific Research on Brand). Beijing: China Market Publishing House.

Ma Qingguo 馬慶國. 2002. *Guanli tongji* 管理統計 (Statistics for Management). Beijing: Beijing Science Publishing House.

Sang Baichuan 桑百川, and Li Yumei 李玉梅. 2005. "Suzao zhongguo zizhu pinpai lu zai he fang" 塑造中國自主品牌路在何方 ("Shaping the Way out for Chinese Proprietary Brands"). *Jiangji daoyin* 經濟導引 (Economic Herald), (12).

Wang Tao 汪濤, and Cao Zixia 曹子夏. 2006. "Woguo qiye yanzhong de zizhu pinpai" 我國企業眼中的自主品牌 ("Proprietary Brand in the Eyes of Chinese Enterprises"). *Jingji guanli* 經濟管理 (Economic Management), (17).

Wang Tengning 王騰寧. 2005. "Zizhu pinpai dui chukou maoyi de yingxiang" 自主品牌對出口貿易的影響 ("Influence of Proprietary Brand on Export Trade"). *Shangye shidai* 商業時代 (Commercial Times) , (9).

Wen Decheng 溫德成. 2006. "Nanfu konggu zhuanyi yu woguo zizhu pinpai jianshe wenti" 南孚控股轉移與我國自主品牌建設問題 ("Digititan Controlling Interest Transfer and Chinese Proprietary Brand Building"). *Qiusuo* 求索 (Investigation), (9).

Wu Yongyi 1. 2006. "Qingpi yanjing jingcheng duijue" 青啤燕京京城對決 ("Competition between Tsingtao Beer and Yanjing Beer"). *Jingying guanli zhe* 經營管理者 Managers and Administrators, (3).

Yan Hengyuan 嚴恒元. 2003. "Yanjing jie Yao Ming xiaoying daru meiguo" 燕京借姚明效應打入美國 ("Yanjing Leverages on Yao Ming to Enter into the U.S. Market"). *Jingji ribao* 經濟日報 Economic Daily, (April 4).

Yu Fang 于方, and Zhuang Hui 莊暉. 2006. "Zhongguo qiche zizhu pinpai fazhan moshi tansuo" 中國汽車自主品牌發展模式探索 ("Exploration of the Development Patterns of Chinese Proprietary Automobile Brands"). *Shanghai guanli kexue* 上海管理科學 (Shanghai Management Science), (1).

Zhang Guangling 張廣玲, and Zou Jie 鄒捷. 2006. "Zizhu pinpai chuangxin de ruogan sikao" 自主品牌創新的若干思考 ("Some Thinking on Proprietary Brand Innovation"). *Zhongguo gongshang guanli yanjiu* 中國工商管理研究 (Study on China Administration for Industry and Commerce), (10).

Zheng Jianling 鄭建玲. 2002. "Pinpai chongzhu zhongwai hudong" 品牌重組中外互動("Interaction between China and Foreign countries on Brand Restructing"). *Zhongguo zhiliang bao* 中國質量報 (China Quality Paper), (July 3).

Chapter 9

English materials:

Aghion, P., and M. Schankerman 2000. "Competition, entry and the social returns to infrastructure in transition economies." *Economics of Transition*, (7).

Atkeson, Andrew, and Patrick J. Kehoe. 2002. "Measuring organization capital." NBER working paper, 8722.

References

Barro, Robert J., and X. Sala-I-Martin. 1995. *Economic growth*. New York: McGraw-Hill.

Burgelman, R., M. A. Maidique, and S. C. Wheelwright. 1996. *Strategic management of technology and innovation*. NewYork: Mc Graw-Hill .

Christensen, J. F. 1995. "Asset profiles for technological innovation." *Research Policy*, (24).

Groning, T. 1988. "POSIT (policies for sustainable technological innovation in the 21Century): states, advantages and difficulties regarding the planned project." Presentation to the ESST Conference, Strasbourg, Sep.12–13. http://www.essst.uio.no/posti/.

Organisation for Economic Cooperation and Development. 1996. "Proposed guidelines for collecting and integrating technological innovation data." OSLO Manual. Paris: OECD.

Schumpeter, Joseph. 1942. *Capitalism, socialism, and democracy*. New York: Harper and Row.

Segerstrom, P., and J. Zolnierek. 1999. "The R&D incentives of industry leaders." *International Economic Review*, (40).

Simon, H. A. 1985. "What we know about the creative process." *Frontiers in creative and innovation management*. Ed. R. L. Kuhn. Cambridge, MA: Ballinger.

Tang, H. K. 1998. "An integrative model of innovation in organizations." *Technovation* 18: 297–309.

Translated material:

Daft, Richard. L. 2003. *Zuzhi lilun yu sheji jingyao* 組織理論與設計精要 (Essentials of Organization Theory and Design). Beijing: Mechanic Industry Press.

Chinese materials:

Chen Jin 陳勁, and Wang Fanghua 王芳華. 2002. *Shehui ziben yu jishu chuangxin* 社會資本與技術創新 (Social Capital and Technological Innovation). Hangzhou: Zhejiang University Press.

Chen Zhongbo 陳仲伯. 2003. "Qiye chixu jishu chuangxin de neihan ji tezheng" 企業持續技術創新的內涵及特徵 ("Connotation and Characteristics of Enterprise Sustainable Technological Innovation"). *Zhongnan daxue xuebao (shehui kexue ban)* 中南大學學報（社會科學版）[Journal of Central South University (Social Science Edition)], (4).

Fu Jiaji 傅家驥, and Jiang Yanfu 姜彥福 et al. 1991. *Jishu chuangxin—*

zhongguo qiye fazhan zhilu 技術創新—中國企業發展之路 (Technological Innovation—Chinese Enterprises Development Path). Beijing: Enterprise Management Press.

Fu Jiaji 傅家驥. 1998. *Jishu chuangxinxue* 技術創新學 (Technological Innovation). Beijing: Tsinghua University Press.

Guan Jiancheng 官建成, and Shi Xiaomin 史曉敏. 2004. "Jishu chuangxin nengli he chuangxin jixiao guanxi yanjiu" 技術創新能力和創新績效關係研究 ("Study on the Relationship between Technological Innovation Capability and Innovation Performance"). *Zhongguo jixie gongcheng* 中國機械工程 (China Mechanical Engineering), (6).

Hu Shousong 胡壽松, and He Yaqun 何亞群. 2006. *Chuchao juece lilun yu yingyong* 粗糙決策理論與應用 (Rough Decision Theory and Application). Beijing: Beijing University of Aeronautics and Astronautics Press.

Li Yuan 李垣. 1994. *Qiye jishu chuangxin jizhi lun* 企業技術創新機制論 (On Enterprise Technological Innovation Mechanism). Xi'an: Xi'an Jiaotong University Press.

Liu Wei 劉偉, and Xiang Gang 向剛. 2001. "Shilun qiye chixu fazhan, chixu chuangxin yu chixu xuexi jizhi jianshe" 試論企業持續發展、持續創新與持續學習機制建設 ("On Building Enterprise Sustainable Development, Sustainable Innovation and Sustainable Learning Mechanism"). *Jingji wenti tansuo* 經濟問題探索 (Inquiry into Economic Issues), (10).

Mao Ning 茅寧, and Wang Chen 王晨. 2005. *Ruan caiwu* 軟財務 (Soft Finance). Beijing: China Economic Publishing House.

Ouyang Xinnian 歐陽新年. 2004. "Qiye jishu chuangxin dongli yu liyi jili jizhi" 企業技術創新動力與利益激勵機制 ("The Drive and Benefit Motivation Mechanism for Enterprise Technology Innovation"). *Kexue guanli yanjiu* 科學管理研究 (Scientific Management Research), (3).

Sun Xiaofeng 孫曉峰. 2007. "Qiye chixu jishu chuangxin nengli jiqi pingjia zhibiao tixi de goujian" 企業持續技術創新能力及其評價指標體系的構建 ("Building of Enterprise Sustainable Technological Innovation Capability and Its Evaluation Index System"). *Gongye jishu jingji* 工業技術經濟 (Industrial Technology and Economy), (1).

Wang Guoyin 王國胤. 2001. *Rough ji lilun yu zhishi huoqu Rough* 集理論與知識獲取 (Rough Set Theory and Knowledge Acquisition). Xi'an: Xi'an Jiaotong University Press.

Wang Xuerong 王雪榮. 2006. "Yizhong jiyu zhengju lilun de dongtai zonghe jixiao pingjia shiyong fangfa" 一種基於證據理論的動態綜合績效評價使用方法 ("The Practical Method of Dynamically Evaluating the Comprehensive Performance Based on Evidential Reasoning"). *Zhongguo guanli kexue* 中國管理科學 (Chinese Journal of Management Science), (8).

Wu Guisheng 吳貴生. 2000. *Jishu chuangxin guanli* 技術創新管理 (Technological Innovation Management). Beijing: Tsinghua University Press.

Xia Baohua 夏保華. 2001. *Qiye chixu jishu chuangxin jiegou* 企業持續技術創新結構 (Structure of Sustainable Enterprise Technological Innovation). Shenyang: Northeastern University Press.

Xia Baohua 夏保華. 2002. "Lun qiye chixu jishu chuangxin de jiegou" 論企業持續技術創新的結構 ("On the Structure of Sustainable Enterprise Technological Innovation"). *Kexue xue yanjiu* 科學學研究 (Studies in Science of Science), (5).

Xia Baohua 夏保華. 2003. "Qiye chixu jishu chuangxin: benzhi, dongying he guanli" 企業持續技術創新：本質、動因和管理 ("Sustainable Enterprise Technological Innovation: Nature, Driving Force and Management"). *Kexue jishu yu bianzhengfa* 科學技術與辯證法 (Science Technology and Dialectics), (2).

Xiang Gang 向剛. 2006. *Qiye chixu chuangxin* 企業持續創新 (Sustainable Enterprise Innovation). Beijing: Science Publishing House.

Xu Jun 徐軍, and Zhao Huifang 趙惠芳 et al. 2006. "Qiye chixu jishu chuangxin de pingjia zhibiao tixi" 企業持續技術創新的評價指標體系 ("Continuous Technological Innovation Evaluation Index System"). *Tongji yu juece* 統計與決策 (Statistics and Decision), (7).

Yang Delin 楊德林, Liu Fang 劉芳 et al. 2004. "Zhongguo hangkong gongye qiye jishu chuangxin nengli pingjia" 中國航空工業企業技術創新能力評價 ("China Aviation Industrial Enterprise Technological Innovation Capability Evaluation"). *Qinghua daxue xueba (zhexue shehui kexue ban)* 清華大學學報（哲學社會科學版） [Journal of Tsinghua University (Philosophy and Social Science Edition)], (4).

Zheng Chundong 鄭春東, and He Jinsheng 和金生 et al. 1999. "Qiye jishu chuangxin nengli pingjia yanjiu" 企業技術創新能力評價研究 ("Study on the Evaluation of Sustainable Enterprise Technological Innovation Capability"). *Zhongguo ruan kexue* 中國軟科學 China Soft Science, (10).

Zheng Qinpu 鄭勤樸. 2001. "Qiantan dingliang pingjia qiye chixu chuangxin nengli" 淺談定量評價企業持續創新能力 ("Brief Analysis of Quantitative Evaluation of Sustainable Enterprise Innovation Capability"). *Gongye qiye guanli* 工業企業管理 (Industrial Business Management), (11).

Zhuang Ziyin 莊子銀. 2005. "Qiyejia jingshen, chixu jishu chuangxin he changqi jingji zengzhang de weiguan jizhi" 企業家精神、持續技術創新和長期經濟增長的微觀機制 ("Entrepreneurship, Continuous Technological Innovation and Long-term Economic Growth Micro Mechanism"). *Shijie jingji* 世界經濟 (World Economy), (12).

Chapter 10

English materials:

Ahuja, Gautam. "Collaboration networks, structural holes, and innovation: A longitudinal study." *Administrative Science Quarterly*. September, 2000: 425–454.

Allison, P. D. 1984. *Event history analysis: Regression for longitudinal event data*. CA: Sage.

Astley, W. Graham, and Andrew H. Van de Ven. 1983. "Central Perspectives and debates in organization theory." *Administrative Science Quarterly* 28: 245–273.

Astley, W. Graham. 1985. "The two ecologies: Population and community perspectives on organizational evolution." *Administrative Science Quarterly* 30: 224–241.

Barney, Jay B. 1986. "Strategic factor markets: expectations, luck, and business strategy." *Management Science* 32(10): 1231–1241.

Bartlett, C. A., and S. Ghoshal. 1989. *Managing across borders: the transnational solution*. Boston: Harvard Business School Press.

Cassiman, B, R. and Veugelers. 2002. "R&D Cooperation and spillovers: some empirical evidence from Belgium." *American Economic Review* 92(3):1169–1197.

Castellacci, Fulvio, Stine Grodal, Sandro Mendonca, and Mona Wibe. 2005. "Advances and challenges in innovation studies." *Journal of Economic Issues* Vol. XXXIX (no. 1).

Child, J. 1972. "Organizational structure, environment, and performance: The role of strategic choice." *Sociology* 6:1–22.

Child, J. 1997. "Strategic choice in the analysis on action, structure, organizations, and environment: retrospect and prospect." *Organization Studies* 18:43–76.

Cohen, W. M., and D. A. Levinthal. 1990. "Absorptive capacity: A new perspective on learning and innovation." *Administrative Science Quarterly* 35:128–152.

D'Aunno, Thomas, Robert I. Sutton, and Richard H. Price. 1991. "Isomorphism and external support in conflicting institutional environments: A study of drug abuse treatment Units." *Academy of Management Journal* 14: 636–661.

Daellenbach, Urs S., Anne M. McCarthy, and Timothy S. Schoenecker. 1999. "Commitment to innovation: the impact of top management team characteristics." *R&D Management* 29(3):198–208.

References

Damanpour, Fariborz. 1987. "The adoption of technological, administrative, and ancillary innovations: impact of organizational factors." *Journal of Management* 13(4):675–688.

Davis, Kingsley. 1949. *Human society*, New York: Macmillan.

Davis, Lance E., and Douglass C. North. 1971. *Institutional change and American economic growth*. Cambridge: Cambridge University Press.

Dierickx, I., and K. Cool. 1989. Asset stock accumulation and sustainability of competitive advantage. *Management Science* 35:1504–1511.

DiMaggio, Paul J., and Walter W. Powell. 1991. "Introduction." *The New Institutionalism in Organizational Analysis*. Eds. Walter W. Powell, and Paul J. DiMaggio. Chicago: University of Chicago Press: 1–38.

Eisenmann, T. R. 2002. *Internet Business Models*. New York: McGraw-Hill.

Fan, J. P. H., Wong T. J., and Zhang T. 2005. "The emergence of corporate pyramids in China." Working paper, Chinese University of Hong Kong.

Faulkner, W. 1994. "Conceptualizing knowledge used in innovation: A second look at the science technology distinction and industrial innovation." *Science, Technology, and Human Values* 19(4):425–458.

Fujita, S. 1997. *A Strategy for Corporate Innovation*.Tokyo: Asian Productivity Organization.

Gibbons, M., and R. Johnston. 1974. "The roles of science in technological innovation." *Research Policy* 3(3):220–242.

Hambrick, D. C. 1983a. "An empirical typology of mature industrial product environments." *Academy of Management Journal* 26:213–230.

Hassink, Robert, and Michelle Wood. 1998. "Geographic 'clustering' in the German opto–electronics industry: its impact on R&D collaboration and innovation." *Entrepreneurship and Regional Development*, (10):277–296.

Hill, C. W. L., and S. A. Snell. 1989. "Effects of ownership structure and control on corporate productivity." *Academy of Management Journal* 32(1):25–46.

Hoffman, Andrew J. 1999. "Institutional evolution and change: environmentalism and the U. S. chemical industry." *The Academy of Management Journal* 42(4):351–371.

Hoffman, Richard C., and W. Harvey Hegarty. 1993. "Top management influence on innovations: effects of executive characteristics and social culture." *Journal of Management* 19(3):549–574.

Hoskisson, Robert E., Lorraine Eden, Chung Ming Lau, and Mike Wright. 2000. "Strategy in emerging economies." *The Academy of Management Journal* 43(3):249–267.

Kim, Linsu, and Youngbae Kim. 1985. "Innovation in a newly industrializing country: a multiple discriminant." *Management Science* (pre-1986), ABI/INFORM Global 31(3): 312.

King, John Leslie, Vijay Gurbaxani, F. Warren McFarlan, K. S. Raman, and C. S. Yap. 1994. "Institutional Factors in Information Technology Innovation." *Information Systems Research* 5(2):139–170.

Lee, P. M. 2005. "A comparison of ownership structures and innovations of US and Japanese firms." *Managerial and Decision Economics*, 26:39–50.

Lu Y. 1996. *Managerial decision making in Chinese enterprises*. London: McMillan.

Lunn, John. 1985. "The roles of property rights and market power in appropriation innovative output." *The Journal of Legal Studies* 14(2):423–433.

Ma Xufei, Yao Xiaotao, and Xi Youmin. 2006. "Business group affiliation and firm performance in a transition economy: A focus on ownership voids." *Asia Pacific J Manage*, 23:467–483.

Magretta, J. 1998. "Governing the family-owned enterprise: An interview with Finland's Krister Ahistrom." *Harvard Business Review*, 76(1):112–123.

Marinova, Detelina. 2004. "Actualizing innovation effort: the impact of market knowledge diffusion in a dynamic system of competition." *Journal of Marketing* 68(7):1–20.

Meyer, John W., and W. Richard Scott. 1992. *Organizational environments: ritual and rationality*, with the assistance of Brian Rowan and Terrence E. Deal [1983b]. Beverly Hills, CA: Sage.

Meyer, John W., and Brian Rowan. 1977. "Institutionalized organizations: formal structure as myth and ceremony." *American Journal of Sociology*, 83:340–363.

Meyer, M., and Lu Xiaohui. 2005. "Managing indefinite boundaries: the strategy and structure of a Chinese business firm." *Management and Organization Review*, 1/1:57–86.

Miller, C. Chet, Linda M. Burke, and William H. Glick. 1998. "Cognitive diversity among upper echelon executives: implications for strategic decision processes." *Strategic Management Journal* 19:39–58.

Nee, V. 1992. "Organizational dynamics of market transition: hybrid forms, property rights and mixed economy in China." *Administrative Science Quarterly* 37(1):1–27.

North, and Douglass C. 1990. *Institutions, Institutional change, and economic performance*. Cambridge, UK: Cambridge University Press.

Oliver, C. 1991. "Strategic responses to institutional processes." *Academy of Management Review* 16(1):145–179.

Oliver, C. 1992. "The antecedents of deinstitutionalization." *Organization Studies*, 13:563–588.

References

Oliver, C. 1997. "Sustainable competitive advantage: combining institutional and resource based views." *Strategic Management Journal*, 18(9):697–713.

Peng, M. W. 2000. *Business strategies in transition economies*. Thousand Oaks, CA: Sage.

Peng, Mike W. 2003. "Institutional transitions and strategic choices." *Academy of Management Review* 28(2):275–296.

Peteraf, M. A. 1993. "The cornerstones of competitive advantage: a resource-based view." *Strategic Management Journal* 14:179–191.

Pfeffer, Jeffrey, and Gerald Salancik. 1978. *The external control of organizations*. New York: Harper and Row.

Powell, and Walter W. 1990. "Neither market nor hierarchy: network forms of organization." in. *Research In Organizational Behavior*. Eds. L. L. Cummings, and B. M. Stam. Greenwich, CT: JAI Press, 12:295–336.

Prahalad, C. K., and G. Hamel. 1990. "The core competence of the corporation." *Harvard Business Review*, (May/Jun):79–91.

Richard, Scott W., and John W. Meyer. 1983. *The organization of societal sectors, Organizational Environments: Ritual and Rationality*. Ed. John W. Meyer, and Scott W. Richard. Chicago: University of Chicago Press:129–153.

Storey, John. 2000. "The Management of Innovation Problem." *International Journal of Innovation Management* 4(3):3.

Teece, David J., Gary Pisano, and Amy Shuen. 1997. "Dynamic Capabilities and Strategic Management." *Strategic Management Journal* 18(7).

Tolbert, P. S., and L. G. Zucker. 1983. "Institutional sources of change in the formal structure of organizations: the diffusion of civil service reforms, 1880–1935." *Administrative Science Quarterly*, 23:22–39.

Wernerfelt, and Birger. 1984. "A resource-based view of the firm." *Strategic Management Journal* 5(2):171–180.

Xu Erming, and Zhang Han. 2007. "The Impact of State ownership on Firm's Technological Innovational Modes." *Transition Economy of China*, 2007ICSM, 4.

Yamaguchi, K. 1991. *Event History Analysis*. Newbury Park, CA: Sage.

Yuan Li, Sun Yongfeng, and Yi Liu. 2006. "An empirical study of SOEs' market orientation in transitional China." *Asia Pacific J Manage*, 23:93–113.

Translated materials:

Agresi, Alan, and Barbara Finley. 2002. *Shehui tongji xue* 社會統計學 (Social Statistics). Taibei: Wunan Book Publishing Co., Ltd.

Drucker, Peter. 2000. *Chuangxin yu chuangye jingshen* 創新與創業精神 (Innovation and Entrepreneurship). Haikou: Hainan Publishing House.

Masahiko Aoki, and Okuno Masahiro. 1999. *Jingji tizhi de bijiao zhidu fenxi* 經濟體制的比較制度分析 (Comparative Institutional Analysis of Economic System). Beijing: China Development Publishing Ltd.

Porter, Michael. 1990. *Guojia jingzheng li* 國家競爭力 (National Competitiveness). Beijing: Hua Xia Publishing House.

Schumpeter, Joseph Alois. 1993. *Jingji fazhan lilun* 經濟發展理論 (The Theory of Economic Development). Beijing: The Commercial Press.

Chinese materials:

Feng Xiaoli 馮曉莉. 2005. "Woguo qiye chuangxin dongli jizhi yanjiu" 我國企業創新動力機制研究 ("Researches on the Driving Force of Chinese Enterprise Innovation"). *Xibei daxue boshi lunwen* 西北大學博士論文 (Doctoral dissertation of Northwest University).

Hua Jinyang 華錦陽. 2003. "Lun gongsi zhili de gongneng tixi ji dui woguo shangshi gongsi de shizheng fenxi" 論公司治理的功能體系及對我國上市公司的實證分析 ("Empirical Analysis on the Functional System of Corporate Governance and Chinese Listed Companies"). *Guanli shijie* 管理世界 (Management World), (1).

Li Fan 李凡, and Nie Ming 聶鳴. 1999. "Qiye zuzhi jiegou dui jishu chuangxin ji kuosan de yingxiang" 企業組織結構對技術創新及擴散的影響 ("Impact of Corporate Organization Structure on Technical Innovation and Dissemination"). *Keji jinbu yu duice* 科技進步與對策 (Scientific and Technological Progress and Countermeasure), 16(3).

Wang Zhihui 王智慧. 2002. *Shangshi gongsi zhili jiegou yu zhanlue xiaoji yanjiu* 上市公司治理結構與戰略效績研究 (Researches on the Governance Structure and Strategic Performance of Listed Companies). Beijing: China International Business and Trade Publishing House.

Xia Dong 夏冬. 2005. *Qiye zhili yu qiye chuangxin—suoyouquan, jingyingzhe yu qiye chuangxin xianghu guanxi de yanjiu* 企業治理與企業創新—所有權、經營者與企業創新相互關係的研究 (Corporate Governance and Corporate Innovation—Researches on the Relations between Ownership, Operators and Business Innovation). Beijing: Economy and Management Publishing House.

Xu Erming 徐二明, and Zhang Han 張晗. 2007. "Zhongguo shangshi gongsi jiyu zhanlue de gongsi zhili yanjiu" 中國上市公司給予戰略的公司治理研究 ("Researches on Strategy-based Corporate Governance of Chinese Listed Companies"). *Dangdai caijing* 當代財經 (Contemporary Finance and Economics), (1).

References

Chapter 11

English materials:

Adler, P. S., B. Goldoftas, and D. I. Levine. 1999. "Flexibility versus efficiency? A case study of model changeovers in the Toyota production system." *Organization Science* 10(1):43–68.

Benner, M. J., and M. L. Tushman. 2003. "Exploitation exploration and process management: the productivity dilemma revisited." *Academy of Management Review* 28(2): 238–256.

Bouchikhi, H. 1998. "Living with and building on complexity." *Organization* 5(2):217–232.

Brown, S. L., and K. M. 1997. Eisenhardt. "The art of continuous change: linking complexity theory and time paced evolution in relentlessly shifting organizations." *Administrative Science Quarterly* 42(1):1–34.

Christensen, C. M. 1997. *The innovator's dilemma*. Boston: Harvard Business School Press.

Duncan, R. B. 1976. "The ambidextrous organization: designing dual structure for innovation." *The Management of Organization*, (1):167–188.

Dutton, J. M., and A. Thomas. 1985. "Relating technological change and learning by doing research on technological innovation." *Management and Policy*, (2).

Eisenhardt, K. M., and M. E. Graebner. 2007. "Theory building from cases: opportunities and challenges." *Academy of Management Journal* 50(1):25–32.

Gassmann, O., P. Sandmeier, and C. H. Wecht. 2006. "Extreme customer innovation in the front-end: learning from a new software paradigm." *International Journal of Technology Management* 33(1):46–66.

Gharajedaghi ,J. 1999. *Systems thinking: managing chaos and complexity: a platform for designing business architecture*. Amsterdam: Elsvier/ Butterworth Heinemann.

Gibson, C. B., and J. Rirbinshaw. 2004. "Antecedents, consequences, and mediating role of organizational ambidexterity." *Academy of Management Journal* 47(2):209–226.

Gupta, A. K., K. G. Smith, and C. E. Shalley. 2006. "The interplay between exploration and exploitation." *Academy of Management Review* 49(4):693–706.

Handy, C. 1994. *The age of paradox*. Boston: Harvard Business School Press.

He, Z. L., and P. K. Wong. 2004. "Exploration vs.exploitation: an empirical test of the ambidexterity hypothesis." *Organization Science* 15(1):481–494.

Hoyt, J., and E. A. Gerloff. 1999. "Organizational environment changing economic conditions, and the effective supervision of technical personnel: a management challenge." *Journal of High Technology Management Research* 10(2):275–293.

Kang, S. C., S. S. Morris, and S. C. Snell. 2007. "Relational archetypes, organizational learning and value creation." *AMR* 32(1):236–256.

Kedia, B. L., R. T. Keller, and S. D. Jullan. 1992. "Dimensions of national culture and the productivity of R&D units." *Journal of High Technology Management Research* 3(1):1–18.

March, J. G. 1991. "Exploration and exploitation in organizational learning." *Organization Science* 2(1):71–87.

McDonough, III E. F., and R. Leifer. 1983. "Using simultaneous structures to cope with uncertainty." *Academy of Management Journal* (26):727–735.

McGrath, R. G. 2001. "Exploratory learning, innovative capability and managerial oversight." *Academic of Management Journal,* (44):118–131.

Mitroff, I. 1995. "Review of the age of paradox." *Academic of Management Review* 20(3):748–750.

Nonaka, I. 1994. A dynamic theory of organizational knowledge creation. *Organization Science,* (5-1):14–37.

Nonaka, I., and H. Takeuchi. 1995. *The knowledge creating company.* New York:Oxford University Press.

O'Reilly, C. A. III, and M. Tushman. 2004. "The ambidextrous organization." *Harvard Business Review* 82(4):74–82.

Perez-Freije, J., and E. Enkel. 2007. "Creative tension in the innovation process: how to support the right capabilities." *European Management Journal* 25(1):11–24.

Poole, M. S., and A. H. Van de Ven. 1989. "Using paradox to build management and organization theories." *Academy of Management Review* 14(4):562–587.

Reagans, R., and E. W. Zuckerman. 2001. "Networks, diversity, and productivity: the social capital of corporate R&D teams." *Organization Science* 12(4):502–516.

Rirbinshaw, J., and C. B. Gibson. 2004. "Building ambidexterity into an organization." *MIT Sloan Management Review* (Summer):47–55.

Siggelkow, N., and D. A. Levinthal. 2003. "Temporarily divide to conquer: centralized, decentrolized, and reintegrated organizational approaches to exploration and adaptation." *Organization Science* 14(6):650–669.

Siggelkow, N., and J. Rivkin. 2006. "When exploration backfires: unintended consequences of multilevel organizational search." *Academy of Management Review* 49(4):779–795.

Smith, W. K., and M. L. Tushman. 2005. "Managing strategic contradictions: a top management model for managing innovation streams." *Organization Science* 16(5):522–536.

Taylor, A., and H. R. Greve. 2006. "Superman or the fantastic four? knowledge combination and experience in innovative teams." *Academy of Management Review* 49(4):723–740.

Tushman, M., C. A. O'Reilly III. 1996. "Evolution and revolution: mastering the dynamics of innovation and change." *California Management Review* 38(4):8–30.

Tushman, M., C. A. O'Reilly III. 1997. *Winning through innovation: a practical guild to leasing organizational change and renewal*. Boston: Harvard Business School Press.

Translated materials:

Daft, Richard. 2003. *Zuzhi lilun yu sheji* 組織理論與設計 (Organization Theory and Design). Beijing: China Renmin University Press.

Dopfer, Kurt. 2004. *Yanhua jingjixue: gangyao yu fanwei* 演化經濟學：綱要與範圍 (Evolutionary Economics: Outline and Scope). Beijing: Higher Education Press.

Druck, Peter. 1987. *Guanli—renwu, zeren, shijian* 管理—任務、責任、實踐 (Management—task, responsibility and practice). Beijing: China Social Science Press.

Iansiti, Marco, and Roy Levien. *Gongyin: shangye shengtai xitong dui qiye zhanlue, chuangxin he ke chixuxing de yingxiang* 共贏：商業生態系統對企業戰略、創新和可持續性的影響 (Win win: What the New Dynamics of. Business Ecosystems Mean for Strategy, Innovation, and Sustainability). Beijing: the Commercial press, 2006.

Chinese materials:

Cohen 科恩. 2007. "Lingdao de zhongji mubiao—zhuanfang haier jituan shouxi zhixingguan Zhang Ruimin" 領導的終極目標—專訪海爾集團首席執行官張瑞敏 (Ultimate goal of leadership—exclusive interview with Zhang Ruimin, CEO of Haier Group). *Hafo shangye pinglun (zhongwen ban)* 哈佛商業評論（中文版）[Harvard Business Review (Chinese version)], (1).

Ju Qiang 鞠強. 2003. "Eryuan xiangdui pingheng guanli lilun" 二元相對平衡管理理論 ("Theory for Relative balanced management of duality"). *Qiye guanli* 企業管理 (Business Management), (11).

Ju Qiang 鞠強. 2007. "Zoujin hexie guanli—eryuan xiangdui pingheng guanli zhexue zongshu" 走進和諧管理—二元相對平衡管理哲學綜述 ("Embark on harmonious management—a summary on the philosophy of relative balanced management of duality"). *Qiye yanjiu* 企業研究 (Business Research), (2).

Li Yuling 李玉玲. 2006. "Eryuan xing zuzhi sheji yu chuangxin liu yanjiu" 二元性組織設計與創新流研究 ("Dual organization design and innovation process research"). *Jilin daxue shuoshi xuewei lunwen* 吉林大學碩士學位論文 Master dissertation of Jilin University.

Li Zhanxiang 李占祥, Yang Du 楊杜, and Xie Peicai 謝培才. 2000. *Maodun guanli xue* 矛盾管理學 (Paradox management). Beijing: Economy and management publishing house.

Shi Jiangtao 史江濤, and Cao Bing 曹兵. 2006. "Jianduan shi jishu chuangxin de zuzhi kunjing ji tupo" 間斷式技術創新的組織困境及突破 ("Organizational dilemma and break-through of interrupted technical innovation"). *Jishu jingji* 技術經濟 (Technical economy), (2).

Yuan Yongzhi 袁勇志. 2001. "Qiye chuangxin yu qiye eryuan zuzhi jiegou" 企業創新與企業二元組織結構 ("Enterprise innovation and the dual organizational structure of enterprise"). *Nanjing nongye daxue xuebao (shehui kexue ban)* 南京農業大學學報（社會科學版） [Journal of Nanjing Agricultural University (social sciences edition)], (1-3).

Zhang Hongshi 張洪石, and Chen Jin 陳勁. 2005. "Tupoxing chuangxin de zuzhi moshi yanjiu" 突破性創新的組織模式研究 ("Studies on the organizational model of groundbreaking innovation"). *Kexue xue yanjiu* 科學學研究 (Studies in Science of Science), (4).

Zhang Xi 張璽. 2006. "Jishu chuangxin de liangnan beilun yu wangluo hua jiqun shi chuangxin yanjiu" 技術創新的兩難悖論與網絡化集群式創新研究 ("Studies on the dilemma paradox of technical innovation and networked, schooling innovation"). *Kexue guanli yanjiu* 科學管理研究 (Scientific Management Research), (24-1).

Zheng Ping 鄭平. 2007. "Jiyu gongsi chuangye de pohuai xing chuangxin yanjiu" 基於公司創業的破壞性創新研究 ("Studies on the disruptive innovation based on entrepreneurship"). *Zhongguo renmin daxue boshi xuewei lunwen* 中國人民大學博士學位論文 (PhD dissertation of Renming University of China).

Zhu Ling 朱凌, Xu Qingrui 徐慶瑞, and Wang Fangrui 王方瑞. 2006. "Cong yanfan-yingxiao de zhenghe dao jishu chuangxin: shichang chuangxin de xietong" 從研發-營銷的整合到技術創新：市場創新的協同 ("From the Integration between R&D and marketing to technical innovation: coordination of market innovation"). *Keyan guanli* 科研管理 (Science Research Management), 27(5).

Index